Primate Ecology and Social Structure

Volume 1: Lorises, Lemurs and Tarsiers

D1073703

Robert W. Sussman

Department of Anthropology Washington University

PEARSON CUSTOM PUBLISHING

Cover photograph by Robert W. Sussman.

Copyright © 1999 by Pearson Custom Publishing.
All rights reserved.

This copyright covers material written expressly for this volume by the editor/s as well as the compilation itself. It does not cover the individual selections herein that first appeared elsewhere. Permission to reprint these has been obtained by Pearson Custom Publishing for this edition only. Further reproduction by any means, electronic or mechanical, including photocopying and recording, or by any information storage or retrieval system, must be arranged with the individual copyright holders noted.

Printed in the United States of America

10 9 8 7 6 5 4 3 2 1

Please visit our website at www.pearsoncustom.com

ISBN 0–536–02256–9
BA 990017

PEARSON CUSTOM PUBLISHING
160 Gould Street/Needham Heights, MA 02494
A Pearson Education Company

THIS BOOK IS DEDICATED TO MY STUDENTS,
PAST, PRESENT AND FUTURE.

CONTENTS

Preface

In this book I review the literature on free-ranging nonhuman primates. Initially I attempt to characterize these animals as they exist in their least disturbed state. When relevant, behavior in disturbed situations and captivity is compared to that occurring in more undisturbed contexts. By doing this, it is possible to gain a better understanding of the reasons primates behave as they do and the ways they fit into the communities in which they live. I realize this is an idealistic approach, since few localities inhabited by primates escape some level of human disturbance.

Research in field primatology during the late 1950s and early 1960s was mainly descriptive natural history, with few comparative and quantitative, or problem-oriented studies. However, by the 1970s, primatology moved into a problem-oriented phase (see Sussman 1979, Smuts et al. 1987, Smuts and Fedigan in press). Problem-oriented studies, focus mainly on determining relationships between behavior and morphology, ecology and social structure, and community interactions, and were a major component of field primatology during the 1970s and 1980s. More recently, many field primatologists are attempting to formulate and test theories that have developed out of classical sociobiology; such as those related to kin selection, inclusive fitness, reciprocal altruism, and sexual selection (see below). In many cases, the theories and studies are elegant and elaborate but the data are meager (e.g., see Casti 1989, Allen 1993, Sussman in press).

The philosopher of science, F.S.C. Northrop (1965) suggests that any healthy scientific discipline goes through three stages during its development. The first stage involves the analysis of the problem. The second is a descriptive natural history phase. Finally, there is a stage of postulationally prescribed theory, in which fundamental theories are tested. Although there is a movement in field primatology towards this final stage of inquiry, in fact, even today many primate species have not been studied in detail, and the most studied species normally are known only from a few localities. The range of variation in the behavior and ecology of most free-ranging primates is still unknown, and basic natural history remains a necessary component of the subdiscipline. As stated by Northrup (1965:37-38):

> In fact, if one proceeds immediately to the deductively formulated type of scientific theory which is appropriate to the third stage of inquiry, before one has passed through the natural history type of science with its inductive Baconian method appropriate to the second stage, the result inevitably is immature, half-baked, dogmatic and for the most part worthless theory.

Although, in the following chapters, I will discuss many of the theories currently popular in primatology, these cautions must be born in mind.

Primates make up only a small part of the communities in which they are found. In fact, plants and insects typically are the most diverse and numerous organisms in these communities. In order to understand primate behavior it is necessary to study the ecological communities and ecosystems in which they live, the broad biogeographical context in which they are found, and their phylogenetic history (Schoener 1988, Brooks and Mclennan 1991, Losos 1996). In the first chapter, I discuss the concept of an ecosystem and describe some of the general types of interactions that occur between the various components of ecosystems. Primates are highly social animals and the way animals space themselves in the environment and reproduce is related to other aspects of their ecology. The study of these relationships is referred to as socioecology. In the final pages of Chapter 1, I discuss some of the terms and theories used in socioecology. Many of these concepts are further explored in later chapters.

In Chapter 2, I discuss the taxonomy of primates, their morphological characteristics, and the origins and evolutionary history of the earliest primates. After these two introductory chapters, I review the literature on the galagos and lorises (Chapter 3), nocturnal Malagasy lemurs (Chapter 4), diurnal Malagasy lemurs (Chapter 5), and tarsiers (Chapter 6). The general organization of each of these review chapters is similar in order to facilitate comparisons. In the final chapter, cross-taxonomic comparisons are made of each of the topics discussed in the review chapters. In a second volume, I will review the literature on New World monkeys, and in a third volume that on Old World monkeys and apes. I have used earlier versions of this book for the past 20 years in teaching courses on primate ecology and social behavior. Revisions of earlier versions have benefited from the input of many students over this time. A few of my colleagues have read various versions of the book in its entirety and I appreciate their comments and suggestions. These include John Buettner-Janusch, Paul Garber, Terry Gleason, Charles Hildebolt, Jane Phillips-Conroy, Alison Richard, Ian Tattersall, and Mildred Trotter. Besides the above, for specific chapters, specialists in the topics covered have generously agreed to read the volume and offer their suggestions. Jonathon Losos and Peter Raven offered their input on Chapter 1, Glenn Conroy, Tab Rasmussen and Peter Raven on Chapter 2, Leanne Nash, Kimberly Nekaris and Tab Rasmussen on Chapter 3, Sylvia Atsalis on Chapter 4, and Myron Shekelle on Chapter 6. I thank all of these people and 8 anonymous reviewers for the assistance although I take responsibility for the final product. I also thank those who provided photographs and Terry Gleason for his assistance in getting this volume into final form. Finally, I thank Brett Nachman for copy editing and Wayne Spohr and Jerry Brennan of Pearson Education for their assistance.

REFERENCES

Allen, G.E. 1993. In Search of the Grail: Biology, Social Science, and the Meaning of Human Nature. *Contemp. Psych.* 38:455–458

Brooks, D.R., Mclennan, D.A. 1991. Historical Ecology: Examining Phylogenetic Components of Community Evolution. Pp. 267–280. In *Species Diversity in Ecological Communities: Historical and Geographic Perspectives.* R.E. Ricklefs; D. Schluter, Eds., Chicago: University of Chicago.

Casti, J.L. 1989. *Paradigms Lost.* New York, William Morrow.

Losos, J.B. 1996. Phylogenetic Perspectives on Community Ecology. *Ecology* 77:1344–1354.

Northrop, F.S.C. 1965. *The Logic of the Sciences and Humanities.* Cleveland, Meridian.

Schoener, T.W. 1988. Ecological Interactions. Pp. 255–297. In *Analytical Biogeography: An Integrated Approach to the Study of Animal and Plant Distribution.* A.A. Myers; P.S. Giller, Eds., London, Chapman and Hall.

Smuts, B.B., Cheney, D.L., Seyfarth, R.M., Wrangham, R.W., Struhsaker, T.T., Eds. 1987. *Primate Societies.* Chicago, University of Chicago.

Smuts, B.B., Fedigan, L.M., Eds. In press. *Primate Encounters: Animals, Scientists and Science.* Chicago, University of Chicago.

Sussman, R.W. 1979. *Primate Ecology: Problem Oriented Field Studies.* New York, Wiley.

Sussman, R.W. In press. The Piltdown Man: Father of American Field Primatology. In *Primate Encounters: Animals, Scientists and Science.* B.B. Smuts; L.M. Fedigan, Eds., Chicago, University of Chicago.

PART 1

CHAPTER 1

Ecology: General Principles

THE ECOSYSTEM

A *system,* according to Webster's dictionary, is a whole that is compounded of many interrelated parts or any phenomenon describable by a large number of variables. Many ecologists believe it useful to view ecosystems as particular types of systems, feedback or self-controlling systems. A feedback system is one that is composed of a group of interrelated parts that function to maintain themselves in a steady state or a state of equilibrium. Such processes always involve a simple principle: some of the output of the system is fed back to help control the input so there is always a loop of interrelated or interdependent events which operate to keep the system in equilibrium.

A good example of a self-controlling system is the heating unit in a building (Figure 1-1). If the temperature drops below 70°, when a thermostat is set at 70°, the sensory component of the system senses this change and automatically switches on, allowing electrical energy to be transformed into chemical energy and heat. This heat is then fed back into the system and the temperature is returned to 70°. The reaction of one component of the system, the sensory part, affects other, responding parts that in turn are fed back to the sensory component.

All feedback systems have the following points in common. (1) They must have *energy sources.* In the case of the thermostat and furnace, electricity is the initial source of energy; in an ecosystem the source of energy is the sun. (2) All systems must have *means of transforming energy* into usable form. This involves transforming energy from one type to another. With these transformations energy is lost. Systems are never perfectly efficient. (3) Systems must have built-in *abilities to vary with conditions.* This allows time for reactions to stimuli. The temperature in your home is never maintained at exactly 70°, for example, but usually fluctuates just above and below this temperature. If it is cold outside and someone opens a window, the temperature will drop below seventy degrees. If there are freezing temperatures and the window is left open, in time the heating system will probably fail. This is the fourth aspect that all systems share. (4) Systems must

OPEN SEQUENCE of control is illustrated by a system for regulating the temperature of a room. T_o is a variation in the temperature outdoors. Th is the variation of a thermometer. F is the fuel control of a furnace. T is the variation of the temperature in the room. In such a system of control there is no feedback.

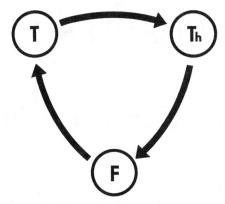

CLOSED SEQUENCE of control is illustrated by a system for regulating the temperature of a room by means of a thermostat. Here Th is a thermostat rather than a thermometer. In such a system there is feedback.

Figure 1-1 Simplified diagram of a self-controlling or feedback system. [Adapted from, "Feedback" by Arnold Tustin. From *Scientific American,* September, 1952. Pp. 48–55. Scientific American, Inc., New York.]

have *inherent limits of efficiency.* Overloads cause erratic behaviors and eventually systems break down.

Systems are usually arbitrarily chosen segments of larger systems. Thus, we can study restricted, self-perpetuating portions of the whole universe. We can extract a cell from a tissue and study the cell as a system in itself, or we can study the tissue as a system. We can, in fact, study systematic interactions or interrelationships of systems at many different levels: at the level of molecules, cells, tissues, organs, organisms, ecosystems, planets, solar systems, galaxies, or at the level of the universe. Thus, as segments of larger systems, ecosystems are open to continual inflow and outflow of materials, organisms, and energy. They are open or dynamic systems subject to constant change in time and space (Schoener 1988).

The fact that ecosystems are dynamic and subject to change has led to some problems with defining what is meant by equilibrium in ecology (Kikkawa and Anderson 1986, Yodzis 1989). Classically, an ecosystem is thought to be in equilibrium if species abundance and population densities remain stable over time (Chesson and Case 1986). However, the notion of constant stability is no doubt an illusion and species and populations in real ecosystems vary in time and space (Ricklefs and Schluter 1993, Maurer 1998). In fact, few ecosystems are in equilibrium over long periods of time. As stated by Pianka (1994:348): "Obviously, the real world is not unchanging—therefore, a fundamental question of interest is 'How do systems respond to various sorts of perturbations?'" Currently, most theories of community ecology assume that systems show a predictable tendency to recover following perturbations and thus can be considered equilibrium theories (Chesson and Case 1986).

There is another problem related to stability in systems. The major function of the components of an ecosystem is *not* to maintain stability, as it is in a heating system. The different organisms in an ecological community are selfish and any stability observed is due to millenia of trial and error interactions. "Communities are not necessarily assembled for orderly and efficient function . . . but rather each species may behave and evolve agonistically toward the other members of its community" (Pianka 1994:365).

Thus, the model of an ecosytem developed here is an abstract, conceptual model of a process. As stated by Watt (1966:12): This model "does not necessarily correspond to any real situation in nature, but rather is built out of components, some of which are found in some situations, and others of which are found in other situations, but some of which are found in all situations."

THE COMPONENTS OF ECOSYSTEMS

An ecosystem is composed of *abiotic* (non-living) and *biotic* (living) components. The abiotic components are solar energy, which provides heat and light, and chemicals. The six most important chemicals in living organisms are oxygen, carbon and hydrogen (which are used in photosynthesis), nitrogen, phosphorous, and potassium. The conditions in which the chemicals and solar energy are available vary according to altitude, longitude, latitude, temperature, moisture, light and atmospheric conditions. The effects of different interactions among these components produce various *biomes* or *plant formations* (see below).

A number of ecologists have illustrated the regularity of the various systems throughout the earth by developing classifications of world life zones (Figure 1-2). These are complex classifications which include such variables as altitude, latitude, longitude, humidity, temperature, and population. Because of the regularity of interactions among these factors, we can often predict which life zones will occur at various localities. Similar conditions produce similar types of plant formations. For example, the rain forests of South America are much like those in Africa, and although the plant species might differ taxonomically, the structure of the plant communities are very similar (Richards 1966, Terborgh 1992, Golblatt 1993). The morphology of the primates and other animals utilizing these plants are also quite similar, as are the behavioral interactions between such plants and animals.

The biotic components of an ecosystem are referred to as producers, consumers, and decomposers. The producers are green plants. Green plants utilize light energy and simple substances to build complex compounds, through the process of photosynthesis. The plants are the only living organisms that use energy directly from the sun for growth, development and reproduction. They transform abiotic components into organic components and use them directly for growth. Plants subsist, for the most part, exclusively on inorganic, nonliving substances.

Consumers *ultimately* obtain all of their nutrition from plants. The minimum nutritional requirements of animals are water, minerals, carbon, nitrogen, vitamins, and essential fatty and amino acids. All of these, except for water and minerals, can be obtained only from plants. A consumer either feeds directly on plants or on animals that feed on plants.

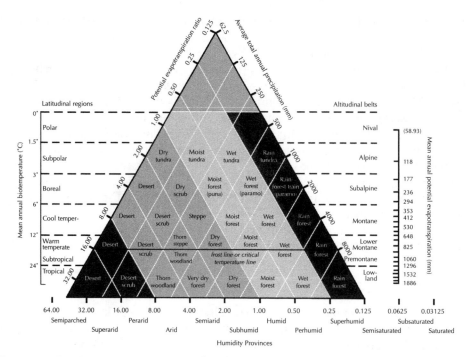

Figure 1-2 The classification scheme developed by U.S. forester Leslie Holdridge for vegetation in Central America. Rainfall increases from left to right on a declining axis; temperature decreases from bottom to top. The maximum evaporation plus the transpiration that could occur in a given environment is termed the potential evapotranspiration. Where rainfall exceeds the potential evapotranspiration, the ratio of the two quantities is less than 1.0, and evergreen vegetation is predominant. Drier climates are represented by ratios of greater than 1.0. Within Central America the cells in the diagram correspond reasonably well to recognized vegetation formations. [Adapted from *Evolutionary Ecology* by Eric Pianka. New York: Harper & Row, 1983, 1994.]

Decomposers are fungi and bacteria. These feed on dead organisms, both plant and animal. They reduce the organic compounds in organisms into simple substances and thus change these compounds back into the simple compounds that can be reused by plants. Although some chemicals are utilized in the process, most chemicals are fed back into the system. Most of the chemicals remain within the system and can be traced through their complex cycles of utilization (Figure 1-3).

ENERGY TRANSFER WITHIN ECOSYSTEMS

The primary source of energy in an ecosystem is the sun. Energy from the sun is transferred through the levels of the system. *This system of energy transfer is referred to as the trophic system.* The first level of the trophic system is the autotrophic, consisting of the plants or producers. Plants utilize solar energy to transform inorganic chemicals into complex organic substances. In the photosynthetic process light from the sun is used as the primary energy source. The second level is the heterotrophic, the animals or consumers. Consumers break down complex organic chemicals to obtain energy. Thus, at the autotrophic level energy is obtained directly from the sun, but at the heterotrophic level energy ultimately is obtained from plants.

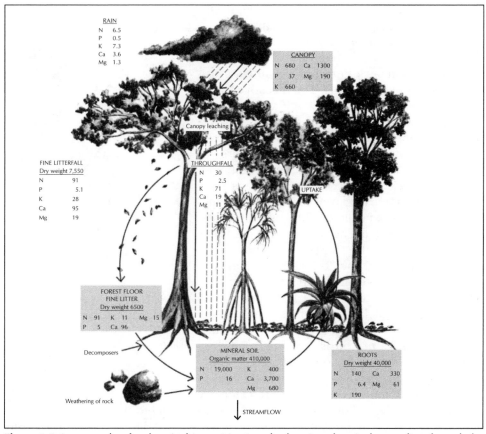

Figure 1-3 An example of a chemical (or nutrient) cycle showing chemicals recycling through the system in a montane rain forest. [Adapted from T. C. Whitmore, 1984.]

Consumers are divided into a number of categories. At a simple level these are primary consumers (herbivores or folivores), secondary consumers (carnivores, scavengers and parasites) and decomposers. Primary consumers eat only plant material. The biomass of consumers is directly related to the biomass of the plants they consume. *Biomass* is the total weight of organic material in a system.

The second level of heterotrophs consists of the carnivores. Carnivores eat herbivores (second level consumers) or other carnivores (third, fourth,....nth level consumers). In oceans there can be as many as twelve or thirteen levels of consumers, with bigger animals eating progressively smaller ones. These levels of gastronomy can be traced in different ecosystems and are referred to as food chains (Figure 1-4). A predator is usually bigger and stronger than its prey. Scavengers eat dead material. Parasites and decomposers feed on everybody, breaking down complex materials into simple inorganic materials. This enables chemicals to be reutilized by the producers and recirculated through the system.

Insects are the major herbivores and probably are responsible for converting something like 80% of plant material into animal material (Raven pers. comm.). The first mammals to evolve were insectivorous taking advantage of animal protein that was easier to digest than plant material. As will be discussed in the next chapter, it seems likely that the earliest primates were omnivores, eating both plants and insects (Harding and Teleki 1981, Martin 1990, Crompton 1995, see chapter 2).

Primary consumers are adapted to break down various types of plant substances, and they have a number of morphological features that enable them to do this. The digestive tract of a herbivore is enlarged and adapted for breaking down the cellulose in plants, detoxifying toxic plant materials, and extracting nutrients from plants (Montgomery 1978, Chivers et al. 1984, Kay and Davies 1994). Some taxonomic groups of primates are highly folivorous and possess adaptations for processing plant cellulose. Only one sub-family of Old World monkey, the Colobinae, has an enlarged foregut like that of ruminants such as cows, goats and antelopes. In fact, these "leaf-eating" monkeys have a complex 4-chambered stomach very similar to that of a cow (Figure 1-5). Some common predators of larger animals are Felidae (cats), Viverridae (similar to cats), Canidae (dogs) and predatory birds (e.g., owls, hawks, and eagles). There is only one taxonomic group

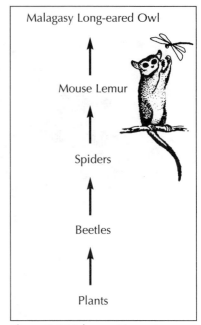

Figure 1-4 Malagasy Mouse Lemur *(Microcebus rufus)* food chain.

of primates, the tarsiers, that is wholly carnivorous. Coyotes, buzzards, crows, hyenas and, to some extent, lions are extensive scavengers. As much as twenty percent of the diet of lions is made up of scavenged material (Schaller 1972). It is possible that some of the diet of early hominids was obtained by scavenging (Blumenshire 1987, Rose and Marshall 1996).

In most systems a simple food chain does not occur but rather one finds a very complex set of feeding interactions. The food web illustrated in Figure 1-6 is an example of this complexity. As can be seen in this food web, some animals feed at more than one level of the trophic system.

Omnivores feed on both plants and other animals. These animals need adaptations which allow them to capture and digest this material, but their digestive tracts and feeding adaptations usually are not as specialized as those of obligate herbivores or carnivores. Many primates, including man, are omnivorous (Harding 1981, Sussman 1987, Martin 1990). Baboons and chimpanzees have been observed hunting and eating meat. This does not mean that they are carnivores, nor is it at all surprising. Animal material is included in the diet of most primate species and some small primates are highly insectivorous. Consumers are usually classified according to the food that makes up the majority of their diets (Kay 1984, see Table 1-1); using these criteria, most primates are frugivores

THE PYRAMID OF ENERGY TRANSFER

Every energy transformation through the levels of the trophic system involves some loss of usable energy. The total amount of energy available for use at each level is less than that available at the preceding level. This is expressed schematically by what is called an *Eltonian pyramid* (Figure 1-7). The Eltonian pyramid illustrates the amount of energy available to and utilized by the various components of

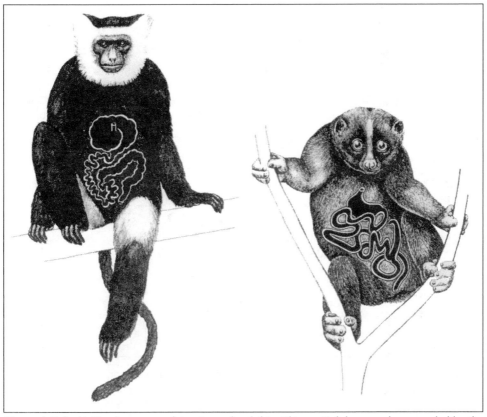

Figure 1-5 (Left) The large sacculate stomach of this African *Colobus* monkey, much like the rumen of a cow, aids in fermenting a leafy diet. Food passes slowly through the long convoluted intestine as microbes assist in the release and transformation of nutrients. (Right) In contrast, the Angwantibo *(Arctocebus calabarensis)*, a nocturnal resident of the same African forests, is largely insectivorous and possesses a much simpler digestive system that passes food relatively quickly. [Adapted from *Diversity and the Rainforest* by John Terborgh. Scientific American Library, New York. 1992.]

the trophic system. The example used in Figure 1-7 is measured in kilogram calories per square meter per year, and represents an oceanic ecosystem. Studies done on terrestrial trophic systems indicate that energy loss at each level in these systems is of a similar order of magnitude. Some ecologists believe that it is reasonable to assume that on average 10-15% of the energy at one given trophic level is available to the next higher trophic level (Slobodkin 1960, Pianka 1994).

Although the above example is expressed in units of energy, the Eltonian pyramid may be expressed in biomass, in numbers of animals, or in density of animals. The biomass of plant material will be greater than that of herbivores, and that of herbivores greater than that of carnivores. Usually the number of animals and the density of each type also will be smaller as one progresses through the system. The density of carnivores is always much lower than that of the animals they feed upon, and the areas they must cover in order to capture sufficient prey are very large (Jewell 1966). Because of this, as ecosystems are altered, carnivores are often the first to suffer. Today, many large carnivores and birds of prey are in danger of extinction. Such relationships are important in planning nature reserves:

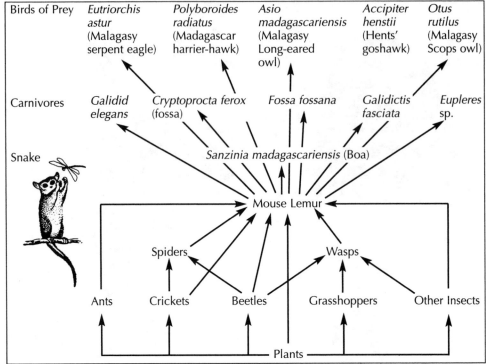

Figure 1-6 Food web of an eastern rainforest mouse lemur. Although the mouse lemur is the smallest primate and eats a large quantity of insects, it relies heavily on fruit. [Atsalis 1998.]

establishing one large reserve rather than several small ones may produce different results (Gilpin and Diamond 1980, Lovejoy 1986, Shafer 1990).

ECOLOGICAL INTERACTIONS

Now that I have provided an outline of the components of ecosystems, I shall describe three types of interactions involved among these components. (1) Those among abiotic components and producers; (2) those among organisms at different trophic levels; and (3) those among animals occupying the same trophic level. I also will provide a brief discussion of coevolutionary interactions.

This is not meant to be a comprehensive discussion of ecological interactions. Rather, I shall introduce some of the general dynamics of ecosystems and some of the ecological rules or laws that are pertinent to many primate field studies. I discuss interactions as they relate to primates living in their natural habitats. I hope this will help to illuminate some of the basic principles underlying many of the field studies that are presented in later chapters.

INTERACTIONS BETWEEN ABIOTIC COMPONENTS AND PRODUCERS

The structure of producers is related to major climatic factors and other abiotic conditions that exist where the plants are found. Thus, plant communities in different parts of the world are quite often similar in structure because they exist in areas

Table 1-1 Dietary terms—term indicates that the species either has a diet consisting mainly of the listed type of food or is morphologically adapted to exploit such food.

carnivorous/faunivorous	feeding on animals
exudativorous	feeding on plant exudates-i.e. saps, resins and gums (see Bearder and Martin 1980).
folivorous	feeding on foliage, or leaves,of trees
frugivorous	feeding on fruit
graminivorous	feeding on grasses
granivorous	feeding on grains
insectivorous	feeding on insects
herbivorous	feeding on plants
nectarivorous	feeding on nectar
omnivorous	feeding on plants and animals

having similar climatic and lithospheric (ground) conditions. There are three major aspects of the lithosphere that affect the type of plant community found in a particular area: (1) the chemical nature and physical texture of the soil (e.g., sandy, rocky, calcareous); (2) the topography (e.g., flat, hilly, mountainous); and (3) the altitude.

Major factors of the climate that affect plant communities are temperature, moisture, wind, and sunlight. The temperature on the earth varies from below freezing to 60° Celsius (140° Fahrenheit) on some desert areas. In many areas temperatures fluctuate widely throughout the day and seasonally, whereas in other areas e.g., the tropics, there are only slight variations. The amount of precipitation varies from more than 500 inches per year in a few tropical regions to only traces in some deserts. Seventy-three percent of the earth's surface is water, which is the major component of the atmosphere and hydrosphere, and of living organisms. Water is also the exclusive source of hydrogen for the biotic

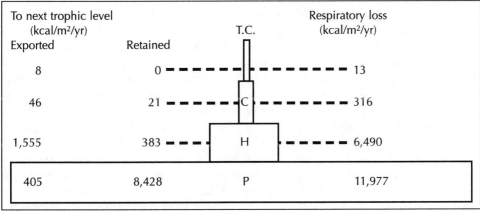

Figure 1-7 Pyramid of energy for Silver Springs, Florida. P = plants, H = herbivores, C = carnivores, TC = top carnivores. [After Odum, 1971.]

world. Wind and sunlight affect both the temperature and the moisture content of the air and ground.

Producers compete for these resources and those better fit or *adapted* to obtain resources are more successful. These organisms are able to pass on their adaptations to the next generation. This is the process of *natural selection.* Because there are a limited number of possible ways to obtain similar resources, the structure of a producer is related directly to climatic and microclimatic factors and to the interactions between other producers competing for the same resources (Horn 1974, Halle et al. 1978, Terborgh 1992, Losos et al. 1998). The same is true for consumers. Thus, animals occupying the same trophic level in similar types of plant communities in various parts of the world often are similar in structure and behavior. This is because they feed on, and compete with other animals that feed on plants of similar structure and/or they are chased or chase similarly adapted animals. Thus, *evolutionary convergence*—the theory that life forms in similar environments independently evolve similar adaptations—is ultimately the result of interactions between abiotic factors and producers.

However, within each, we must always be aware that the phylogenetic history of the particular species involved will effect the resulting ecological community (Ricklefs and Schluter 1993, Losos 1996, 1998).

BIOMES AND THE PROCESS OF SUCCESSION

Because climatic and lithospheric conditions in any area are fairly constant, the type of *plant community* that will develop in a particular place is broadly predictable (Pianka 1994). Furthermore, the various stages this community will go through before it reaches its final or climatic condition are also somewhat predictable, passing from pioneer stages to the climax stage. This predictable series of changes is referred to as the *process of succession,* and is defined as the orderly change in the environment from one community to another until a stable equilibrium is obtained (Clements 1916). This process is illustrated in Figure 1-8. Recently, researchers have suggested that there is a continuity of different climaxes in any region varying along environmental gradients and controlled by such factors as soil moisture, nutrients, herbivores, fires, etc (Krebs 1994).

During each stage of a succession most organisms (both plants and animals), if not all, are different. Cyclic changes, dictated by life cycles of the dominant organisms, are repeated over and over again as part of the internal dynamics of the community. With age plants decline in vigor and treefall gaps can create a mosaic of patches undergoing cyclic changes within a climax community. If mortality, for any reason, increases tremendously, the community returns to an earlier stage in the succession. Because of short-term climatic changes and cyclic changes of growth and decay, particular communities are not stable for long periods in nature (Krebs 1994).

Climatically delineated assemblages of organisms that have characteristic appearance, referred to as biomes, are distributed over wide areas of land. The community of plants and animals found in any area depends upon the particular combination of abiotic features mentioned earlier. Although biomes have been classified in several ways, I will group them into the following broad categories (following Raven and Johnson 1991): deserts, savannahs, tropical rain forests, temperate grasslands, temperate deciduous forests, taiga, and tundra (see Figure 1-9). The first of these distinct habitat zones are found mainly in the tropics (near the equator, +/- 20 degrees latitude). They

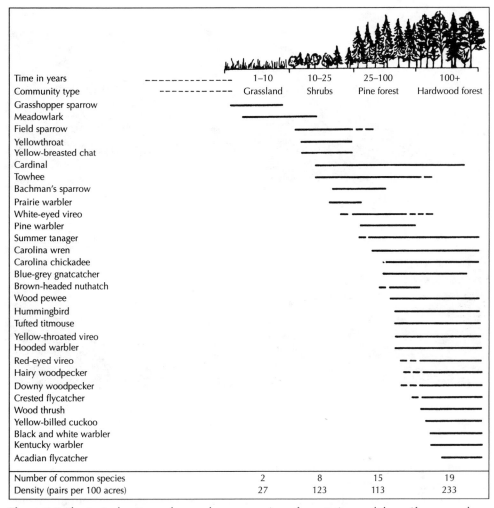

Time in years	1–10	10–25	25–100	100+
Community type	Grassland	Shrubs	Pine forest	Hardwood forest
Grasshopper sparrow				
Meadowlark				
Field sparrow				
Yellowthroat				
Yellow-breasted chat				
Cardinal				
Towhee				
Bachman's sparrow				
Prairie warbler				
White-eyed vireo				
Pine warbler				
Summer tanager				
Carolina wren				
Carolina chickadee				
Blue-grey gnatcatcher				
Brown-headed nuthatch				
Wood pewee				
Hummingbird				
Tufted titmouse				
Yellow-throated vireo				
Hooded warbler				
Red-eyed vireo				
Hairy woodpecker				
Downy woodpecker				
Crested flycatcher				
Wood thrush				
Yellow-billed cuckoo				
Black and white warbler				
Kentucky warbler				
Acadian flycatcher				
Number of common species	2	8	15	19
Density (pairs per 100 acres)	27	123	113	233

Figure 1.8 The typical pattern of secondary succession of vegetation and the avifauna on abandoned farmland in the southeastern United States. Number of bird species increases markedly with increased vertical structural complexity of the vegetation. [Odum 1971]

are characterized by high annual temperatures and daily temperature variation which is greater than seasonal variation. As can be seen in Table 1-2, the nature of the three tropical habitats is determined mainly by the amount of rainfall. Table 1-2 also summarizes the distribution of non-human primates in the biomes of the earth.

SOME FEATURES OF A CLIMAX COMMUNITY

The structure of a plant community is determined by competition for particular resources. Because of this, plants found in climax communities often share structural features, and different communities have particular features (see Horn 1971). *Vertical stratification* is characteristic of some communities. For example, many tropical forests in different parts of the world have similar stratification (Richards 1966, Whitmore 1984, Gentry 1990, Terborgh 1992). Figure 1-10 illustrates the plant levels of some tropical forests. The microclimate at each stratum is different. More

Desert

Deciduous Forest

Savannah

Rain Forest

Figure 1-9 Examples of biomes (all photos taken in Madagascar). [Photos by Robert W. Sussman]

sunlight reaches the upper layers than the ground and this affects both the temperature and humidity at these levels.

The structure of a plant stratum determines its utilization by animal species. Stratification is also present in many animal populations. For example, Dowdy (1947) collected 240 species of insects in an oak-hickory forest with five major strata. One hundred eighty-one of these species (78%) used one level, 32 used two levels, 19 used 3 levels, and only 3–5 used 4 or 5 levels. In the Caribbean, Anolis lizard ecomorphs also are distributed by plant strata (Williams 1983). Many nonhuman primates are restricted in their use of the forest strata, as I discuss later.

The various strata found in a forest are interdependent, although they may have different microclimates and animal populations. Thus, they are considered as subdivisions of an ecosystem and not separate communities. Stratification increases the number of habitats in a given community, and reduces interspecific competition, enabling a larger number of species to utilize the total community (Richards 1966, Schoener 1988, Terborgh 1992, Pianka 1994). In Table 1-3, I give definitions of a number of botanical terms related to plant community structure.

LEIBIG'S "LAW" OF THE MINIMUM

The material (chemical, water, heat, energy, space) which is most limited in supply tends to be the factor limiting the size of the population. One might say an

Table 1-2 Characteristics of Some of the Major Terrestrial

Biome	Geographical Distribution	Amount of Rainfall/year	Annual Temperature	Temperature Variation	Characteristic Plants	Characteristic Animals
Desert	Tropical (± 20° latitude)	<10"	High	Daily variation > seasonal variation	Succulents and cactus—reduced surface area of leaves to minimize water loss.	Small, cold blooded (reptiles). Mammals and birds relatively rare or absent. A few primates live on fringes of desert areas of Africa (i.e. *Papio hamadras*) and in semi-deserts of Madagascar (i.e., *Lepilemur, Propithecus*).
Savanna and Grassland	Tropical (± 20° latitude)	10–40" intermittent, erratic rainfall	High	Daily variation > seasonal variation	Grasses and bamboos—particularly adapted to alternating precipitation and dryness.	More ground-living species than any other habitat. Many herbivores and carnivores. Some species of primates, especially *Papio Erythrocebus, Cerocipithecus aethiops, Macaca, Lemur catta, Galago senegalensis.*
Tropical Rain Forest	Tropical (±20° latitude)	Torrential rainfall practically daily (short dry season)	High	Daily variation > seasonal variation	Plant growth year round (perennial). Continuous canopy (sometimes 3 layers) cuts off practically all sunlight, much rainwater and a good deal of wind. Greater diversity of plant species than all other	Greatest diversity of species. Most species of primates.

Table 1-2 (continued)

Biome	Geographical Distribution	Amount of Rainfall/year	Annual Temperature	Temperature Variation	Characteristic Plants	Characteristic Animals
Deciduous Forest	Temperate Zone	30–40″	High in summer low in Winter	Daily variation < seasonal variation	Growing season discontinuous—large plants shed leaves and hibernate. Trees further apart and fewer species than tropical forest.	Many species of mammals and birds, both arboreal and terrestrial (e.g. deer, boars, raccoons, foxes, squirrels, woodpeckers are typical.) Many species of primates, including the most cold-adapted species (e.g, *Macaca mulata* of the Himalayas, the snow monkeys of Japan *Macaca fuscata*, and *Rhinopithecus roxellanae* of China).
Taiga	40°–50° N. latitude		Low—long severe winters few months of summer. Persistent winter snow cover.	Daily variation < seasonal variation	Hardy conifers—often a single species of tree. Perennial herbs are common.	Moose, bear, wolves are characteristic species. No non-human primates.
Tundra	Arctic Circle (1/5 of earth's land surface)	< 10″	Freezing, except for 1–2 months/year. Ground is frozen.	Daily variation < seasonal variation Winter—continuous night.	Perennial plants low to the ground. Almost no annuals and few woody plants. Large woody plants are absent.	Caribou, arctic hares, lemmings, foxes, musk oxen, polar bears, migratory birds. No non-human primates.

Figure 1-10 Profile diagram of primary mixed forest, Moraballi Creek, Guyana. [From Richards 1966.]

organism is no stronger than the weakest link in its ecological chain of requirements. For example, during periodic droughts that last one or two years out of every five to six, food supplies may run low. This may cause densities of animal populations also to remain low, even if the food supply is plentiful for the other four to five years. In fact, Birdsell (1953) found a direct correlation in different areas of Australia between amount of rainfall and density of human hunting and gathering populations. In this case water would be considered the limiting factor in a broad sense. However, the density of the population was probably limited more directly by the availability of certain nutrients obtained from the plants and animals eaten, the density of such resources being directly dependent upon the amount of rainfall. Thus, the biomass of plants and of the animals that eat them, including the human population, and the amount of available water are all interrelated. Ultimately, in desert communities such as that of central Australia, the amount of water is the limiting factor.

Production of fruit and other plant resources in rain forests also is often unpredictable and irregular. Alternating seasons of scarcity and abundance might be the key to understanding much about the ecology of animal inhabitants of tropical forests (Terborgh 1986). For example, in a number of rain forests, fruit production is far in excess of consumption by vertebrates for 8–9 months of the year. For the remaining 3–4 months consumption is precariously close to production. For the higher vertebrate community, including primates, frugivory is the overriding prevalent way of life in tropical rain forests.

Table 1-3 Some Botanical Terms

Dominant	The species of plant that has the greatest total biomass in a plant community or that intercepts the greatest amount of sunlight (i.e., provides the greatest amount of cover in a vertical projection to the ground).
Co-dominant	When more than one species are dominant in a plant community, those species are co-dominant.
Consociation	Vegetation units characterized mainly by dominance in the component layers. Thus, a *Tamarindus indica* consociation would be one dominated and characterized by the tamarind tree.
Association	Vegetation units characterized mainly by two or more co-dominant species. A *Picea glauca-Larix laricina* association would be one characterized by white spruce and tamarack trees.
Stand*	An area in which one species or type of plant is more densely distributed within a larger area dominated by different species. Thus, there can be a stand of *Ficus* within a *Tamarindus* consociation.
Synusae	A group of plants of similar life form, filling the same niche and playing a similar role, in the community of which it forms a part.
Ecotone	The border between two communities—usually contains species of both plus some unique species.
Edge effect	The tendency for an increase in variety and density at community junction points.
Emergent	Isolated trees whose crowns rise above the general canopy. These trees may be considered collectively as forming an "extra" (discontinuous) layer above the general tree canopy.

*Also referred to as a "society."

(see Richards 1966 for more detailed descriptions.)

In a study of five species of primates in Peru, Terborgh (1983, 1986) found that certain plant resources play a prominent role in sustaining frugivores through these periods of general food scarcity. Palm nuts, figs, and nectar were found to set the carrying capacity of the community of frugivores although they represented less than one percent of the local plant diversity. These plants, therefore, were limiting resources and, because of their key role in sustaining frugivores, are referred to as *keystone plant resources* (see also Sauther 1998).

Limiting factors are not always food resources. Animal and plant species have different space requirements. The density of a species in a given area is related to a number of factors. Clearly, there must be enough food and water, but in addition each species has spatial requirements and must maintain adequate social interactions. Even if the social organization of a species varies in different environmental conditions, the amount of variation possible for each species does have limitations. Early studies of rodents (Elton 1942, Calhoun 1952), tree shrews (Von Holst 1974) and primates (Rudran 1973) illustrated that lack of adequate space can be a limiting factor causing physiological and behavioral changes that keep birth rates low and rates of infant mortality high. Disease and predation also are important factors limiting the density of populations.

The population of a species can be maintained at or near a steady state in a given area due to "internal" spacing mechanisms, e.g., species specific patterns of social spacing and of birth spacing. For example, birth spacing is usually physiologically determined and ultimately can be related to the availability or lack of nutrients. An example of these interrelationships can be found in the history of the human population.

For the first million or two million years of its existence, the human population remained relatively stable in numbers. About 9–10,000 years ago, the growth rate increased tremendously (Cohen 1989, Landers 1992). This increase correlates with the advent of agriculture in some human groups. Many scientists believe that this pattern of subsistence changed the carrying capacity in some areas and this led to a direct increase in the size of the human population. Another explanation, however, is that population increase actually brought about a need to increase carrying capacity (Carneiro 1968, Spooner 1972, Cohen 1977). Before 10,000 years ago, most human groups were gatherers and hunters who were constantly moving because their resources were widely and patchily distributed. In modern gathering populations, women constantly carry their infants while moving, and only one infant can be managed at a time (Figure 1-11). Birth spacing averages about 4 years (Sussman 1972, 1981, Lee 1972, 1979). When a group becomes sedentary, for whatever reason, a woman does not have to constantly carry her offspring and birth spacing *can* be decreased (Sussman 1972, 1981, Dumond 1979, Landers 1992).

Thus resource distribution, population density, population growth, subsistence techniques, individual behavior, social behavior, and reproductive biology are all interrelated and the specific factors limiting population growth may be quite complex and numerous.

Factors limiting population size have not been extensively studied in nonhuman primates. Some interesting questions related to the "law of the minimum" are as follows: Is it possible to relate the population density of some savannah or ecotone-living monkeys to rainfall? Can the biomass of specific resources extensively or periodically utilized by primates be related to the density of these populations in parts of their range? Can the geographical distribution of certain species be shown to be related to the presence or absence of specific limiting or keystone resources? Some of these questions are considered in the following chapters.

Leibig's "law" of the minimum is related to Shelford's law of tolerance (Shelford 1913). Not only can a minimum of some factor limit the size of a population but too much of the same materials (e.g., heat, light, water) also can be limiting. Species have ecological minima and maxima for certain conditions and these are called the limits of tolerance of the species, though determining tolerance zones for any organism is often difficult (Krebs 1994, Pianka

Figure 1-11 San woman. [From *!The Kung San: Men, Women, and Work in a Foraging Society* by Richard B. Lee. Cambridge University Press, Cambridge. 1977.]

1994). Animals with narrow limits of tolerance may be forms that are specialized and have greater efficiency in specific conditions at the expense of adaptability. The populations of these forms can become very abundant when conditions are favorable and stable (Odum 1971).

INTERACTIONS BETWEEN POPULATIONS OF CONSUMERS

Six possible ways two species affect each other and eight possible types of interaction are listed in Table 1-4. In this section, I concentrate on interactions between species at different trophic levels, e.g., predator-prey interactions, and those between species at the same trophic level, e.g., coexisting or competing species. I also briefly discuss coevolution.

INTERACTIONS BETWEEN ANIMALS OCCUPYING DIFFERENT TROPHIC LEVELS: PREDATOR-PREY INTERACTIONS

In a stable ecological system predator and prey populations are usually in equilibrium (Krebs 1994). In actuality, they often are mutually dependent. If a predator over kills its prey this could lead to the extinction of both populations. In some cases, a predator may limit the population of its prey. For example, when fences are used to prevent the dingo, a dog-like predator, from entering sheep country in Australia, the abundance of red kangaroos and of emus, large flightless birds, increases tremendously (Caughley et al. 1980). However, predators normally are neither the only nor the most important factor limiting the population of the prey although availability of prey in many systems limits the predator population. In the Serengeti Plains of eastern Africa, there are a number of large ungulates and their predators—lions, leopards, cheetahs, wild dogs, and spotted

Table 1-4 Population Interaction—
Six Basic Ways Two Species Can Interact (Eight Actual Types of Interaction)

Species			Nature of the Interaction
1	2		
0	0	(1)	Neutralism—neither population is affected by the interaction
-	-	(2)	Competition—each population adversely affects the other in struggle for food, nutrients, space, etc.
+	+	(3)	Mutualism—growth and survival of both populations are benefitted and neither can survive without the other
+	+	(4)	Protocooperation—both populations benefit but the relationship is not obligatory
+	0	(5)	Commensalism—one population benefits, the other is not affected
-	0	(6)	Amensalism—one population is inhibited and the other is not affected
+	-	(7)	Parasitism—one population adversely affects the other by direct attacks
+	-	(8)	Predation—one population, the predator, kills and consumes the other population, the prey

[Adapted from Odum 1971]

hyenas. Predation may be an important factor limiting the population of zebra and Thompson's gazelle, but it has little impact on the populations of wildebeest and buffalo (Schaller 1972, Bertram 1978). The latter two ungulate populations may have reached the carrying capacity of the Serengeti, and disease and malnutrition affect the size of the population more than predation.

Klopfer (1962) identified the following behavioral mechanisms that protect an organism from excessive predation: (1) *The prey exploits the limits of the predator's motor or sensory capacity.* Speed, climbing ability, agility, camouflage, dispersal or cryptic behavior are all mechanisms by which prey escape their predators. Most primates are arboreal (living mainly in the trees), quick, and agile; they use these traits to minimize predation. It is not surprising that the most terrestrially adapted primate, the patas monkey, is the fastest extant primate, clocked at 55 km/hr over long distances (Hall 1965). Lorises, on the other hand, are exceedingly slow primates, moving much like chameleons (Figure 1-12). These nocturnal primates move quietly through the dense foliage and rarely vocalize, thus avoiding excessive predation by their cryptic behavior. Most primates are diurnal and live in social groups. When approached by a predator, a group of forest living primates often disperses and individuals are difficult to locate. I have seen groups of 18–24 ringtailed lemurs, and an entire group of over 80 long-tailed macaques, disappear into a thicket in less than a minute after being frightened by a potential predator.

(2) *The prey exploits the learning abilities of the predator.* In this case, the predator is disinclined to take a potential prey species because it *looks like* a species that the predator has learned is toxic, unpalatable or otherwise inedible or unapproachable. Often this involves mimicry: the edible species looks like, or mimics, the toxic or poisonous model. Many insects and snakes mimic distasteful or deadly species.

Another device that might fall into this category is the use of displays. Many primates, when faced with a predator, will enter into a display, or the group will mob the predator. Mobbing may involve loud vocalizing, vigorous jumping, tail-waving, displaying canines, or even branch breaking and throwing. Although the displaying animals are generally weaker than the predator and would most likely flee before attack, the predator often will look for prey that appear less threatening.

(3) *Finally, prey may protect themselves from predation by direct attack.* Horns, scales, vertebral spines, large canine teeth, and group defense are used by various species to ward off predation. Schaller (1972) found that buffalo and rhinoceros readily attack lions when their young are threatened, and that wildebeest and zebra sometimes defend their offspring

Figure 1-12 *Loris* locomotion. [Photo by Kimberly Nekaris]

against wild dogs and hyenas. He states that "if the prey outweighs a predator by a ratio of at least 3:1, then it may feel secure enough to attack" (Schaller 1972:38).

Primates are generally smaller than their predators and often use group defense mechanisms for protection. The senses of sight, hearing and smell are all well developed among primates, and living in groups adds eyes, ears and noses to better detect predators. As already stated, mobbing by the group will often discourage a predator. Under certain conditions, the presence of many large males with long canines is a direct deterrent to predation and, under provocation, these males do attack.

A number of investigators have observed direct attacks on predators by baboons. For example, the Altmanns observed the following reaction to a predator: "The baboons sprang away, then turned on the leopard, barking loudly as several members of the group ran at the leopard . . . at one moment, the dominant male was closest to the leopard. Faced with this mass attack, the leopard turned and ran" (Altmann and Altmann 1970:176). The Altmanns also observed the baboons threaten jackals, a cheetah and a tawny eagle.

Many species, especially those under the most severe predator pressures, use a combination of these three methods of predator avoidance. The patas monkey and the slow-moving loris are probably among the most vulnerable species of primates. Both of these taxa use a combination of defense mechanisms. The male patas monkey displays before a predator, trying to gain its attention, while the females and younger animals disperse and hide, their coloration helping them blend into the tall grass. When the male has the full attention of the predator, he uses his tremendous speed to divert the predator from the group and then to outrun it. As a last resort, however, the male patas will attack a predator. "In one case, a patas male chased a jackal which carried an infant patas in its fangs. The jackal soon dropped its prey, and the infant, apparently unharmed, was retrieved by a patas female" (Kummer 1971:55).

The slow moving locomotion of the lorises also necessitates a very specialized and complex defensive behavior. Besides concealment and other cryptic behaviors, lorises have specific defense postures and methods of attack. The potto faces the predator with its head beneath its arms and thrusts its body forward, using a scapular shield (Figure 1-13) to butt the attacker (Charles-Dominique 1977). There also is evidence that slow and pygmy lorises exude body chemicals that may be toxic to predators (Alterman 1995).

INTERACTIONS BETWEEN ANIMALS AT THE SAME TROPHIC LEVEL: COMPETITION AND COEXISTENCE

In 1934, Gause performed an experiment in which he placed populations of two closely related protozoans, *Paramecium aurelia* and *P. caudatum,* into culture mediams that were maintained under identical conditions. The populations of each species exhibited normal growth and reached equilibrium. When both species were placed in the same culture, however, only *P. aurelia* survived after sixteen days. Neither organism attacked the other or secreted harmful substances. *P. aurelia* simply had a higher intrinsic rate of increase (reproduced faster) and consumed more of the limited supply of food. Thus, *P. aurelia* had a competitive advantage over *P. caudatum*. In these experiments, Gause illustrated that two species cannot coexist in the same environment if they compete for the same resources if these resources are in limited supply. Since these classic experiments were performed, a number of laboratory studies have confirmed Gause's observation, and the above theorem has been referred to as Gause's law or the principle of competitive exclusion.

In further experiments Gause placed *P. caudatum* with a third species, *P. bursaria,* in a similar medium. Although they utilized the same food source, *P. bursaria* occupied a different part of the culture where it could feed without competing with *P. caudatum.* As a result of this, Gause suggested that "the intensity of competition is determined not by the systematic likeness, but by the similarity of the demands of the competitors upon the environment" (Gause 1934:19). In fact, in their natural environments, many closely related species coexist in the same environment and do not compete. For exam-

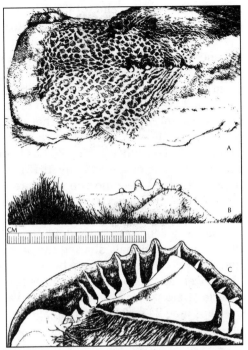

Figure 1-13 Scapular shield of Potto. [From Charles-Dominique 1977]

ple, in many areas of Africa, five or more monkey species of the genus *Cercopithecus* are found to coexist in the same tropical forest (Gautier-Hion and Gautier 1988).

The principle of competitive exclusion has often been stated as follows: no two closely related species can coexist in the same environment if they share a common *niche.* However, this is actually a misinterpretation of the results of Gause's original experiments and leads to a number of problems concerning the processes of competition. In the first place, as Gause illustrated, competition between two species is not determined by their taxonomic relationship. Second, the concept of niche is not easily defined nor is it necessary or useful in a discussion of competition.

The concept of niche is extremely important, especially to community ecology. Yet, it is a concept that has been defined in a number of ways. Pianka (1994:268) states:

> *The concept of niche pervades all of ecology; indeed, were it not for the fact that the ecological niche has been used in so many different ways, ecology might be defined as the study of niches (p. 268).*

Niche has been defined behaviorally as what animals do, functional or behavioral niche, and spatially as where they do it, place niche or habitat niche. Perhaps the most widely adopted use is that of Hutchinson (1957, 1965). His definition is based on the idea that there are many resources available in any area and that all those resources relevant to the life of the organism make up the organism's niche. In this sense, the niche is a set of measurable variables; it is that portion of the resources available in the environment that are actually, realized niche, or potentially, fundamental niche, utilized by the organism. It is possible to conceive of unfilled or "empty" niches, using this definition.

The patterns of resource allocation among species and patterns of their spatial and temporal abundance constitute the structure of the community. Convergence in community structure between similar habitats or resources in different geographical areas leads to similarities in niche utilization by different species (Cody and Diamond 1975). A group of species that exploits similar resources in a similar way or, in other words, that overlaps significantly in their niche requirements, regardless of taxonomic relationship or position is referred to as a *guild* (Root 1967). The concept of guild has been utilized a great deal in the study of bird species (see for example Cody 1975). Terborgh and Robinson (1986:90) believe that "guilds will become the standard currency of ecologists in their effort to understand community relationships of many kinds." Guilds have not been studied extensively in primates (see Sussman 1979).

Having given these basic definitions, I now return to the description of competition between animals sharing the same trophic level. But first I pose two questions. What is the relationship between competition and niche overlap? How much niche overlap can coexisting species tolerate? In a study of eight coexisting South American monkeys, Mittermeier and Van Roosmalen (1981) found dietary niche overlap to be 24 to 97%, with an average of 69%, between the twenty-eight possible species pairs. Pianka states:

> . . . in nature, niches often do overlap yet competitive exclusion does not take place. Niche overlap in itself obviously need not necessitate competition. . . . In fact, extensive niche overlap may often be correlated with **reduced** competition (1994:274).

Some species can coexist and others do not, and the extent to which two species compete does not necessarily depend on their degree of relationship nor on niche overlap *per se*. Under what conditions does competition occur and how can it be reduced or avoided? Mayr (1963:67) defines competition as follows:

> No unanimity has yet been achieved concerning the precise definition of competition, yet it always means that two species seek simultaneously an essential resource of the environment (such as food, or a place to live, to hide, or to breed) that is in limited supply. Consequently competition becomes more acute as the population of either species increases. Any factor the effect of which becomes more severe as the density of the population increases (and this may be true for all causes of mortality and fecundity) is called a density-dependent (=controlling) factor.

Thus, using Mayr's definition and the experiments of Gause, we can say that two species are in competition when they have a density-dependent or controlling factor in common and when the supply of this factor is limited. Given this definition, two or more species can coexist if they exploit resources that are not in limited supply or if they differentially exploit them, and thereby reduce the risk that specific resources become rare.

There are species that are able to coexist and share the same resources because their densities are low and fixed by intraspecific mechanisms of control, e.g. territoriality, by a high rate of predation or disease, or by extreme climate. However, the most frequently cited examples of coexistence are those in which two or more species persist together in the same area and are adapted to exploit different resources, or have different habitat preferences. This is often referred to as *resource partitioning* (Schoener 1986). Resource partitioning may be the result of divergence caused directly by interaction between sympatric populations, or the result of long-term, genetically fixed adaptations to different environmental conditions. Migration of species into and out of a community also can be a factor mitigating competition (Schoener 1988), as can predation (Losos pers. comm.). If predators feed on the best competitors, they can reverse the outcome of competition.

Differences in the ways in which species use the environment are usually related to differences in morphology and physiology. The study of resource partitioning among interacting species may give us indications as to the nature of the forces that produce differentiation among species. Differences in behavior and ecology offer clues to the significance of subtle morphological variations. Morphological differences often can be related to specific behavioral and ecological adaptations.

In any case, a correspondence exists among morphology, behavior, and ecology, and hypotheses concerning the existence of particular adaptations, and mechanisms involved in the evolution of species can be formulated and tested by the study of habitat selection in coexisting species. These studies also give us an overview of the structure of communities within an ecological system. A number of studies of resource partitioning among coexisting primate species will be described in the following chapters.

COEVOLUTION

Coevolution refers to the long-term, mutual evolution of two or more taxa that have close ecological relationships, and in which reciprocal selective pressures operate to make the evolution of each taxon partially dependent upon that of the other; yet no genes are exchanged (Futuyma and Slatkin 1983, Raven and Johnson 1991, Krebs 1994, Pianka 1994). The term was first used by Ehrlich and Raven (1964) in a more restricted sense "to refer primarily to the interdependent evolutionary interactions between plants and animals, mainly herbivores" (Pianka 1994:329).

Since plants are immobile they must develop clever means of protecting themselves from predators. Cross-breeding plants usually entice animals to aid in reproduction and in dispersal of fertilized seeds. Many plants have developed morphological (e.g., thorns, corky bark, hard seed coats) or behavioral (e.g., timing of leaf growth, sensitive leaves) defenses against predation. Chemical defenses (secondary compounds) are of paramount importance and are well developed in plants (Janzen 1978). The methods used by animals to detoxify these secondary compounds and strategies of folivory by herbivores have been the subject of numerous studies (e.g., Freeland and Janzen 1974, Montgomery 1978, Stamp 1992). There also have been numerous studies of the adaptive strategies of primates in relation to leaf eating (e.g., Hladik 1969, Oates et al. 1977, Waterman 1984, Glander et al. 1989, Davies and Oates 1994).

The use of animals as agents for pollination by plants has been studied extensively (Faegri and van der Pijl 1971; Heinrich and Raven 1972; Gilbert and Raven 1975, Regal 1977, Fleming 1988). The most frequent pollinators are flying animals such as insects, birds and bats. Different plants have developed characteristics specifically adapted to attract particular pollinators, and bird pollinated plants usually are different from bat pollinated ones. In turn, bat and bird pollinators have coevolved traits that facilitate feeding on floral nectar (see Table 1-5). In

Table 1-5 The Syndrome of Ornithophily (Bird Pollination)

Bird Flowers	Flower Birds
1. Diurnal anthesis	Diurnal
2. Vivid colors, often scarlet or with contrasting parrot-colors	Visual with sensitivity for red, not for u.v.
3. Lip or margin absent or curved back, flower tubate and/or hanging, zygomorphy unnecessary	Too large to alight on the flower itself
4. Hard flower wall, filaments still or united, stiped or otherwise; protected ovary, nectar stowed away	Hard bill
5. Absence of odor	Scarcely any sense of smell
6. Nectar abundant	Large—and great consumers
7. Capillary system bringing nectar up or preventing its flowing out	
8. Possibly deep tube or spur, wider than in butterfly flowers	Long bill and tongue
9. Distance nectar—sexual sphere may be large	Large, long bill; large body
10. Nectar-guide absent or plain	Intelligent in finding an entrance

The Syndrome of Chiropterophily (Bat Pollination)

Bat Flower	Flower Bat
1. Nocturnal anthesis, mostly only one night	Nocturnal
2. Sometimes whitish or creamy	Good eyes, probably for near orientation
3. Often drab color, greenish or purplish, rarely pink	Color-blind
4. Strong odor at night	Good sense of smell for far orientation
5. Stale smell reminiscent of fermentation	Glands with stale odor as attraction
6. Large mouthed and-strong single flowers, often strong (brush) inflorescences of small flowers	Large animals, clinging with thumb claws
7. Exceedingly large quantity of nectar	Large, with strong metabolism
8. Large quantity of pollen, large or many anthers	Pollen as sole source of protein
9. Peculiar position outside the foliage, flagelliflory, cauliflory	Sonar system less developed, flying inside foliage difficult

[Adapted from Faegri and van der Pijl 1971]

these interrelationships, nectar provides important nutrients for the animal, and reproduction is enhanced for the plant.

I have mentioned the fact that a predator and its prey are, at least in one sense, mutually dependent and that species at the same trophic level can coexist by dividing up available resources. Species also may aid each other by exchanging and reacting to one another's warning calls when a predator is sighted. In a general sense, all of these interactions involve coevolution. When several species are involved in reciprocal evolutionary interactions, the term *diffuse coevolution* is often used (Herrera 1981). Thus, coevolution shapes the characteristics of coevolving pairs of species, whereas diffuse coevolution might occur in communities of many species and may be an important factor in determining community structure (Johnson and Raven 1991, Krebs 1994).

The importance of coevolution as a dynamic process affecting present interactions between primates and the plants that they utilize, and the possible effects of this process on the evolution of primate communities recently have been recognized (e.g., Sussman and Raven 1978, Sussman 1979, Garber 1986, Howe 1990, Fleming and Estrada 1993, Kress 1993). The study of this process offers a promising field of research in primate ecology.

SOCIAL GROUPS, SOCIAL STRUCTURE AND SOCIAL ORGANIZATION

The distribution of individuals within a species is related to such factors as the distribution and abundance of resources, predation pressure, and genetically determined behavioral propensities. In some species, individuals spend most of their time alone, whereas in others most or all of their time is spent in groups. With one exception, all *diurnal* primates live in social groups. In many animal taxa temporary aggregations form under certain conditions (e.g., schools of fish, flocks of birds, herds of ungulates) but in most of these cases, aggregations are made up of different individuals each time the animals congregate. One of the most common features of primate groups is the relative constancy of group membership. There are many theories attempting to explain why diurnal primates live in groups and the reasons for different types of groups (e.g., van Schaik 1983, Terborgh and Janson 1986, Wrangham 1987, Rowell 1991, Chan 1992, Runciman et al. 1996). A number of these theories will be explored in later chapters. Here I offer some definitions.

If one searches the primate literature for a definition of the term "social group" it is quite difficult to find. It seems that primatologists know a social group when they see one but are reluctant to define the term generally. I have been able to find only three attempts to formally define the primate "social group" (Altmann 1962, Kummer 1967, 1971 and Struhsaker 1969) and three critical evaluations of these definitions (Richard 1978, Dunbar 1988, and Fuentes 1999). Altmann (1962) defines a group as consisting of conspecific, intercommunicating individuals bounded by frontiers of far less frequent communication. Kummer (1967:378) considers a group as a "spatial aggregation of primates that travel and rest together and at the same time avoid the proximity of other such aggregations is, at closer inspection, also a functional and reproductive unit". These are both broad definitions that emphasize social proximity, social interaction and synchrony of activity. As we shall see, all of these criteria are not equally useful in defining the social group.

Reproduction was once thought to be a major *raison d'etre* of primate groups (Zuckerman 1932) but a number of factors indicate that it is not a major common denominator of the primate group. For example, in many species breeding is seasonal but the group is stable throughout the year. In fact, in some species, behavior during the breeding season actually disrupts the social group. In *Propithecus,* groups break up during the short mating season and reformulate after breeding takes place (Richard 1974). Most species of primates have mechanisms that ensure continuous gene flow between groups. Groups exist within a breeding population (or deme) and genetic exchange is common between groups, at higher or lower frequencies, depending on the species.

Struhsaker (1969) offers a more detailed definition of the social group. He includes in his definition the following factors: (1) membership in the same social network; (2) the majority (at least 80%) of non-aggressive social interactions take place within this social network; (3) membership is temporally stable; (4) group members occupy the same home range; (5) group members have distinct social roles; (6) progressions between members are synchronized; (7) group members exhibit different behavior toward non-group conspecifics; and (8) social groups are relatively closed to new members.

Using some, but not all of the above criteria, I define a "social group" in primates as follows: *A number of individuals who interact socially more frequently among themselves than with other individuals; group members exhibit different behavior towards non-group members and occupy the same home range.* Group membership is relatively stable over time and group members have distinct social roles but, although these two factors are characteristics of the group, the particulars of group stability and of social roles are extremely variable from species to species and are not necessary components of my definition.

It is also true that most social groups are relatively closed to new membership but this is also highly variable and would be difficult to include in a general definition. In fact, the only social groups that seem to be closed to new membership are monogamous pairs, and this may not be as true as was previously believed (Bartlett 1999); individual exchange between groups is a common occurrence in most diurnal primate species. Synchronized progressions, mentioned in the definitions of both Kummer and Struhsaker, are not common to all primate groups. A number of groups divide into subgroups for various periods of time and these subgroups do not necessarily synchronize movements nor activity while separated. Because of the dynamics of individual social relationships, Dunbar (1988) prefers to view primate groups or social systems in terms of their constituent relationships: "Where the sets of relationships of several individuals coincide and are reciprocated, a formal group emerges as a stable cohesive unit. . . . These clusterings are sufficiently distinctive in their cohesion in time and space that we can define them as levels of groupings" (p. 12).

There are many different types of social group found among the diurnal primates (Table 1-6). At a very general level these include: monogamous pairs (i.e., one adult male, one adult female and their immature offspring); multi-male groups (groups containing more than one adult male); one-male groups (groups containing only one adult male); and fission-fusion groups (groups which form dispersing and coalescing subgroups). Of course, within each of these superficial categories we are not really dealing with groups of identical social structure nor social organization. For example, multi-male groups of lemurs are quite different from those of baboons in structure, function, evolutionary history and ecological advantage. As we shall see, this is even more obvious when we compare species with "fission-fusion" group structures.

I have now introduced the terms *social structure* and *social organization*. These terms are often used interchangeably, however, it is useful to distinguish between them. I will follow definitions similar to those proposed by Rowell (1972, 1979). *The social structure of a species can best be divided into two components: group structure and population structure. Group structure is the size and composition (as to age and sex) of a primate group. Population structure is the demographic character of the population, population density, and the density and dispersal of groups within the breeding population or deme.* One measures social structure using censusing

TABLE 1-6 Nonhuman Primate Social Groups

1. *Solitary but Social*—found in most nocturnal prosimians *plus* orang utans. Characteristics: Some individuals may sleep together in nesting groups, but forage separately—home ranges of females overlap; those of adult males usually do not. Adult males avoid each other. Each adult male interacts with more than one adult female. Adult females form *nesting groups* consisting of females and young. Male offspring forced out of group at puberty.

2. *One-Male Group*—found in Black-and-White colobus monkeys *(Colobus guereza)*, Patas monkeys *(Erythrocebus patas)*, Gelada baboons *(Theropithecus gelada)*, Hamadryas baboons *(Papio hamadryas)*, and most members of the genus *Cercopithecus*. Types of one-male groups can vary slightly, but the basic elements are the same: One adult male interacts with more than one adult female and immature offspring. Male offspring are forced to leave the group at puberty.

3. *Fission-Fusion*—characterized by the existence of a larger group that breaks into subgroups and reforms again over a period of time. Three types:

    (a) Spider monkeys (genus *Ateles*)—fairly small group breaks up into smaller subunits for feeding.

    (b) Gelada baboons *(Theropithecus gelada)* and Hamadryas (Papio hamadryas)—In this case the subunits are one-male units that remain constant from day to day. Comosition of the larger group varies when different one male units come together.

    (c) Chimpanzees (genus *Pan*)—the large group, composed of up to 100 animals is called a regional population. Subgroups vary.

4. *Pair bonds* ("Family groups")—found in *Callicebus*, Hylobatidae, *Indri,* and other species that are characterized by exclusive use of space. One adult male and one adult female live together for life with their various *immature* offspring. Offspring of both sexes are forced out of the group at puberty.

5. *Cooperative polyandrous group* (communal breeding)—group composed of more than one unrelated male and female but only one female is reproductively active. Females, however, mate with more than one male and all animals (especially males) carry the young. Found only in the Callitrichidae.

6. *Multi-Male Group*—most common type of nonhuman primate social group—found widely throughout the order Primates, e.g., most Cebidae, savanna baboons (except those above), diurnal lemurs, langurs (subfamily Colobinae), gorillas. Size varies from approximately 10 to more than 100 animals. Group consists of more than one adult male, more than one adult female, and offspring, of all ages. Larger groups sometimes called "troops."

techniques. *Social organization is the pattern of social interactions which occurs between individuals within and between groups: it is a description of social behavior.*

Thus social structure refers to the components of the social system whereas social organization refers to the patterns of interaction between the components. Both patas monkeys and Hamadryas baboons live in one-male groups that can have essentially similar age and sexual composition. However, in Hamadryas one-male groups the male is the center of attention of the group and directs group travel. In the patas monkey, the male takes a peripheral position in the group and the females initiate and direct group travel. Thus the patas monkey and the Hamadryas baboon have groups with similar structure but the social organization of the two species is very different. Furthermore, differences in degrees of tolerance between adult males result in very different distribution patterns of one-male groups. Hamadryas one-male groups often sleep and move together, whereas patas males are intolerant of each other and one-male groups are usually widely dispersed. Although the group structure is similar, the structure of the populations of these two species is radically different.

When dealing with animals living in groups, we must consider two levels of social spacing: between groups of the same species and between individuals living in the same group. Each species has developed complex communication patterns for both intergroup and intragroup spacing. These patterns are part of the social organization of a species and they help to maintain the species-specific social structure. In group-living animals, such as diurnal primates, the terms home range, core area, area of exclusive use and territory relate to intergroup spacing, whereas dominance hierarchies, roles, kinship lines and peer groups usually are pertinent to spacing patterns within a group.

In the above definition of a "social group," I stated that group members occupy the same home range. *Home range* can be defined as "the area over which a group normally travels in pursuit of its routine activities." Other terms such as *day range* and *monthly range* are used to denote the area over which the group travels in a given time period. If a group uses a particular part of its home range more frequently, or with greater regularity, than other parts, this zone is referred to as a *core area* (see Burt 1943, Kaufman 1962 and Jewell 1966 for a discussion of these terms). The core area often contains many of the major resources of the group. In some species of primates, the boundaries of different groups overlap only slightly, whereas in others home ranges overlap a great deal and each group may have a very small area that is not shared by other conspecific groups (referred to as an *area of exclusive use*). In a number of primate species, home ranges of groups may overlap completely.

Mechanisms used for intergroup spacing also vary between species. In some species, spacing between groups involves the maintenance of exclusive use of a particular area but, in others, it seems to involve the maintenance of intragroup integrity and a patterned use, by two or more groups, of the same space. The term *territory* was first developed in studies of birds (Noble 1939). The essence of the concept, namely "any defended area," was then accepted by Burt (1943, 1949) to apply to mammals. However, a unitary definition of territoriality and a simple dichotomy between territorial and nonterritorial species might do more to confuse the issue than to aid in an understanding of ecological parameters of spatial relationships between individuals and groups of conspecifics (see Waser and Wiley 1980). For example, active, ritualized, agonistic defense behaviors are used by many species of primates to maintain the integrity of home range boundaries or areas of exclusive use. However, in other species, spatial integrity is maintained between groups by exchange of vocalizations or with no apparent outward behavioral spacing mechanisms. In fact, very few species of primates could be considered territorial in the strict sense of the term.

A large number of species which do not maintain exclusive areas employ ritualized postures and gestures, and/or vocalizations, to maintain group integrity and to regulate spatial relationships within a shared home range. Some of these species were thought to be highly territorial until recent studies have shown this not to be the case. Thus, we are dealing with two separate phenomena: (1) the behavioral mechanisms that control spacing and (2) the degree of isolation or overlap between groups. The use of the term territoriality often confounds these two factors and in this book it will only refer to *the active defense of individual or group home range boundaries by actual or ritualized agonistic encounters, thereby maintaining essentially exclusive use of the home range.* This, then, describes a very specific set of behaviors that control spacing and a particular pattern of spacing. However, even using this strict definition may conceal important variation in conditions of spacing between two species labeled as territorial.

Besides the above, there are a number of terms and concepts currently in vogue in socioecology and in sociobiological theory that will be discussed throughout the book.

BIBLIOGRAPHY

Alterman, L. 1995. Toxins and Toothcombs: Potential Allospecific Chemical Defenses in *Nycticeb* and *Perodicticus*. Pp.413-424 In *Creatures of the Dark: the Nocturnal Prosimians*. Alterman, L., Doyle, G. and Izard, K. New York, Aldine. Eds.,

Altmann, S.A. 1962. A Field Study of the Sociobiology of Rhesus Monkeys, *Macaca mulatta*. *Ann. N.Y. Acad. Sci.* 102:338–435.

Altmann, S.A. and Altmann, J. 1970. *Baboon Ecology*. Chicago, University of Chicago Press.

Bartlett, T.Q. 1999. *Socio-Ecology of the White-Handed Gibbon in Khao Yai National Park, Thailand*. Ph.D. Thesis. Washington University, St. Louis.

Bertram, B.C.R. 1978. Living in Groups: Predators and Prey. In *Behavioural Ecology*. Eds., Krebs, J.R. and Davies, N.B. Sunderland, Sinauer Associates.

Birdsell, J.B. 1953. Some Environmental and Cultural Factors Influencing the Structuring of Australian Aboriginal Populations. *Am. Natur.* 87:171–207.

Blumenshine, R.J. 1987. Characteristics of an Early Hominid Scavenging Niche. *Current Anthropol.* 28:383–407.

Burt, W.H. 1943. Territoriality and Home Range Concepts as Applied to Mammals. *J. Mammal.* 27:346–352.

Burt, W.H. 1949. Territoriality. *J. Mammal.* 30:25–27.

Calhoun, J.B. 1952. *The Social Aspects of Population Dynamics*. London, Oxford University Press.

Carneiro, R. 1968. The Transition From Hunting to Horticulture in the Amazon Basin. *Proc. VIII Congr. Anthropol. Ethnol. Sci.* Pp. 224–248.

Caughley, G., Grice, D., Barker R. and Brown, B. 1980. Does Dingo Predation Control the Densities of Kangaroos and Emus? *Australian Wildlife Research* 7:1–12.

Chan, L.K. 1992. Problems with Socioecological Explanations of Primate Social Diversity. Pp. 1–30. In *Social Processes and Mental Abilities in Non-Human Primates*. Burton, F.D. ed. Lewiston, Edwin Mellen Press.

Charles-Dominique, P. 1977. *Ecology and Behaviour of Nocturnal Primates*. New York, Columbia University Press.

Chesson, P.L. and Case, T.J. 1986. Overview: Nonequilibrium Community Theories: Chance, Variability, History and Coexistence. Pp. 229–239. In *Community Ecology*. Eds., Diamond, J. and Case, T.J. New York, Harper and Row.

Chivers, D.J., Wood, B.A. and Bilsborough, A., Eds., 1984. *Food Acquisition and Processing in Primates*. New York, Plenum.

Clements, F.E. 1916. *Plant Succession: An Analysis of the Development of Vegetation*. Washington, D.C., Carnegie Institute Publication No. 242.

Cody, M.L. and Diamond, J.M. Eds., 1975. *Ecology and Evolution of Communities.* Cambridge, Belknap Press.

Cody, M.L. 1975. Towards a Theory of Continental Species Diversity. Pp. 214–257. In *Ecology and Evolution of Communities.* Cody, M.L. and Diamond, J.M. Eds., Cambridge, Belknap Press

Cohen, M.N. 1989. *Health and the Rise of Civilization.* New Haven, Yale University Press.

Cohen, M.N. 1977. *The Food Crisis in Prehistory: Overpopulation and the Origins of Agriculture.* New Haven, Yale University Press.

Crompton, R.H. 1995. "Visual Predation," Habitat Structure, and the Ancestral Primate Niche. Pp. 11–30. In *Creatures of the Dark: The Nocturnal Prosimians.* Eds., Altman, L., Doyle, G.A. , and Izard, M.K. New York, Plenum Press.

Davies, A.G. and Oates, J.F. Eds., 1994. *Colobine Monkeys: Their Ecology, Behaviour and Evolution.* Cambridge, Cambridge University Press.

Dowdy, W.W. 1947. An Ecological Study of the Arthropoda of an Oak-Hickory Forest with Reference to Stratification. *Ecol.* 28:418–439.

Dumond, D.E. 1979. The Limitation of Human Population: A Natural History. *Science* 187:713–721.

Dunbar, R.I.M. 1988. *Primate Social Systems.* Ithaca, Comstock.

Ehrlich, P.R. and Raven P.H. 1964. Butterflies and Plants: A Study in Coevolution. *Evol.* 18:586–608.

Elton, C.S. 1942. *Voles, Mice and Lemmings: Problems in Population Dynamics.* London, Oxford University Press.

Faegri, K. and Vander Pijl, L. 1971. *The Principles of Pollination Ecology.* (2nd edition). London, Oxford.

Fleming, T.H. 1988. *The Short-Tailed Fruit Bat.* Chicago, University of Chicago Press.

Fleming, T.H. and Estrada, A. Eds., 1993. *Frugivory and Seed Dispersal: Ecological and Evolutionary Aspects.* Dordrecht, Kluwer Academic Publishers.

Freeland, W.J. and Janzen, D.H. 1974. Strategies in Herbivory by Mammals: The Role of Plant Secondary Compounds. *Am. Natur.* 108:269–289.

Fuentes, A. (1999) Variable Social Organization in Primates: What Can Looking at Primate Groups Tell Us About the Evolution of Plasticity in Primate Societies? pp. 183-188. In *The Non Human Primates.* Eds., Dolhinow, P. Fuentes, A. and Mountain View, Mayfield.

Futuyma, D.J. and Slatkin, M. Eds., 1983. *Coevolution.* Sunderland, Sinauer.

Garber, P.A. 1986. The Ecology of Seed Dispersal in Two Species of Callitrichid Primates (*Saguinus mystax* and *Saguinus fuscicollis*). *Am. J. Primatol.* 10:155–177.

Gause, G.F. 1934. *The Struggle for Existence.* New York, Hafner.

Gautier-Hion, A. and Gautier, J.-P. 1988. The Diet and Dietary Habits of Forest Guenons. In *A Primate Radiation: Evolutionary Biology of the African Guenons.* Eds., Gautier-Hion, A., Bouliere, F., Gautier, J.-P. and Kingdon, J. Pp. 257–283. New York, Cambridge University Press.

Gentry, A.H. 1990. *Four Neotropical Rainforests.* New Haven, Yale University Press.

Gilbert, L.E. and Raven, P.H. 1975. *Coevolution of Animals and Plants.* Austin, University of Texas Press.

Gilpin, M.E. and Diamond, J.M. 1980. Subdivision of Nature Reserves and the Maintenance of Species Diversity. *Nature* 285: 567–568.

Glander, K., Wright, P., Seigler, D., Randrianasolo, V. and Randrianasolo, B. 1989. Consumption of Cyanogenic Bamboo by a Newly Discovered Species of Bamboo Lemur. *Am. J. Primatol.* 19:119–124.

Goldblatt, L.P., ed. 1993. *Biological Relationships Between Africa and South America.* New Haven, Yale University Press.

Hall, K.R.L.H. 1965. Behaviour and Ecology of the Wild Patas Monkey, *Erythrocebus patas,* in Uganda. *J. Zoology* 148:15–87.

Halle, F., Oldemann, R.A. and Tomlinson, P.B. 1978. *Tropical Trees and Forests: and Architectural Analysis.* Berlin, Springer-Verlag.

Harding, R.S.O. and Teleki, G., Eds., 1981. *Omnivorous Primates.* New York, Columbia University Press.

Heinrich, B. and Raven, P.H. 1972. Energetics and Pollination Ecology. *Science* 176:597–602.

Herrera, C.M. 1981. Determinants of Plant-Animal Coevolution: The Case of Mutualistic Dispersal of Seeds by Vertebrates. *Oikos* 44:132–144.

Hladik, C.M. 1969. Surface Relative du Tractus Digestif de Quelques Primates, Morphologie des Villosités Intestinales et Corrélations avec Régime Alimentaire. *Mammalia* 31:120–147.

Horn, H.S. 1971. *The Adaptive Geometry of Trees. Monographs In Population Biology 3.* Princeton, Princeton University Press.

Horn, H.S. 1974. The Ecology of Secondary Succession. *Ann. Rev. Ecol. Syst.* 5:25–37.

Howe, H.F. 1990. Seed Dispersal by Birds and Mammals: Implications for Seedling Demography. Pp. 191–218. In *Reproductive Ecology of Rain Forest Plants.* Eds., Bawa, K.S. and Hadley, M. Paris, Parthenon.

Hutchinson, G.E. 1957. Concluding Remarks. *Cold Spring Harbor Symp. Quant. Biol.* 22:415–427.

Hutchinson, G.E. 1965. *The Ecological Theater and the Evolutionary Play.* New Haven, Yale University Press.

Janzen, D.H. 1978. Complications in Interpreting the Chemical Defenses of Trees Against Tropical Arboreal Plant-Eating Vertebrates. Pp. 73–84. In *The Ecology of Arboreal Folivores.* Montgomery, G.G., Ed. Washington, D.C., Smithsonian Institution Press.

Jewell, P.A. 1966. The Concept of Home Range in Mammals. Pp. 85–109. In *Play, Exploration and Territory in Mammals.* Eds., Jewell, P.A. and Loizos, C. New York, Academic Press.

Kaufmann, J.H. 1962. Ecology and Behaviour of the Coati, *Nasua narica,* on Barro Colorado Island, Panama. *Univ. Calif. Publ. Zool.* 60:95–222.

Kay, R.F. 1984. The Use of Anatomical Features to Infer Foraging Behavior in Extinct Primates. Pp. 21–53. In *Adaptations for Foraging in Nonhuman Primates.* Eds., Rodman, P.S. and Cant, J.G. New York, Columbia University Press.

Kay, R.N.B. and Davies, A.G. 1994. Digestive Physiology. PP. 229-249. In *Colobine Monkeys: Their Ecology, Behaviour and Evolution.* Eds., Davies, A.G. and Oates, J.F. Cambridge, Cambridge University Press.

Kikkawa, J. and Anderson, D.J. 1986. *Community Ecology: Pattern and Process.* Melbourne, Blackwell Scientific Publications.

Klopfer, P.H. 1962. *Behavioral Aspects of Ecology.* Englewood Cliffs, Prentice Hall.

Krebs, C.J. 1994. *Ecology, Fourth Edition.* New York, Harper Collins.

Kress, R.K. 1993. Coevolution of Plants and Animals: Pollination of Flowers by Primates in Madagascar. *Current Science* 65:253–257.

Kummer, H. 1967. Dimensions of a Comparative Biology of Primate Groups. *Am. J. Phys. Anthropol.* 27:357–366.

Kummer, H. 1971. *Primate Societies: Group Techniques of Ecological Adaptation.* Arlington Heights, Harlan Davidson.

Landers, J. 1992. Reconstructing Ancient Populations. Pp. 402–405. In *The Cambridge Encyclopedia of Human Evolution.* Eds., Jones, S., Martin, R.D. and D. Pilbeam. Cambridge, Cambridge University Press.

Lee, R.B. 1972. Population Growth and the Beginning of Sedentary Life Among the !Kung Bushman. In *Population Growth: Anthropological Implications.* ed. Spooner, B. Pp. 329–342. Cambridge, MIT Press.

Lee, R.B. 1979. *The !Kung San.* London, Cambridge University Press.

Losos, J.B. 1996. Phylogenetic Perspectives on Community Ecology. *Ecol.* 77:1344–1354.

Losos, J.B., Jackman, T.R., Larson, A., de Queiroz,K. and Rodriguez-Schettino, L. 1998. Contingency and Determinism in Replicated Adaptive Radiations of Island Lizards. *Science* 279:2115–2118.

Lovejoy, T.E. 1986. Species Leave the Ark One by One. In *Preservation of Species.* Norton, B.G. ed. Pp. 13–27. Princeton, Princeton University Press.

Martin, R.D. 1990. *Primate Origins and Evolution: A Phylogenetic Reconstruction.* Princeton, Princeton University Press.

Maurer, B.A. 1998. Ecological Science and Statistical Paradigms, At the Threshold. *Science* 279:502–503.

Mayr, E. 1963. *Animal Species and Evolution.* Cambridge, Belknap.

Mittermeier, R.A. and van Roosmalen, M.G.M. 1981. Preliminary Observations on Habitat Utilization and Diet in Eight Surinam Monkeys. *Folia Primatol.* 36:1–39.

Montgomery, G.G., ed. 1978. *The Ecology of Arboreal Folivores.* Washington, D.C., Smithsonian Institution Press.

Noble, G.K. 1939. The Role of Dominance in the Social Life of Birds. *Auk* 56:263–273.

Oates, J.F., Swain, T. and Zantovska, J. 1977. Secondary Compounds and Food Selection by Colobus Monkeys. *Biochem. Syst. Ecol.* 5:317–321.

Odum, E.P. 1971. *Fundamentals of Ecology. Third Edition.* Philadelphia, Saunders.

Pianka, E.R. 1994. *Evolutionary Ecology, Fifth Edition.* New York, Harper Collins.

Raven, P.H. and Johnson, G.B. 1991. *Biology.* St. Louis, Mosby.

Regal, P.J. 1977. Ecology and Evolution of Flowering Plant Dominance. *Science* 196:622–629.

Richard, A.F. 1978. *Behavioral Variation: Case Study of a Malagasy lemur.* Lewisburg, Bucknell University Press.

Richards, P.W. 1966. *The Tropical Rain Forest.* Cambridge, Cambridge University Press.

Ricklefs, R.E. and Schluter, D. 1993. Species Diversity and Historical Influences. Pp. 350–363. In *Species Diversity in Ecological Communities: Historical and Geographic Perspectives.* Eds., Ricklefs, R.E. and Schluter, D. Chicago, University of Chicago Press.

Root, R.B. 1967. The Niche Exploitation Pattern of the Blue-Gray Gnatcatcher. *Ecol. Monogr.* 37:317–350.

Rose, L.M. and Marshall, F. 1996. Meat Eating, Hominid Sociality, and Home Bases Revisited. *Current Anthrop.* 37: 307–338.

Rowell, T.E. 1972. *Social Behaviour of Monkeys.* Middlesex, Penguin Press.

Rowell, T.E. 1979. How Would We Know if Social Organization Were *Not* Adaptive. Pp. 1–22. In *Primate Ecology and Human Origins.* Bernstein, I.S. and Smith, E.O. Eds., New York, Garland.

Rowell, T.E. 1991. What Can We Say About Social Structure? Pp.255–270. In *The Development and Integration of Behaviour: Essays in Honour of Robert Hinde.* Bateson, P. ed. Cambridge, Cambridge University Press.

Rudran, R. 1973. Adult Male Replacement in One-Male Groups of Purple-Faced Langurs *(Presbytis senex senex)* and Its Effect Population Structure. *Folia Primatol.* 19:166–192.

Runciman, W.G., Smith, J.M. and Dunbar, R.I.M. 1996. *Evolution of Social Behaviour Patterns in Primates and Man.* Oxford, Oxford University Press.

Sauther, M.L. 1998. Interplay of Phenology and Reproduction in Ringtailed Lemurs: Implication for Ringtailed Lemur Conservation. *Folia Primatol.* 69 Supplement 1:309–320.

Schaller, G.B. 1972. *The Serengeti Lion, A Study of Predator-Prey Relations.* Chicago, University of Chicago Press.

Schoener, T.W. 1986. Resource Partitioning. Pp. 91–126. In *Community Ecology: Pattern and Process.* Eds., Kikkawa, J. and Anderson, D.J. Melbourne, Blackwell Scientific Publications.

Schoener, T.W. 1988. Ecological Interactions. Pp. 255–297. In *Analytical Biogeography: An Integrated Approach to the Study of Animal and Plant Distributions.* Eds., Myers, A.A. and Giller,P.S. London, Chapman and Hall.

Shafer, C.L. 1990. *Nature Reserves: Island Theory and Conservation Practice.* Washington, D.C., Smithsonian Institution Press.

Shelford, V.E. 1913. *Animal Communities in Temperate America.* Chicago, University of Chicago Press.

Slobodkin, L.B. 1960. Ecological Energy Relationships at the Population Level. *Am. Natur.* 94:213–236.

Spooner, B. (ed.) *Population Growth: Anthropological Implications.* Cambridge, MIT Press.

Stamp, N.E. 1992. Theory of Plant-Insect Herbivore Interactions on the Inevitable Brink of Re-synthesis. *Bull. Ecol. Soc. Am.* 73:28–34.

Struhsaker, T.T. 1969. Correlates of Ecology and Social Organization Among African Cercopithecines. *Folia Primatol.* 11:80–118.

Sussman, R.W. 1972. Child Transport, Family Size and Increase in Human Population During the Neolithic. *Current Anthropol.* 13:258–259.

Sussman, R.W. 1979. Nectar-Feeding by Prosimians and Its Evolutionary and Ecological Implications. Pp. 119–125. In *Recent Advances in Primatology: Volume 3, Evolution.* Eds., Chivers, D.J. and Joysey, K.A. London, Academic Press.

Sussman, R.W. 1981. Preagricultural Mobility—A Factor Limiting Growth in Human Populations. Pp.199–212. In *The Perception of Evolution.* Eds., Mai, L.L., Shanklin, E. and Sussman, R.W. Los Angeles, U.C.L.A. Publication Services.

Sussman, R.W. 1987. Morpho-Physiological Analysis of Diets: Species-Specific Dietary Patterns in Primates and Human Dietary Adaptations. Pp. 151–179. In *The Evolution of Human Behavior: Primate Models.* ed. Kinzey, W.G. Albany, State University of New York Press.

Sussman, R.W. and Raven, P.H. 1978. Pollination by Lemurs and Marsupials: An Archaic Coevolutionary System. *Science* 200:731–736.

Terborgh, J. 1983. *Five New World Primates: A Study in Comparative Ecology.* Princeton, Princeton University Press.

Terborgh, J. 1986. Keystone Plant Resources in the Tropical Forest. In *Conservation Biology: The Science of Scarcity and Diversity.* ed. Soulé, M.E. Pp.330–344. Sunderland, Sinauer.

Terborgh, J. 1992. *Diversity and the Tropical Rain Forest.* New York, Scientific American Scientific Publications.

Terborgh, J. and Janson, C.H. 1986. The Socioecology of Primates. *Ann. Rev. Ecol. Syst.* 17: 111.

Terborgh, J. and Robinson, S. 1986. Guilds and Their Utility in Ecology. Pp. 65–90. In *Community Ecology: Pattern and Process.* Eds., Kikkawa, J. and Anderson, D.J. Melbourne, Blackwell Scientific Publications.

van Schaik, C.P. 1983. Why Are Diurnal Primates Living in Groups? *Behaviour* 87:120–144.

von Holst, D. 1974. Social Stress in the Tree-Shrew: Its Causes and Physiological and Ethological Consequences. Pp. 389–411. In *Prosimian Biology.* Eds., Martin, R.C., Doyle, G.A. and Walker, A.C. London, Duckworth.

Waser, P.M. and Wiley, R.H. 1980. Mechanisms and Evolution of Spacing in Animals. Pp. 159–233. In *Handbook of Behavioral Neurobiology. Vol. 3.* Marler, P. and Vanderbergh, J.G. Eds., New York, Plenum.

Waterman, P.G. 1984. Food Acquisition and Processing as a Function of Plant Chemistry. Pp. 177–211. In *Food Acquisition and Processing in Primates.* Eds., Chivers, D.J., Wood, B.A. and Bilsborough, A. New York, Plenum.

Watt, D. (ed.) 1966. The Nature of Systems Analysis. Pp. 1–14. In *Systems Analysis in Ecology.* ed. Watt, D. New York, Academic Press.

Whitmore, T.C. 1984. *Tropical Rainforests of the Far East. Second Edition.* Oxford, Clarendon.

Williams, E.E. 1983. Ecomorphs, Faunas, Island Size, and Diverse End Points in Island Radiations of *Anolis*. In *Lizard Ecology: Studies of a Model Organism.* Eds., Huey, R.B., Pianka, E.R. and Schoener, T.W. Pp. 326–370. Massachusetts, Harvard University Press.

Wrangham, R.W. 1987. Evolution of Social Structure. Pp. 282–296. In *Primate Societies.* Smuts, B.B., Cheney, D.L., Seyfarth, R.M., Wrangham, R.W. and Struhsaker, T.T. Eds., Chicago, University of Chicago Press.

Yodzis, P. 1989. *Introduction to Theoretical Ecology.* New York, Harper and Row.

Zuckerman, S. 1932. *The Social Life of Monkeys and Apes.* New York, Harcourt, Brace.

CHAPTER 2

The Taxonomy and Evolution of Primates

PRINCIPLES OF ANIMAL TAXONOMY

The necessity of aggregating things into categories is a general characteristic of all living things. The ability to classify is an absolute and minimal characteristic of being or staying alive. Even one-celled organisms must discriminate classes of objects. Animals group objects into classes but not necessarily the same objects into the same classes. For example, what may be dangerous, food, an enemy, a sleeping site, or a sex partner for one organism may not be so for another. The ways in which an ant, tick, bat, primate, and a human sees the world and classifies it are quite different (von Uexkull 1957, reprinted from 1934 original), and these ways are related to the senses of the particular organism, its learning experiences, and, at least in a human, his or her cultural background (Figure 2-1).

According to Simpson (1961:3), there are two major ways to order our perceptions: "association by contiguity" and "association by similarity."

> *Association by contiguity (for our purposes) is a structural and functional relationship among things that, in a different psychological terminology, enter into a single Gestalt. The things involved may be quite dissimilar, or in any event their similarity is irrelevant. Such, for instance, is the relationship between a plant and the soil in which it grows, between a rabbit and the fox that pursues it, between the separate organs that compose an organism, among all the trees of a forest, or among all the descendents of a given population. Things in this relationship to each other belong both structurally and functionally to what may be defined in a broad but technical sense as a single system. Objects associated by similarity, on the other hand, are classed together simply because they share one or more common characteristics.*

Figure 2-1a Forester and oak tree. von Uexkull's views. Adapted from *Instinctive Behavior: The Development of a Modern Concept* (ed. C.H. Schiller). International Universities Press, New York, 1957.

Figure 2-1b Little girl and oak tree. von Uexkull's views. Adapted from *Instinctive Behavior: The Development of a Modern Concept* (ed. C.H. Schiller). International Universities Press, New York, 1957.

Table 2-1 Definitions

Systematics	The scientific study of the kinds of diversity of organisms and any and all relationships among them.
Zoological Classification	The ordering of animals into groups (or sets) on the basis of their relationship.
Zoological nomenclature	The application of distinctive names to each of the groups recognized in any given zoological classification.
Taxonomy	The theoretical study of classification, including its basis, principles, procedures and rules.
Taxon	A group of real organisms recognized as a formal unit at any level of hierarchic classification.
Species	A population or group of populations of actually or potentially inter-breeding animals that are reproductively isolated from other such groups.
Sympatric species	Two *non*-interbreeding populations living in the same environment.
Allochronic species	Two groups of animals in the same lineage but living in different time periods
Allopatric species	Two groups of animals in the same or different species but isolated geographically.

They may be all yellow, or all smooth, or all with wings, or all ten feet high. They may, for that matter, all be seats with four legs and a back, in which case we call them chairs (Simpson 1961:4).

If there were no means of classification or of ordering nature, natural phenomena would appear chaotic. Indeed, the whole aim of theoretical science, it might be said, is to carry to the highest possible and conscious degree the perceptual reduction of chaos in natural phenomena (Simpson 1961). Is the order so achieved, however, an objective characteristic of the phenomena or an artifact constructed by the western scientist? In classifying potatoes, the western scientist would use only one term, *Solanum tuberosum*. The local Bolivian farmer has over 200 terms for potatoes. The botanist and the Bolivian farmer use different criteria in making their classifications. In fact, the usefulness of the classification depends upon the specific reasons or needs of the classifier.

The systematic study of biology, from the viewpoint of a western scientist, requires the construction of a formal system in which organisms are grouped into categories *(zoological classification)* and of names for these categories *(zoological nomenclature)* (Table 2-1). The present classificatory scheme used in biology is based on Linnaeus' *Systema Naturae* (10th edition, 1758). Since the mid-eighteenth century, however, the philosophy underlying biological classification has changed radically. Linnaeus based his scheme on the typological or archetypal concept, in which all species are unchanging or immutable. The living animals in a class (or taxon) were considered to be imperfect copies of an ideal or perfect type which was created by God: humans were considered the imperfect copy of God—western European man bearing the greatest likeness! Each class was considered to be fixed and distinct from all others. Thus, animals were classed into groups using association by similarity.

There have been two major changes in the philosophy of biology since the time of Linnaeus. The first grows from the discovery of evolutionary mechanisms and processes by Darwin, who demonstrated that species are not immutable and that they diverged from a single common ancestor. The branching tree of all organisms, linked by their increasingly ancient common ancestors, is called a phylogeny. Darwin's discovery of phylogeny provided the biological basis for classifying organisms according to their degree of relatedness. This remains the most important criterion for biological classification above the species level.

The second major change is typology as the basis for species has been replaced by the concept of breeding populations. Linnaeus classified animals by similarity. Modern biological systematists recognize the enormous variability that exists within breeding populations. The potential ability to interbreed, not just similarity, is now the major basis for classification at the species level.

In principle, animal classification in modern biology is based upon phylogeny and on genealogical relationships among organisms in a population. In practice, however, much of biological classification is pragmatic. Phylogeny and capacity to interbreed are difficult to ascertain and, thus, classification still sometimes is based on similarity, whatever the theory may be. For example, even modern DNA hybridization techniques gauge overall similarity in the genetic material.

Modern zoological classification, the basis for most zoological nomenclature, is referred to as the Linnaean hierarchy. The major categories of this classificatory system are listed in Table 2-2. The customary indentation of subsequent categories indicates the decreasing inclusiveness of the various levels. Figure 2-2 represents the hierarchical classification of the major taxa of the Order Primates.

The basic unit of the Linnaean hierarchy is the *species*. As stated above, natural populations are not fixed, static qualities. Populations change through time

Table 2-2 Categories in the Current Classification (The Linnaen Hierarchy)

Kingdom*	Animal	
Phylum*	Chordate	
Subphylum		
Superclass		
Class*	Mammalian	
Infraclass		
Cohort		
Superorder		
Order*	Primates	
Suborder		
Infraorder		
Superfamily		
Family*	Hominidae	
Subfamily		
Tribe		
Subtribe		
Genus*		*Homo*
Subgenus		
Species*		*sapiens*
Subspecies		

Adapted from Buettner-Janusch (1966)
*designates seven major categories of the Linnaen Hierarchy

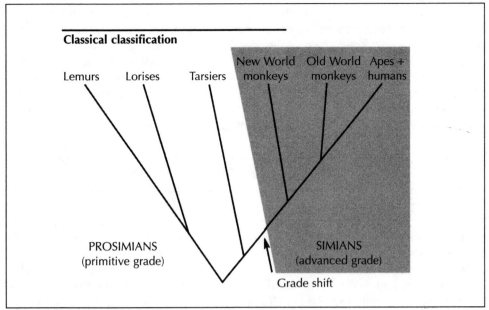

Classical classification

Lemurs   Lorises   Tarsiers   New World   Old World   Apes +
                                 monkeys   monkeys   humans

PROSIMIANS   SIMIANS
(primitive grade)   (advanced grade)

Grade shift

Figure 2-2 Primate phylogenetic tree.

and vary morphologically, genetically and geographically. Naming them often involves creating discrete units out of continuous phenomena. Most biologists follow Mayr's (1963, Mayr and Ashlock 1991) definition of a species as *a population or group of populations of actually or potentially interbreeding animals that are reproductively isolated from other such groups.* It is considered to be a real, natural phenomenon and not an arbitrary, analytical category. However, there is often a small amount of gene flow between populations that do not regularly interbreed. In these cases of morphologically distinct populations in which hybridization occurs; how much genetic leakage should be tolerated before recognizing an independent species? This has led to considerable debate (e.g., papers in Vrba 1985, Otte and Endler 1989, Kimbel and Martin 1993). These problems have led some biologists to disagree with the biospecies concept (Templeton 1989, Raven et al. 1992). Problems also arise when attempting to identify fossil species in an evolving lineage, where breeding is moot (see Rose and Bown 1986), and when comparing morphologically similar extant populations that are allopatric (geographically isolated) and, thus, have no chance to naturally express their potential for interbreeding.

Recently, Templeton (1989, 1994) has introduced a new definition of species, *the cohesion concept of species.* Rather than looking at mechanisms that isolate populations, Templeton defines species by the mechanisms that maintain cohesion within a population and speciation is defined as the evolution of cohesive mechanisms. Using this definition *species are groups of animals that share a suite of genetic, phenotypic, demographic, ecological, and behavioral characteristics or cohesive mechanisms that help maintain the group as a cohesive unit.* These characteristics allow what Templeton refers to as genetic and demographic exchangeability within the species. Similarly, Paterson (1985, see also Templeton 1987) defines sexually reproducing species as the most inclusive population of individual biparenting organisms which share a common fertilization system. In many animals, an important component of this system is the means by which the animals attract and recognize each other, what Paterson calls *the specific-mate recognition system (SMRS)*. By using Paterson's species recognition system, primatologists have tested the validity of current taxonomic divisions among the nocturnal galagos and a number of new species have been identified (e.g., Bearder 1995, see also Masters 1998). I will discuss this further in the next chapter. In any case, as emphasized by Mayr (1963) and reiterated by Jolly (1993:104): ". . . taxonomic organization inevitably forces an oversimplified structure upon a complex, evolutionarily dynamic situation."

Following the rules of nomenclature each species is scientifically referred to by its genus and species names (binomen). Thus the scientific name of humans is *Homo sapiens.* That of the rhesus monkey is *Macaca mulatta. Homo* and *Macaca* are names for the genera and *H. sapiens* and *M. mulatta* are names for the species. There is only one living species in the genus *Homo;* the genus is monotypic. The genus *Macaca* is polytypic for it contains 12 species. The genus name is capitalized and underlined; the species name is underlined. Other conventions have been devised for the purpose of forming names of some of the higher categories in the Linnaean hierarchy. Some of these are found in Table 2-3. It should be noted that for each Linnean name (e.g. family Hominidae) it is possible to generate an informal English version that lacks capitilization and lacks the formal suffix (in this case, "hominid"). Students should be careful never to use the Linnean name as an adjective (Hominidae evolution), while, in contrast, the English version is acceptable (hominid evolution).

The higher categories in zoological classification are those above the level of species. Groups of animals included in a higher taxon should be descended from a common ancestor, in which case the group is called *monophyletic*. Although it is not always possible, the classification of animals into higher categories should reflect valid phylogenetic entities. The criteria used to discover phylogeny have been the subject of many books and articles (see for example Simpson 1945, 1961, 1975, Mayr 1969, Hennig 1966, Sneath and Sokal 1973, Kluge 1984, Eldridge 1989, Martin 1990, Mayr and Ashlock 1991). Even more subjectivity enters the process when moving from a phylogeny to a classification, because phylogeny itself does not dictate the hierarchical level at which organisms that are related to each other should

Table 2-3 Names of Higher Categories

Category	Suffix	Genus	Stem*	Name of Higher Category
Infraorder	-IFORMES	*Lemur*	LEMUR-	LEMURIFORMES
		Tarsius	TARSI-	TARSIIFORMES
Superfamily	-OIDEA	*Lemur*	LEMUR-	LEMUROIDEA
		Cercopithecus	CERCOPITHEC-	CERCOPITHECOIDEA
Family	-IDAE	*Lemur*	LEMUR-	LEMURIDAE
		Cercopithecus	CERCOPITHEC-	CERCOPITHECIDAE
Subfamily	-INAE	*Lemur*	LEMUR-	LEMURINAE
		Alouatta	ALOUATT-	ALOUATTINAE

*The stem is usually taken from the name of a genus within the higher category.
Adapted from Buettner-Janusch (1966)

be classified. In a discussion of the origin of the Order Primates, to which I will return shortly, Cartmill (1972:97–98) summarizes these criteria as follows:

> *Taxonomic boundaries are expected to correspond as much as possible to major adaptive shifts, represented in the fossil record by phases of rapid morphological change that led to the fixation of some new complex of traits underlying a subsequent evolutionary radiation. Such a complex of traits can be used as a diagnosis of the derived higher taxon. . . . It is not always possible to identify a major adaptive shift accompanying the appearance of a new grade of organization. . . . In a situation like this—when, in Simpson's (1944) terms, there is no quantum evolution from one adaptive zone to another at the base of an adaptive radiation–placement of any taxonomic boundary between ancestral and descendent groups becomes largely a matter of caprice.*

In Table 2-4, I present the classification of the Order Primates that will be used in this book. Because of the difficulties of dividing a continuously evolving group into discrete units, and of other difficulties inherent in interpreting taxonomic relationships, there are still many debates concerning the classification of primates. For example, until the mid-1960's, tree shrews (Tupaiiformes) were considered to be primates, now it is generally agreed that they are not (Van Valen 1965, Goodman 1966, Martin 1968, 1990, Campbell 1974, Luckett 1980).

In addition, the Order Primates commonly is divided into two suborders, Prosimii (lemurs, lorises, and tarsiers) and Anthropoidea (monkeys, apes, and humans). The prosimians are the more primitive of the two suborders because they preserve many morphological characteristics similar to those of primates living in the Eocene epoch, 40 to 50 million years ago (Fleagle 1999). However, tarsiers apparently share some anatomical features with anthropoids, which suggests that anthropoids and tarsiers may be derived from a common ancestor (see relevant papers in Luckett and Szalay 1975 and Chivers and Joysey 1978, Aiello 1986, Martin 1990, Fleagle and Kay 1994).

For some taxonomists, who believe all higher taxa should be *holophyletic*—a strict form of monophyly in which not only must the members of a group share an ancestor but *all* the descendents of the ancestor must be classified in the taxon—the taxon Prosimii may be invalid, even though it is monophyletic. For these classifiers, Primates should be divided into the suborders Haplorhini (including tarsiers, monkeys, apes and humans) and the Strepsirhini (including lemurs and lorises), which are considered to be separate holophyletic assemblages. However, this classification is scientifically controversial and also impractical to apply to fossil primates (Rasmussen 1990a, 1994). A number of other taxonomic debates are currently in progress but are of little concern to this discussion. I mention these debates to make the reader aware of two facts: (1) phylogeny is often difficult to establish with certainty; and (2) the names given to taxa by various authors may differ, even in cases when the phylogeny is agreed upon.

WHAT IS A PRIMATE?

Using the terms discussed in the first section of this chapter, I can now say that the Order Primates is a monophyletic group of animals bound together by common

Table 2-4 Present Classification of the Order Primates

Genus	Subfamily	Family	Superfamily	Intraorder	Suborder	Order
Indri		Indriidae				
Propithecus						
Avahi						
Daubentonia		Daubentoniidae				
Lepilemur		Lepilemuridae				
Hapalemur		Lemuridae		Lemuriformes		
Eulemur						
Lemur						
Varecia						
Microcebus		Cheirogaleidae				
Mirza						
Cheirogaleus						
Allocebus					Prosimil	
Phaner						
Galago	Galaginae	Lorisidae		Lorisiformes		
Perodicticus	Lorisinae					
Arctocebus						
Nycticebus						
Loris						
Tarsius		Tarsiidae		Tarsiiformes		Primates
Cebuella	Callitrichinae	Callitrichidae				
Callithrix						
Saguinus						
Leontopithecus						
Callimico						
Saimiri	Cebinae	Cebidae	Ceboidea	Platyrrhini	Anthropoidea	
Cebus						
Aotus	Aotinae					
Callicebus						
Pithecia	Pitheciinae	Atelidae				
Chiropotes						
Cacajao						
Alouatta	Atelinae					
Lagothrix						
Brachyteles						
Ateles						
Allenopithecus	Cercopithecinae	Cercopithecidae	Cercopithecoidea			
Erythrocebus						
Cercopithecus						
Macaca						
Cercocebus						
Papio						
Mandrillus						
Theropithecus						
Colobus	Colobinae					
Presbytis						
Simias				Catarrhini	Anthropoidea	
Nasalis						
Pygathrix						
Rhinopithecus						
Hylobates		Hylobatidae	Hominoidea			
Pongo		Pongidae				
Gorilla						
Pan						
Homo		Hominidae				

ancestry. The question still remains, however, as to whether this group can be described by a complex of traits related to a major adaptive shift, which accompanied the origin of the order. This is an ecological, as well as a taxonomic, question and will be the focus of the remainder of this chapter.

An early, classic definition of primates by Mivart (1873) was expanded by Le Gros Clark (1963). The resultant list of anatomical characters and evolutionary trends (Table 2-5) is still used by most primatologists to define primates. Because these features, for the most part, involve the retention and elaboration of generalized mammalian features rather than a list of specialized anatomical traits, primates are often distinguished from other mammalian orders by their lack of specialization. As stated by Le Gros Clark (1963:42):

> While many other mammalian orders can be defined by conspicuous specializations of a positive kind which readily mark them off from one another, the Primates as a whole have preserved rather a generalized anatomy and, if anything, are to be mainly distinguished from other orders by a negative feature—their lack of specialization.

This definition, however, does not clearly identify probable *derived* (= evolutionarily specialized) features of primates that would provide strong support for their ancestry. By including likely primitive features of placental mammals, it is difficult to distinguish primates from other mammals in either a phylogenetic or an ecological sense. Furthermore, the trends included in this definition refer not to universally shared primate features, but to developments found only in some members of the group. Because of these problems, Martin (1986a, 1990) devised a list of derived characters shared by primates that distinguish them from other mammals, and that should aid in any reconstruction of primate evolution.

Martin's resulting new definition of living primates, with some slight modifications, is as follows: Primates are (1) essentially *arboreal inhabitants of tropical/subtropical forest ecosytems*. This is related to a number of interrelated

Table 2-5 Traits Often Used to Define Primates Developed from LeGros Clark and Others

(1) Preservation of a generalized limb structure with primitive pentadactyly—five fingers or toes on each of the extremities.

(2) Enhancement of free mobility of the digits, especially the thumb (pollex) and big toe (hallux) (which are used for grasping).

(3) Sharp, compressed claws are replaced by flattened nails, and this is associated with the development of highly sensitive tactile pads on the digits.

(4) Reduction of olfaction and associated areas of the brain, and abbreviation of the snout.

(5) Elaboration of the visual apparatus, with development of binocular vision.

(6) The loss of certain elements of the primitive mammalian dentition, but the preservation of a simple cusp pattern of the molar teeth and the retention of teeth which are regionally differentiated in form (heterodonty).

(7) Expansion and elaboration of the brain, especially the cerebral cortex.

(8) Increase and elaboration of the development of the processes of gestation and of the uterine and placental membranes.

locomotor adaptations, including (2) extremities adapted for *prehension of branches* rather than clinging with claws. This is enhanced by possession of: (3) *flat nails and sensitive tactile pads* with cutaneous ridges (dermatoglyphs); (4) a widely divergent hallux allowing for a *powerful grasping action of the foot;* (5) locomotion is *hindlimb dominated;* and (6) the foot has a unique orientation at the point of thrust *(tarsi-fulcrumating* type of foot) and relative *elongation of the distal segment of the calcaneus* (heel bone).

There are also features of the sense organs and skull. (7) The *visual sense is greatly emphasized.* The (8) *eyes are relatively large* in relation to skull length, and the orbits possess a (9) *postorbital bar* (orbit completely encircled with a rim of bone). Primates have a pronounced degree of (10) forward rotation of the orbits (viz. *orbital convergence*) which ensures a large degree of (11) *binocular vision.* (12) Ipsilateral (from the same side eye) and contralateral (from opposite side eye) nerve *fibres passing to the optic tectum are approximately balanced* in primates, whereas in non-primates contralateral fibres predominate—this allows (13) effective *stereoscopic vision* because single areas of the brain integrate information from two different visual perspectives. In primates, the bony enclosure of the middle ear cavity (the auditory bulla) is formed from a bone called the petrosal, (14) *(petrosal bulla),* a unique condition among mammals. (15) *The brain is typically moderately enlarged in relation to body size,* and possesses some (16) *unique sulcal patterns.* Furthermore, (17) the *brain constitutes a significantly larger proportion of body weight at all stages of gestation.*

Unique reproductive features include: (18) *very early descent of the testes* into a postpenial scrotum; (19) *absence of a urogenital sinus* and (20) involvement of the *yolk-sac in placentation is suppressed.* Primates are also adapted for (21) *slow reproductive turnover:* they have (22) *long gestation periods* relative to maternal body size, (23) produce *small litters of precocial infants,* and (24) *fetal and postnatal growth are slow* in relation to maternal size. (25) *Sexual maturity is attained late* and (26) *life spans are long* relative to body size.

The (27) *dental formula exhibits a maximum of two incisors in each quadrat of the jaw,* instead of the primitive eutherian number of three incisors. The (28) *cheek teeth (molars and premolars) are typically relatively unspecialized,* (29) *cusps are generally low and rounded,* and the lower molars possess a (30) broad, grinding basin, or *talonid.*

Another feature, not included in Martin's list but which is related to the unique reproductive features listed above, is the tendency of primates to be highly social. All diurnal primates except orangutans live in relatively stable social groups, whereas the nocturnal prosimians generally are "solitary but social." In Table 1-6, I define the types of social structure found among extant primates.

Now I have gone one step further in answering the question "What is a primate?" A primate is a member of a monophyletic order of mammals. All primate taxa are believed to be descended from a common ancestral species, and share a number of traits in common. Can we learn more about the origin of primates by studying the adaptive significance of these traits, or by a study of the fossil record? Can we reconstruct the niche to which the ancestral primate was adapted? It is upon the morphological traits of this ancestral group of primates that all further adaptations have been built, but what was the initial impetus that brought about the adaptive radiation of the primates?

THE ORIGIN OF PRIMATES: THE ARBOREAL THEORY OF PRIMATE EVOLUTION

Traditionally, certain morphological traits of primates were thought to be related to the acquisition of an arboreal way of life by a pre-primate, primitive mammal of the order Insectivora in the late Mesozoic (Table 2-6). Grafton Elliot Smith (1912), an English anatomist who specialized in comparative neurology, explained the reduction in the sense of smell and the elaboration of the senses of vision, touch, and hearing in early primates as adaptations to life amidst the branches of trees.

Wood Jones, an assistant to Smith, elaborated Smith's theories (Jones 1916). He focused on the postcranial features of the primate morphological pattern. Jones emphasized the adaptive significance of the use of the forelimbs for touch and climbing in the trees. This, Jones claimed, led to the "emancipation of the forelimbs" and improved eye-hand coordination. Since these early papers, this theory was elaborated upon and referred to as the "arboreal theory" of primate evolution (Howells 1947, Clark 1963).

Until the 1970s, this theory was generally accepted as the explanation for the adaptive significance of many unique primate morphological traits. These traits were considered to be adaptations of our earliest ancestors to the acquisition of an arboreal way of life. LeGros Clark (1963:43) summarizes as follows:

> *The evolutionary trends associated with the relative lack of structural and functional specialization are a natural consequence of an arboreal habitat, a mode of life which among other things demands or encourages prehensile functions of the limbs, a high degree of visual acuity, and the accurate control and coordination of muscular activity by a well-developed brain.*

Table 2-6 Recent Geologic Time Periods

Eras	Duration of Periods (Millions of Years)	Periods	Epochs	(Millions of Years)
			Recent	0.01
	3	Quarternary	Pleistocene	3
			Pliocene	12–3
CENOZOIC			Miocene	25–12
(65–70	65	Tertiary	Oligocene	34–25
million			Eocene	58–34
years			Paleocene	65–58
duration)				
MESOZOIC	60	Cretaceous		
(130 million	35	Jurassic		
years	35	Triassic		
duration)				
PALEOZOIC				
(300 million				
years duration)				

THE ORIGIN OF PRIMATES: A PALEONTOLOGICAL APPROACH

The arboreal theory of primate evolution was developed by primate anatomists and behavioralists using traits found in taxa of living primates. Paleontologists have discovered late Cretaceous and early Paleocene mammals with some features that are unique and primate-like, and other features that are shared by a number of early mammalian groups. The best classification for these early primate-like mammals remains uncertain. Because they are working with forms close to the point of divergence between primates and other early Tertiary mammals, paleontologists often find it difficult to draw boundaries between higher taxonomic groups at this early stage. In fact, many students of the Primates claimed that no major adaptive shift occurred in the earliest stages of primate evolution (Simpson 1955, 1961, Clark 1963, McKenna 1966). These authors asserted that one would have to rely on specific "taxonomic traits", the adaptive significance of which is not necessarily recognized, to define the Order Primates. One such trait, the petrosal bulla (Figure 2-3), was thought by many to serve as a marker of the order (Van Valen 1965, McKenna 1966, Martin 1968, 1986a, Szalay and Delson 1979, Covert 1986, Szalay et al. 1987).

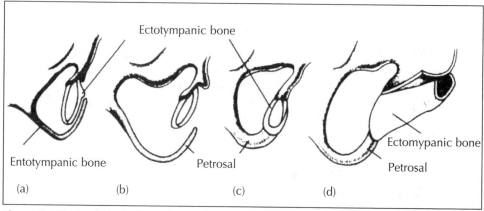

Figure 2-3 The auditory bulla in mammals. The main bone forming the bulla is the entotympanic in tree shrews (a) and the petrosal (part of the temporal bone) in primates: lemurs (b), lorises (c), and in tarsiers, Old World Monkeys, apes and humans (d). [From Conroy 1990]

The early primate-like mammals are usually grouped into five or six families, the Paromomyidae, Picrodontidae, Plesiadapidae, Carpolestidae, Saxonellidae, and for some authors the Microsyopidae, all of which make up the suborder Plesiadapiformes (see reviews by Gunnell 1989, Conroy 1990, Fleagle 1999). The Plesiadapiformes, which may or may not fit in the Order Primates, stand in contrast to later certain primates of the Eocene which are called euprimates (true primates) to avoid confusion with these archaic groups.

The plesiadapiforms are found in various Paleocene and Eocene deposits of Europe and North America, and range in size from that of a mouse to a large domestic cat. They did not look like modern primates (Figure 2-4). They possessed long snouts, relatively small brains, a relatively large olfactory apparatus, minimally convergent orbits, and where postcranials have been discovered, clawed digits similar to those of a squirrel. However, the plesiadapoids (except for microsyopids) presumably possessed a petrosal bulla (but see MacPhee et al. 1983 and Kay et al.

Figure 2-4 Artists rendition of a plesiadapoid. Drawing by Laurie Schlueb.

1990). Furthermore, some authors believed that they shared a number of dental characteristics with genera of Eocene euprimates (e.g. Simpson 1935, Szalay 1968, 1969, Gingerich 1986; but see Wible and Covert 1987, Martin 1990). Thus, despite their obvious dissimilarities it was believed that plesiadiformes and euprimates might be closely related to each other, and in fact, that the former may have been the direct ancestor of the latter. Therefore, this group of Paleocene mammals has been classified by many as the earliest form of primate. Unlike many authors, however, Szalay (1968, 1972) proposed an ecological explanation rather than just a taxonomic one by suggesting that both were the result of an evolutionary shift from mammals that were primarily insectivorous to those including more and more plant material in their diet.

> *It is safe to presume that the various features of the early prosimian dentition reflect a rather important shift in the nature of the whole feeding mechanism. Sporadic finds of primate skulls in the early Tertiary confirm this shift as a change from an insectivorous diet (i.e., a carnivorous diet in a special sense) of the insectivore ancestry to a herbivorous one . . . it is only an increasing occupation of feeding on fruits, leaves, and other herbaceous matter that explains the first radiation of primates* (Szalay 1968:32).

THE ORIGIN OF PRIMATES: TERMINAL BRANCH FEEDING INSECTIVORES

Now I have presented three divergent points of view concerning primate origins and early adaptations: (1) primates are arboreally adapted mammals; (2) no recognizable major adaptive shift occurred at the primate-insectivore border and non-adaptive ordinal traits (e.g., the petrosal bulla) must be used to distinguish early primates; and (3) dental features indicate that the insectivore-primate transition involved a dietary shift from predominantly insect feeding to predominantly plant feeding animals. More recently, Cartmill (1972, 1974, 1992) presented a fourth argument. He rejected the notion that the plesiadapoids are primates and did not agree with the arboreal theory of primate evolution.

Cartmill considered plesiadapoids to be close collateral relatives of the earliest primates with whom they shared many traits (Cartmill 1974, Kay and Cartmill 1977), but he rejected the view that the plesiadapoids shared a common adaptive trend with the primates. Cartmill argued that whereas some of these early mammals possess primate-like features, others do not display any significant adaptation that would justify including them in Primates. He argued further that an insectivore to plant-feeding shift does not define a boundary between insectivores and primates. In a detailed study of plesiadapoid dentitions, Kay and Cartmill (1977:31) stated:

> *Early and Middle Paleocene plesiadapoids appear to have been insectivorous. Specializations for eating plants do not appear among plesiadapoids until as much as 6 million years after the earliest known relatives of the group.*

Cartmill asserted that a marked adaptive shift *can* be postulated to account for the evolution of primates. The earliest primates were those characterized by the traits described by Clark (1963) and Martin (1986a, 1990): e.g., prehensility of the hands and feet with flattened nails on the fingers and toes, orbital convergence and stereoscopic vision, and regression of the snout and olfactory senses. These "primates of modern aspect" (Simons 1972) or "euprimates" (Hoffstetter 1977) first appeared in the Eocene. The arboreal theory of primate evolution was usually used to explain the evolution of these traits: i.e., they signify an adaptive trend in primate evolution towards an arboreal way of life. Cartmill argues otherwise:

> *If the primate evolutionary trends have not been characteristic of other lineages of arboreal mammals, we may conclude that there is something wrong with the arboreal theory in its received form* (1972:103).

Of the fourteen orders of terrestrial mammals, nine have arboreal forms. Many of these animals have highly successful adaptations to life in the trees and do not have the characteristic traits developed by the primates. For example, tree squirrels fill a specific set of niches in the forest canopy and these differ from those filled by primates. Unlike primates, tree squirrels normally range vertically using scansorial locomotion along large branches and tree trunks, and they are the primary and often sole consumers and dispersers of large hard fruits and nuts (Emmons 1980, Garber and Sussman 1984).

In an elegant use of the comparative method, Cartmill compared the function of specific traits shared by primates and other animals in an attempt to determine the precise niche that early primates might have filled. Cartmill (1972) found that,

> *grasping hands and feet are advantageous to animals that habitually forage in terminal branches, since they permit these animals to suspend themselves by their hind limbs (and tail, if prehensile) while using the forelimbs to reach and manipulate food items (1972:107). . . . Prehensile hands and feet are universal among shrub layer insectivores and related herbivorous forms (1972:108).*

He also concluded that,

> *outside of primates, pronounced convergence of the optic axes is largely restricted to predators. Optic convergence is particularly marked in such animals as owls, hawks, and cats which depend on vision for the detection of prey (113).*

Thus, Cartmill argued that the earliest primate adaptation involved nocturnal (because of their size), visually-oriented predation on insects in the lower canopy and undergrowth of tropical forests.

> *This implies that the last common ancestor of the extant primates, like many extant prosimians (for example, Tarsius, Microcebus, Loris, Arctocebus, and the smaller galagines), subsisted to an important*

extent on insects and other prey, which were visually located and manually captured in the insect-rich canopy and undergrowth of tropical forests (Cartmill 1974:441; also Cartmill 1992:11).

WERE THE EARLIEST PRIMATES VISUALLY ORIENTED PREDATORS?

Although Cartmill's argument is indeed an elegant one, I now ask, "Is it correct?" There are two parts to this question. The first is whether or not the plesiadapoids are primates. The second, were the earliest primates visually-oriented predators living in the terminal branches of the trees and shrubs? The first aspect of this question will not greatly concern me here. Whether or not plesiadapoids are considered to be primates often appears to revolve around one's nomenclatural philosophy (Szalay 1975, Kay and Cartmill 1977, Schwartz et al 1978, Szalay and Delson 1979, MacPhee et al. 1983, Gingerich 1986, Martin 1986a, Tattersall 1986, Wible and Covert 1987).

At present, there is no direct phyletic link and no significant adaptive resemblance between the plesiadapoids and euprimates (Martin 1993). In fact, recently discovered postcranials of the genus *Ignacius* have led to the hypothesis that one family of plesiadapoids, the paromomyids, are not related to primates but to the Southeast Asian gliding mammals, the dermopterans (Figure 2-5) (Beard 1990, Kay et al. 1990, Krause 1991). Finally, the relationship of the Plesiadiformes is secondary at best because few if any biologists believe that the euprimates are not a natural phylogenetic group.

The more important aspect of my question, to an ecologist at least, is the second: are the features shared by living primates the result of the adaptive shift to the role of visually-oriented predator that searched for insects in the terminal branches? It is true that many small mammals, including primates, eat insects. However, since Cartmill first presented his theory, the diet of a number of prosimians has been studied in detail. Most small nocturnal primates feed mainly on crawling insects, many of which are captured on the ground (Martin 1972, Doyle 1974, Fogden 1974, Charles-Dominique 1977, Charles-Dominique and Bearder 1979, MacKinnon and MacKinnon 1980, Neimitz 1984, Crompton and Andau 1986, Oxnard et al. 1990, Atsalis 1997).

Very few species of primates (i.e., the smaller galagos and lorises, and the tarsiers) are known to have a diet including a greater proportion of insects than plant material and there is evidence that the primitive (i.e., ancestral) primate

Figure 2-5 Flying lemur. Published by G. Kearsley, 1808, October, Fleet Street, London.

condition is omnivory. For example, the locomotor anatomy and general anatomy of the digestive tract of primates (e.g., the presence of the caecum and the relative size of gut compartments) reflects adaptations for an omnivorous diet (Martin et al 1985, Martin 1990, Crompton 1995). The dietary pattern found among the vast majority of primates (over 95%) is omnivory (Harding and Teleki 1981). Furthermore, according to Cartmill (1974), the trend towards orbital convergence and approximation in primates culminates in the slow-moving lorises. However, lorises rely heavily on scent to detect their prey. Eighty-five to 95% of the prey captured by the lorises are slow-moving but odoriforous. Lorises detect these insects with their highly developed sense of smell (Charles-Dominique 1977). Galagos and tarsiers seem mainly to use hearing in hunting prey.

> *When a moving prey is present, bushbabies* (Galago demidovii) *direct their mobile ears in the appropriate direction, and . . . can localize perfectly sounds emitted by insects. Flying locusts and scurrying crickets can be localized on the other side of a plywood screen, and bushbabies will follow the movements of these insects with their heads, just as if they could actually see the prey* (Charles-Dominique 1977:39).

Tarsiers, the primate most specialized for predation, also seem to use hearing to detect prey. The Bornean tarsiers "locate their prey primarily by sound and only secondarily by sight. . . . *Tarsius bancanus* catch their prey with their eyes closed" (Neimitz 1979: 642). In the larger Sulawesi tarsier, *Tarsius spectrum,* "cryptic or motionless prey are located by sight, moving prey are usually first detected by hearing, then fixated visually. The hunting tarsier's large papery ears are constantly twitching to the tiny sounds of the forest. Tarsiers pounce on insects on the ground or stretch and grab at insects in foliage" (MacKinnon and MacKinnon 1980:370; see also Crompton 1995). Almost identical observations of insect-foraging behavior have been described for *Galago senegalensis* (Doyle 1974, Oxnard et al 1990) and *Microcebus murinus* (Martin 1972, Atsalis 1997).

In a study of the role of vision in prosimian behavior, Pariente (1979:453) concludes:

> *. . . in certain nocturnal prosimian species which are able to hunt mobile animal prey, there has been considerable development of the apparatus of hearing, particularly for the detection of high frequency sound. . . . In fact, the reduction of the sphere of vision imposed by very low light intensities renders nocturnal vision insufficient for this type of diet.*

The visual anatomy of cats and primates are not very similar, and auditory stimuli are generally more compelling to carnivores than are visual stimuli (Raczkowski 1975). "It is well known that cats can catch mice in the dark or when they are hidden under fallen leaves in wood . . . a blindfolded cat can locate a mouse on a table very quickly and, as soon as it touches with its vibrissae, grasps it with a precise nape bite at lightning speed." (Leyhausen 1979: 71).

Finally, Rasmussen (1990b) showed that the arboreal marsupial *Caluromys* hunts for insects visually in the terminal branches of trees, and this species does not have marked orbital convergence. In fact, some animals that have stereoscopy, such as sloths, koalas, and some phalangerid marsupials are strictly plant eaters (Figure 2-6). Many birds that do not have convergent eyes are highly insectivorous (see Cartmill 1992, Sussman 1995). The only mammals that possess a complex visual system similar to that of primates are the fruit bats (Megachiroptera) (Allman 1982, Pettigrew 1986). As stated by Martin (1986b:483):

> *The fact that forward-facing eyes and primatelike organization of the retinotectal system should have evolved in fruit-eating megachiropteran bats, rather than in insect-eating microchiropteran bats, now provides support for the modified suggestion that the primate visual system evolved in connection with feeding on both fruits and arthropods in the 'fine-branch niche' constituted by the terminal branches of trees. It seems likely, therefore, that visual predation per se is not a sufficient explanation of the visual adaptations of the post-Paleocene primates.*

Figure 2-6 Greater Glider, a leaf-eating marsupial. From *Possums and Gliders* (ed. A. Smith and I. Hume). Surrey Beatty and Sons Publishing, Chipping Norton. 1984.

A NEW THEORY OF PRIMATE ORIGINS: COEVOLUTION WITH ANGIOSPERMS

The Paleocene-Eocene boundary was a period of rapid change that involved coincidental adaptive shifts in a number of plant and animal groups, including primates. It is in the context of the interrelationships between these groups that we might find an alternative hypothesis for the origin of primates (Sussman 1991, 1995).

I believe that the uniqueness of the earliest primates of modern aspect (the euprimates) involved a combination of the features described by both Cartmill and Szalay. I suggest that these early primates were *omnivores,* feeding on small-sized objects found in the terminal branches of trees. Thus the novel adaptive shift involved two aspects: (1) becoming well adapted to feed in the small branch milieu and (2) including a high proportion of plant material in the diet. The most important difference is that in this model the ecological resource providing a new basis for exploitation is identified—the fruit, flowers, and attracted insects of the angiosperms or flowering plants.

The evolution of birds and mammals is directly related to that of angiosperms and, in fact, vertebrate herbivores and dispersers have had a powerful influence on angiosperm evolution and visa versa. As stated in Chapter 1, Herrera (1984) has termed this kind of broad evolutionary effect of one set of lineages upon another as "diffuse coevolution." Although the general outline of parallel evolutionary events has been traced between angiosperms and a number of tetropod herbivores (Niklas et al. 1980, Tiffney 1984, Wing and Tiffney 1987a,b, Collinson and Hooker 1987), this only recently has been done specifically for primates.

The major events of angiosperm evolution are outlined in Table 2-7. Angiosperms arose in the early Cretaceous, approximately 120 million years ago (MYA). The earliest flowering plants were pollinated by primitive insects, contained small seeds, and were wind or water dispersed with little or no specializations for animal dispersal (Wing and Tiffney 1987b, Friis and Crepet 1987, Behrensmeyer et al. 1992). They were low status shrubs and weeds located mainly in unstable environments. By the late Cretaceous, some larger tree-like species appeared but most flowering plants remained of low stature, occupying marginal habitats dominated by gymnosperms (Figure 2-7).

During the Cretaceous, major coevolutionary events occurred between flowering plants and their insect pollinators. By the early Paleocene, floral biology had already reached modern form and most modern families of insects had appeared.

Figure 2-7 Late cretaceous. Drawing by Laurie Schlueb.

Although advanced pollination systems were in place by the mid-Paleocene, most fruit and seeds were still of small size and were dispersed abiotically.

In living plants, the size of the diaspore (propagative plant organ, e.g., fruit, seed, spore) is highly correlated with two classes of ecological traits: habit and habitat of the parent plant, and means of dispersal (van der Pijl 1982, Wing and Tiffney 1987b). Large, dominant forest trees and plants of late successional status tend to have large fruits and seeds, whereas herbaceous plants and early successionals usually have small propagules. The large seeds of large canopy trees reflect a higher need for stored nutrients in environments with low light levels. Furthermore, abiotically dispersed diaspores are small, while those that are dispersed by animals are generally larger and possess some reward or attractant, the nature of which depends upon the specific dispersal agent (van der Pijl 1982, Janson 1983, Howe and Westley 1988).

Table 2-7 Outline of the Evolution of Angiosperms and their Dispersers (from Sussman 1991)

	Early Cretaceous	Late Cretaceous	Early Paleocene	Mid-Late Paleocene	Eocene
Angiosperms	Successful appearance and radiation (7, 11)	Period of refinement of adaptations. Rapid increase in diversity and abundance. Established early successional plants. (7, 11)	Diversity continues to rise, but period of relative stability continues throughout Paleocene. (11, 13, 14, 16)	Slow continuous rise in diversity. Period of stability continues. More diversity in size. (11, 13, 16, 17)	Major taxonomic upheaval. Change from archaic to modern taxa. Modernization of world's flora occassioned by initiation of co-evolutionary spiral resulting from establishment of biotic dispersal relationships. (11, 14, 15, 16, 17)
Angiosperm form	Low status shrubs and "weeds." (11, 14)	Some larger arborescent individuals, but mainly as before. (4, 11, 14)	Resembled modern secondary succession, deciduous. Begin to move into closed canopy forest, some sharing of canopy status with dominant gymno-sperms. (11, 15, 16)	Still mainly r-selected, colonizing plants. Some paratropical forest, but low species diversity. (1, 15)	First closed canopy forest of modern aspect. Climax of arborescence. Angiosperm dominated tropical forests. (11, 13, 14, 15, 16, 17)
Habitat of angiosperms	Unstable environments, river edge aquatic and understory (11, 14)	Occupying only open, marginal habitats in gymnosperm dominated vegetation. Possibly some movement into association with dominant gymnosperms. (4, 11, 14, 16, 17)	Still mainly colonizing (r-selected) species with some movement into patches of woodland forest and clearings. No radical jumps into new subzones. (11, 15)	Open disturbed environments with limited forested areas. Some movements into tropical stable forest niches but archaic taxa. Greater abilities to colonize and form canopy vegetation. (1, 16, 17)	Modern taxa replace forms adapted to similar tropical niches. General climatic warming. (11, 12, 14)

Table 2-7 *(continued)*

	Early Cretaceous	Late Cretaceous	Early Paleocene	Mid-Late Paleocene	Eocene
Pollination system	Initially primitive insect pollination. (5, 6, 8)	Experimentation and diversification of pollination systems. Major increase in insect diversity and modernizing of floral morphology. Origin of bees. (5, 8)	Specializations in many major features of floral biology reach a modern level. Most modern families of insects. More specialized biotic pollinators; bees and Lepidoptera. (5, 8)	Modern pollination systems; bee and lepidopteran pollinators diversified. (5, 14)	Modern pollination systems. Major increase in diversity of modern insect taxa. (5, 8, 14)
Dispersal system	Small seeds: wind dispersed. (8, 14, 16)	First appearance of small, fleshy fruit, but mostly still wind dispersed. (8, 14, 16)	Berries first appear. Fleshy fruit relatively common. Large diaspores begin to appear, but still rare. (5, 14)	Diverse array of large and small diaspores. Range of diaspores almost as great as that of modern floras, but large diaspores still not abundant. (5, 14, 16, 17)	Very large diaspores first appear in large numbers and dominate by Mid-Eocene. Bat dispersed fruits appear. Wide range of fruits and seeds adapted to animal dispersal. Dispersal patterns achieve modern form. (14)
Dispersal agents	Mammals and birds appear.	Dispersal agents for angiosperms were few, but mammal dispersion may appear at end of this period. Herbivore community dominated by huge generalist dinosaurs. Mammals and birds relatively insignificant. (14, 16, 17)	Large dinosaurs gone; major radiation of birds and mammals. Radiation of mixed feeding mammals. Some partially arboreal forms. (9, 10, 16, 17)	Still few seeds available for bird and mammal dispersal. Few adaptations for mammalian herbivory had evolved, though adaptations for herbivory and arboreality continue to increase. Beginning of coevolutionary spiral between angiosperm and dispersal agents increases rate of speciation. (1, 2, 3, 17)	Rodents, lagomorphs, artiodactyls, perissodactyls all commence radiation. First appearance of major modern dispersers. Origin of early radiation of bats, vegetarian birds and modern primates. (14, 16, 17)

*(1) Collinson and Hooker, 1987; (2) Conroy, 1990; (3) Covert, 1986; (4) Crane, 1987; (5) Crepet & Friis, 1987; (6) Doyle & Hickey, 1976; (7) Friis, et al., 1987; (8) Friis & Crepet, 1987; (9) Krause, 1982; (10) Krause & Jenkins, 1983; (11) Niklas, et al., 1980; (12) Parrish, 1987; (13) Tiffney, 1981; (14) Tiffney, 1984; (15) Upchurch & Wolfe, 1987; (16) Wing & Tiffney, 1987a; (17) Wing & Tiffney, 1987b.

Only in the latest Cretaceous or early Tertiary did large diaspores begin to appear and did angiosperms evolve to include some dominant trees of stable, climax forests. This is presumed to be related to a change from the dominance of abiotic dispersal mechanisms in the Cretaceous to an increase in the importance of biotic dispersal agents beginning in the Tertiary (Tiffney 1984). Although there appears to be no radical jump into new adaptive zones, the total diversity of angiosperms continued to increase across the Cretaceous-Tertiary boundary and into the early Tertiary (Niklas et al. 1980, Tiffney 1981, Friis et al. 1987). This coincided with the extinction of the dinosaurs (Wing and Tiffney 1987a, Sussman 1995), and with a major radiation of birds (Tiffney 1984, Olson 1985, Feduccia 1996) and mammals (Colbert 1969, Lillegraven et al. 1979).

This radiation included the origin and diversification of the plesiadapoids in Europe and North America during the Paleocene. A similar diversification of marsupials occurred at the same time in South America (Clemens 1968, Clemens et al. 1979). It included the invasion in more or less arboreal mixed feeding adaptive zones by some plesiadapoids and other mammals (Figure 2-8) (Kay and Cartmill 1977, Szalay and Dagosto 1980, Collinson and Hooker 1987, Beard 1990, Kay et al. 1990). As mentioned above, a number of these mammals have dental morphology that suggests convergent feeding adaptations (Szalay 1968).

This new feeding niche was the small branch milieu of the newly radiating angiosperms, which offered an array of previously unexploited resources, e.g., flowers, fruits, floral and leaf buds, gums, nectars, and also the insects that feed upon these items (Rasmussen 1990b). Among the angiosperms, however, the Paleocene was a time of relative stability, during which plant species diversity continued to increase slightly but generally was maintained at a plateau level. Nothing seems to suggest that flowering plants were undergoing any major evolutionary change at this time (Hickey 1981, Tshudy 1977, Tiffney 1981, Niklas et al. 1980).

Not until the late Paleocene did large seeds with large endosperm reserves become common. Thus, fruits and seeds with clear adaptations, (thick walls,

Figure 2-8 Paleocene. Drawing by Laurie Schlueb.

attractive flesh, etc.) for animal dispersal were not a major part of the flora until the end of the Paleocene (Wing and Tiffney 1987a,b). Furthermore, Late Paleocene mammalian faunas contain few arboreal forms, few herbivores and a large proportion of insectivores (Collinson and Hooker 1987). Few adaptations for mammalian plant-feeding had evolved, although the number of animals with adaptations for plant-feeding was undoubtedly continuing to increase throughout the Paleocene (Kay and Cartmill 1977, Beard 1990, Martin 1993).

If it is true that a major impetus for this diversification of mammals was the concurrent radiation of angiosperms and their occupation of new adaptive zones, then we would expect that certain groups of animals would evolve more efficient and competitive means of exploiting these resources. At the same time, angiosperms would evolve better means of protecting themselves from predation and efficient means of exploiting these animals as dispersal agents. These coevolutionary interactions appear to have reached a threshold at the Paleocene-Eocene boundary. By the beginning of the Eocene angiosperms had evolved a range of diaspore sizes almost as great as modern floras (Table 2-7). This dramatic increase in diaspore size is strong evidence for the increased importance of animal dispersal. It also implies that angiosperms would have had a greater ability to colonize and form closed canopy vegetation and, indeed, large angiosperm trunks do not become common until the Eocene (Tiffney 1984, Wing and Tiffney 1987a,b, Upchurch and Wolf 1987).

Thus, the apparent ecological stasis in angiosperm evolution is broken in the late Paleocene and early Eocene when there is a major rearrangement in the taxonomic composition of the angiosperm community across the entire Northern Hemisphere (Tiffney 1981). The Paleocene-Eocene boundary was the time of disappearance of many archaic taxa and of the appearance of a diverse array of modern families and genera of flowering plants. Modern evergreen tropical rain forests appear and become widespread during the Eocene (Niklas et al. 1980, Tiffney 1981, Upchurch and Wolfe 1987). As stated above, by this time a wide range of fruits and seeds adapted to animal dispersal were present (Figure 2-9) (Tiffney 1984).

Paralleling this modernization of the world's flora were major changes in the fauna of the Eocene. The fossil record of birds indicates a major radiation of modern groups beginning in the latest Cretaceous. The number of modern families increased greatly in the Eocene and small, arboreal, fruit-eating birds first appear at that time. Furthermore, although the proportion of omnivorous birds remains steady, the proportion of fruit and seed-eating birds continues to rise, while that of carnivorous families falls (Tiffney 1984, Olsen 1985). Among mammals, the earliest known fossils of the important Orders Rodentia, Lagomorpha (rabbits), Artiodactyla (even-toed ungulates), and Perissodactyla (odd-toed ungulates) appear at the base of the Eocene followed by dramatic evolutionary radiations. Bats also first appeared in the early Eocene. Although the earliest bats were most likely insectivorous, the morphological characters of bat-dispersed fruit and, likely, fruit bats evolved during this time period (Tiffney 1984). Along with these major changes, the first appearance and radiation of our first euprimate ancestors, the adapids and omomyids, occurred in the late Paleocene or early Eocene (Figure 2-10) (for reviews, see Conroy 1990, Martin 1993, Fleagle 1999,).

Figure 2-9 Variety of fruits. Adapted from *Fruits of the Guianan Flora,* by Marc G. M. van Roosmalen. Utrech & University. Inst. of Syst. Botany (1985).

Thus, the establishment of biological interactions between angiosperms and their pollinators and dispersers is reflected in the rapid appearance of modern families and genera in the Eocene. The evolution of modern primates parallels that of other herbivorous mammals, of plant-eating birds, and of modern angiosperms; it also appears that many of these organisms were linked in a tight coevolutionary relationship. At present, frugivorous birds, bats, and primates are the most important seed dispersal agents in the tropics (Howe 1980, 1989, Terborgh 1986, 1992, Stiles 1989). As stated by Niklas et al. (1980:6), "plants and animals are not homologous evolutionary systems and plants, by virtue of their inherent difference from animals will provide new insights into paleontological and evolutionary arguments."

The evolution of primates of modern aspect, therefore, as well as that of fruit bats and fruit-eating birds, may be directly related to the evolution of improved means of exploiting flowering plants. Furthermore, the particular pattern of Paleocene extinctions of the plesiadapoids may be related to the rapid evolution and radiation of the primates, bats, and rodents and their improved ability to exploit insects, fruits and flowers (see Sussman and Raven 1978, Maas et al. 1988).

Some plesiadapoids persisted into the Eocene, thereby overlapping in time with euprimates. We still do not know which mammalian lineage, if any of those known from the fossil record, gave rise to the primates of modern aspect sometime during the Paleocene. The fossil record from this time period is very poor in tropical regions of the world. In contrast to the plesiadapoids, the new euprimates possessed a divergent toe and thumb and flattened nails to produce effective grasping organs. It is generally agreed that these adaptations would have allowed Eocene prosimians far greater access to fruits and flowers, as well as plant-visiting insects, making them much more efficient at locomoting and foraging in the small terminal branches of bushes and trees than were the plesiadapoids (Cartmill 1972, 1992, Martin 1993, Sussman 1995).

Figure 2-10 Eocene. Drawing by Laurie Schlueb.

This hypothesis does not explain the unique visual adaptations of primates and fruit bats, and this subject needs further study. However, these nocturnal animals were feeding on and manipulating items of very small size (e.g., fruits, flowers, and insects) at very close range and under low light conditions. This might require acute powers of discrimination and precise coordination. Tree squirrels cut large fruits or nuts from one set of branches and then move to a large horizontal support to feed (Garber and Sussman 1984). Primates do not transport food items from one support to another but usually consume a large number of items where they are acquired. They are able to venture out on very small supports and then, maintaining a firm grasp with their hindlimbs, detach the food item with their forelimbs and/or mouth (Figure 2-11).

In summary, although they vary greatly in dietary preferences, most nocturnal primates are omnivorous, feeding mainly on plant material. The largest proportion of their prey are crawling insects, many of which are captured on the ground and detected by the senses of smell or hearing. It does not seem, therefore, that the visual adaptations of post-Paleocene primates can be explained by visually-oriented predation. It is more likely that the explanation will be found in adaptations providing the fine discrimination needed to exploit the small food items available on the newly diversifying flowering plants. It is also likely that this improved, generalized ability to feed on a variety of items in the fine terminal branches was the most important impetus for the major adaptive shift seen in the Eocene primates. These first primates of modern aspect, the Eocene adapids and omomyids, were simply a more efficient version of the frugivorous-nectarivorous-insectivorous, i.e., omnivorous, plesiadapoids and other similar Paleocene mammals and they were a product of a diffuse coevolutionary interaction with the angiosperms.

Thus, I have answered the question "what is a primate?" in three ways. First, taxonomically, it is a monophyletic order of mammals containing two suborders,

Figure 2.11 Terminal branch feeding. Photo by Robert W. Sussman.

Prosimii and Anthropoidea, which derived from a common, yet unknown, ancestor. Second, primates share a number of morphological traits, especially of the
locomotor anatomy, skull morphology, dentition, and reproductive biology that
distinguish them from other mammals. Finally, these traits indicate that the adaptive shift accompanying the appearance of the primates was the occupation of
new locomotor and feeding niches made available by the co-evolving angiosperm
tropical rain forests. Although bats and birds can reach the terminal branches of
large rain forest trees by flying to them, primates need their grasping appendages
to obtain the same advantage. In fact, primates are the only major taxonomic
groups of non-flying vertebrates to regularly exploit the terminal branch niche of
the tropical forest.

BIBLIOGRAPHY

Aiello, L.C. 1986. The Relationship of the Tarsiiformes: A Review of the Case for the
    Haplorhini. Pp. 47–65 in *Major Topics In Primate and Human Evolution.* B.A.
    Wood; L.B. Martin; P.J. Andrews, eds. Cambridge, Cambridge University Press.

Allman, J.M. 1982. Reconstructing the Evolution of the Brain in Primates Through
    the Use of Comparative Neurophysiological and Neuroanatomical Data. Pp.
    13–28 in *Primate Brain Evolution.* E. Armstrong; D. Falk, eds. New York, Plenum.

Atsalis, S. 1997. *Ecology and Behavior of Microcebus rufus (Family Cheirogaleidae).*
    Ph.D. Thesis, New York, CUNY.

Beard, K.C. 1990. Gliding Behavior and Palaeoecology of the Alleged Primate Fam-
    ily Paromomyidae (Mammalia, Dermoptera). *Nature* 345:340–341.

Bearder, S.K. 1995. Species Diversity Among Galagos with Special Reference to Mate Recognition. Pp. 331–352 in *Creatures of the Dark: The Nocturnal Prosimians*. L. Alterman, G.A. Doyle, M.K. Izard, eds. New York, Plenum.

Berensmeyer, A.K.; Damuth, J.D.; DiMichele, W.A.; Potts, R.; Sues, H.D.; Wing, S.L. 1992. *Terrestrial Ecosystems Through Time: Evolutionary Paleoecology of Terrestrial Plants and Animals.* Chicago, University of Chicago Press.

Buettner-Janusch, J. 1966. *The Origins of Man.* New York, Wiley.

Campbell, C.B.G. 1974. On the Phyletic Relationships of the Tree Shrews. *Mammal Rev.* 4:125–143.

Cartmill, M. 1972. Arboreal Adaptations and the Origin of the Order Primates. Pp. 97–122 in *The Functional and Evolutionary Biology of Primates.* R. Tuttle, ed. Chicago, Aldine.

Cartmill, M. 1974. Rethinking Primate Origins. *Science* 184:436–443.

Cartmill, M. 1974b. Pads and Claws in Arboreal Locomotion. Pp. 45–83 in *Primate Locomotion.* F.A. Jenkins Jr., ed. New York, Academic.

Cartmill, M. 1992. New Views on Primate Origins. *Evol. Anthropol.* 1:105–111.

Charles-Dominique, P. 1977. *Ecology and Behavior of Nocturnal Primates.* New York, Columbia University Press.

Charles-Dominique, P.; Bearder, S.K. 1979. Field Studies of Lorisid Behavior: Methodological Aspects. Pp. 567–629 in *The Study of Prosimian Behavior*, G.A. Doyle; R.D. Martin, eds. New York, Academic.

Chivers, D.J.; Joysey, K.A., eds. 1978. *Recent Advances in Primatology: Vol. 3: Evol.* New York, Academic.

Clark, W.E. LeGros 1963. *The Antecedents of Man,* New York: Harper and Row.

Clemens Jr., W.A. 1968. Origin and Early Evolution of Marsupials. *Evol.* 22:1–18.

Clemens Jr., W.A.; Lellegraven, J.A.; Lindsay, E.H.; Simpson, G.G. 1979. Where, When, and What—A Survey of Known Mesozoic Mammal Distribution. Pp. 7–58 In *Mesozoic Mammals.* Lillegraven, J.A.; Kielan-Jaworowska, Z.; Clemens, W.A. eds., Berkeley, University of California Press.

Colbert, E.H. 1969. *Evolution of the Vertebrates: History of the Backboned Animals Through Time. Second Edition.* New York, Wiley.

Collinson, M.E.; Hooker, J.J. 1987. Vegetational and Mammalian Faunal Changes in the Early Tertiary of Southern England. Pp. 259–304 in *The Origins of Angiosperms and Their Biological Consequences.* E.M. Friis, W.G. Chaloner, P.R. Crane, eds. Cambridge, Cambridge University Press.

Conroy, G.C. 1990. *Primate Evolution,* New York, Norton.

Covert, H.H. 1986. Biology of Early Cenozoic Primates. Pp. 335–359 in *Comparative Primate Biology. Vol. 1: Systematics, Evolution, and Anatomy,* D.R. Swindler; J. Erwin, eds. New York, Alan R. Liss.

Crompton, R.H.; Andau, P.M. 1986. Locomotion and Habitat Utilization in Free-Ranging *Tarsius bancanus*: A Preliminary Report. *Primates* 27:337–355.

Crompton, R.H. 1995. "Visual predation." Habitat Structure, and the Ancestral Primate Niche. Pp. 11–30. In *Creatures of the Dark: The Nocturnal Prosimian.* L. Alterman, G.A. Doyle, M.K. Izard, eds. New York, Plenum.

Doyle, G.A. 1974. Behavior of Prosimians. Pp. 155–353 in *Behavior of Nonhuman Primates,* A.M. Schrier; F. Stolnitz, eds. New York, Academic.

Eldridge, N. 1989. *Macroevolutionary Systematics.* New York, McGraw-Hill.

Emmons, L.H. 1980. Ecology and Resource Partitioning Among Nine Species of African Rain Forest Squirrels. E*col. Monogr.* 50:31–54.

Fleagle, J.G. 1988. *Primate Adaptation and Evolution*, New York, Academic Press.

Feduccia, A. 1996. *The Origin and Evolution of Birds.* New Haven, Yale University Press.

Fleagle, J.G. 1999. *Primate Adaptation and Evolution,* New York, Academic.

Fleagle, J.G.; Kay, R.F., eds. 1994. *Anthropoid Origins,* New York, Plenum Press.

Fogden, M. 1974. A Preliminary Study of the Western Tarsier, *Tarsius bancanus* Horsfield. Pp. 151–165 In *Prosimian Biology,* R.D. Martin; G.A. Doyle; A.C. Walker, eds. London, Duckworth.

Friis, E.M.; Chaloner, W.G.; Crane, P.R. 1987. Introduction to the Angiosperms. Pg. 1–15 in *The Origins of Angiosperms and Their Biological Consequences.* E.M. Friis, W.G. Chaloner, P.R. Crane, eds. Cambridge, Cambridge University Press.

Friis, E.M.; Crepet, W.L. 1987. Time of Appearance of Floral Features. Pp. 259–304 In *The Origins of Angiosperms and Their Biological Consequences.* E.M. Friis, W.G. Chaloner, P.R. Crane, eds. Cambridge, Cambridge University Press.

Garber, P.A.; Sussman, R.W. 1984. Ecological Distinctions Between Sympatric Species of *Saguinus* and *Sciurus. Am. J. Phys. Anthropol.* 65:135–146.

Gingerich, P.D. 1986. *Plesiadapis* and the delineation of the Order Primates. Pp. 32–46 in *Major Topics in Primate and Human Evolution,* B.A. Wood; L.B. Martin; P. Andrews, eds. Cambridge: Cambridge University Press.

Goodman, M. 1966. Phyletic Position of Tree Shrews. *Science* 153:1550.

Gunnell, G.F. 1989. Evolutionary History of Microsyopoidea and the Relationship Between Plesiadapiformes and Primates. *University of Michigan Papers in Paleontol.* 27:1–154.

Harding, R.S.O. 1981. An Order of Omnivores: Nonhuman Primates Diets in the Wild. Pp. 191–214 In *Omnivorous Primates,* R.S.O. Harding; G. Teleki, eds. York, Columbia University Press.

Hennig, W. 1966. *Phylogenetic Systematics.* Urbana: University of Illinois Press.

Herrera, C.M. 1984. Determinants of Plant-Animal Coevolution: The Case of Mutualistic Dispersal of Seeds by Vertebrates. *Oikos* 44:132–144.

Howe, H.F. 1980. Monkey Dispersal and Waste of a Neotropical Fruit. *Ecol.* 61:944–959.

Howe, H.F. 1989. Scatter- and Clump-Dispersal and Seedling Demography: Hypothesis and Implications. *Oecologia* 79:417–426.

Howe, H.F.; Westley, L.C. 1988. *Ecological Relationships Between Plants and Animals.* Oxford, Oxford University Press.

Hickey, L.J. 1981. Land Plant Evidence Compatible with Gradual, Not Catastrophic, Change at the End of the Cretaceous. *Nature* 292:529–531.

Hoffstetter, R. 1977. Phylogénie des Primates: Confrontation des Résultats Obtenus par les Diverse Voies d'Approches de Problème. *Bull. Mem. Soc. Anthropol.*, Paris 4:327–346.

Howells, W.W. 1947. *Mankind So Far.* Garden City, Doubleday.

Janson, C.H. 1983. Adaptation of Fruit Morphology to Dispersal Agents in a Neotropical Forest. *Science* 219:187–189.

Jurmain, R.; Nelson, H. 1994. *Introduction to Physical Anthropology, Sixth Edition.* St. Paul, West.

Jones, F.W. 1916. A*rboreal Man.* London, Edward Arnold.

Jolly, C.J. 1993. Species, Subspecies, and Baboon Systematics. pp. 67–107 in *Species, Species Concepts, and Primate Evolution.* Kimbel, W.H.; L.B. Martin, eds. New York, Plenum.

Kay, R.F.; Cartmill, M. 1977. Cranial Morphology and Adaptations of *Palaecthon nacimienti* and Other Parmomyidae (Plesiadapoidea, ? Primates), with a description of a New Genus and Species. *J. Hum. Evol.* 6:19–53.

Kay, R.F. 1990. Thorington, R.W.; Houde P.; Eocene Plesiadapiform Shows Affinities with Flying Lemurs Not Primates. *Nature* 345:342–344.

Kimbel, W.H.; L.B. Martin, eds. 1993. *Species, Species Concepts, and Primate Evolution.* New York, Plenum.

Kluge, A.G. 1984. The Relevance of Parsimony to Phylogenetic Inference. Pp. 24–38 In *Cladistics.* T. Duncan; T.F. Stuessy, eds. New York, Columbia University Press.

Krause, D.W. 1991. Were Paromomyids Gliders? Maybe, Maybe Not. *J. Hum. Evol.* 21:177–188.

Leyhausen, P. 1979. *Cat Behavior: The Predatory and Social Behavior of Domestic and Wild Cats.* New York, Garland.

Lillegraven, J.A.; Kielan-Jaworowska, Z.; Clemens, W.A. eds., 1979. *Mesozoic Mammals.* Berkeley, University of California Press.

Luckett, W.P.; Szalay, F.S. eds. 1975. *Phylogeny of the Primates: A Multidisciplinary Approach.* New York, Plenum.

Luckett, W.P. 1980. *Comparative Biology and Evolutionary Relationships of Tree Shrews.* New York, Plenum.

Maas, M.C.; Krause, D.W.; Strait, S.G. 1988. Decline and Extinction of Plesiadapiforms in North America. Paleobiology 14:410–431.

MacKinnon, J; MacKinnon, K. 1980. The Behavior of Wild Spectral Tarsiers. *Intl. J. Primatol.* 1:361–379.

MacPhee, R.D.E.; Cartmill, M.; Gingerich, P.D. 1983. New Palaeogene Primate Basicrania and the Definition of the Order Primates. *Nature* 301:509–511.

Martin, R.D. 1968. Towards a New Definition of Primates. *Man* 3:376–401.

Martin, R.D. 1972. A Preliminary Field-Study of the Lesser Mouse Lemur *(Microcebus murinus* J.F. Miller 1777). *Tierpsychol.* 9:43–89.

Martin, R.D. 1986a. Primates: A Definition. Pp. 1–31 In *Major Topics in Primate and Human Evolution,* B.A. Wood; L.B. Martin; P. Andrews, eds. Cambridge, Cambridge University Press.

Martin, R.D. 1986b. Are Fruit Bats Primates. *Nature* 320:482–483.

Martin, R.D. 1990. *Primate Origins and Evolution: A Phylogenetic Reconstruction.* Princeton, Princeton University Press.

Martin, R.D. 1993. Primate Origins: Plugging the Gaps. *Nature* 363:223–234.

Martin, R.D.; Chivers, D.J.; MacLarnon, M.A.; Hladik, C.M. 1985. Gastrointestinal Allometry in Primates and Other Mammals. Pp. 61–89 In *Size and Scaling in Primate Biology,* W.L. Jungers, ed., New York, Plenum.

Masters, J.C. 1998, Speciation in the Lesser Galagos. *Folia Primatol.* 69: 357–370.

Mayr, E. 1963. *Animal Species and Evolution.* Cambridge, Belknap.

Mayr, E. 1969. *Principles of Systematic Zoology.* New York, McGraw-Hill.

Mayr, E.; Ashlock, P.D. 1991. *Principles of Systematic Zoology. Second Edition.* New York, McGraw-Hill.

McKenna, M.C. 1966. Paleontology and the Origins of Primates. *Folia Primatol.* 4:1–25.

Mivart, St. G.J. 1873. On *Lepilemur* and *Cheirogaleus* and on the Zoological Rank of the Lemuroidea. *Proc. Zool. Soc. Lond.* 1873:484–510.

Neimitz, C. 1979. Outline of the Behavior of *Tarsius bancanus.* Pp. 631–660 in *The Study of Prosimian Behavior,* G.A. Doyle; R.D. Martin, eds. New York, Academic.

Neimitz, C. 1984. Synecological Relationships and Feeding Behavior of the Genus *Tarsius.* Pp. 59–76 In *Biology of Tarsiers,* C. Neimitz, ed. Stuttgart: G. Fischer Verlag.

Niklas, K.J.; Tiffney, B.H.; Knoll, A.H. 1980. Apparent Changes in the Diversity of Fossil Plants: A Preliminary Assessment. Pp. 1–89 In *Evol. Biol., Vol. 12,* M.K. Hecht; W.C. Steere; B. Wallace, eds. New York, Plenum.

Olsen, S.L. 1985. The Fossil Record of Birds. Pp. 79–237 In *Avian Biology,* Vol. III, D. Farner, J. King; J.K. Parkes, eds. New York, Academic Press.

Otte, D.; Endler, J.A. eds. 1989. *Speciation and Its Consequences.* Mass., Sinauer.

Oxnard C.E.; Crompton, R.H.; Lieberman, S.S. 1990. *Animal Lifestyles and Anatomies: The Case of the Prosimian Primates.* Seattle, University of Washington Press.

Pariente, G. 1979. The Role of Vision in Prosimian Behavior. Pp. 411–459 In *The Study of Prosimian Behavior,* G.A. Doyle; R.D. Martin, eds. New York, Academic.

Paterson, H.E.H. 1985. The Recognition Concept of Species. Pp. 21–19. In *Species and Speciation,* E.S. Vrba, ed. Transvaal, Transvaal Museum.

Pettigrew, J.D. 1986. Flying Primates? Megabats Have the Advanced Pathway from Eye to Midbrain. *Science* 231:1304–1306.

Pijl, L. van der, 1982. *Principles of Dispersal in Higher Plants.* Berlin, Springer-Verlag.

Raczkowski, D. 1975. Primate Evolution: Were Traits Selected for Arboreal Locomotion or Visually Directed Predation. *Science* 187:455–456.

Rasmussen, D.T. 1990a. The Phylogenetic Position of *Mahgariba stevensi:* Protoanthropoid or Lemuroid? *Intl. J. Primatol.* 11:439–469.

Rasmussen, D.T. 1990b. Primate Origins: Lessons from a Neotropical Marsupial. *Am. J. Primatol.* 22:263–277.

Rasmussen, D.T. 1994. The Different Meanings of a Tarsioid-Anthropoid Clade and a New Model of Anthropoid Origins. Pp. 335–360 in *Anthropoid Origins*. J.G. Fleagle and R.F. Kay, eds. New York, Plenum.

Raven, P.H.; Evert, R.F.; Curtis, H. 1992. *Biology of Plants, Fifth Edition*. New York, Worth.

Rose, K.D.; Bown, T.M. 1986. Gradual Evolution and Species Discrimination in the Fossil Record. *Contrib. Geol. Univ. Wyoming* 3:119–130.

Schwartz, J.H.; Tattersall, I.; Eldredge, N. 1978. Phylogeny and Classification of the Primates Revisited. Y*rbk. Phys. Anthropol.* 21:95–133.

Simons, E.L. 1972. *Primate Evolution: An Introduction to Man's Place in Nature*. New York, Macmillan.

Simpson, G.G. 1935. The Tiffany Fauna, Upper Paleocene. 2. Structure and Relationships of *Plesiadapis. Am. Mus. Novit.* 816:1–30.

Simpson, G.G. 1944. *Tempo and Mode in Evolution.* New York, Columbia University Press.

Simpson, G.G. 1945. The Principles of Classification and a Classification of Mammals. *Bull. Amer. Mus. Nat. Hist.* 85:1–350.

Simpson, G.G. 1955. The Phenacolemuridae, New Family of Early Primates. *Bull. Amer. Mus. Nat. Hist.* 105:411–442.

Simpson, G.G. 1961. *Principles of Animal Taxonomy*. New York, Columbia University Press.

Simpson, G.G. 1975. Recent Advances in Method of Phylogenetic Inference. Pp. 3–19 In *Phylogeny of the Primates: A Multidisciplinary Approach.* W.P. Luckett; F.S. Szalay, eds. New York, Plenum.

Smith, G.E. 1912. *The Evolution of Man. Smith. Inst. Ann. Rep.*.

Sneath, P.H.A., Sokal, R.R. 1973. *Numerical Taxonomy: The Principles and Practice of Numerical Taxonomy.* San Fransisco, Freeman.

Sussman, R.W. 1991. Primate Origins and the Evolution of Angiosperms. *Am. J. Primatol.* 23:209–223.

Sussman, R.W. 1995. How Primates Invented the Rainforest and Visa Versa. Pp. 1–10 in *Creatures of the Dark: The Nocturnal Prosimians.* L. Alterman; K. Izard; G.A. Doyle, eds. New York, Plenum.

Sussman, R.W.; Raven P.H. 1978. Pollination by Lemurs and Marsupials: An Archaic Coevolutionary System. *Science* 200:731–736.

Stiles, E.W. 1989. Fruits, Seeds, and Dispersal Agents. Pp. 87–122 in *Plant-Animal Interactions.* W.G. Abrahamson, ed. New York, Mcgraw–Hill.

Szalay, F.S. 1968. The Beginnings of Primates. *Evol.* 22:19–36.

Szalay, F.S. 1969. Mixodectidae, Microsyopidae, and the Insectivore-Primate Transition. *Bull. Am. Mus. Nat. Hist.* 140:193–330.

Szalay, F.S. 1972. Paleobiology of the Earliest Primates. Pp. 3–35 in *The Functional and Evolutionary Biology of Primates.* R. Tuttle, ed. Chicago, Aldine.

Szalay, F.S. 1975. Phylogeny of Primate Higher Taxa: the Basicranial Evidence. Pp. 91–125 in *Phylogeny of the Primates: A Multidisciplinary Approach.* W.P. Luckett; F.S. Szalay, eds. New York, Plenum.

Szalay, F.S.; Dagosto, M. 1980. Locomotor Adaptations as Reflected on the Humerous of Paleogene Primates. *Folia Primatol.* 34:1–45.

Szalay, F.S.; Delson E. 1979. *Evolutionary History of the Primates,* New York, Academic Press.

Szalay, F.S.; Rosenberger, A.L.; Dagosto, M. 1987. Diagnosis and Differentiation of the Order Primates. *Yrbk. Phys. Anthropol.* 30:75–105.

Tattersall, I. 1986. Review of: Major Topics in Primate and Human Evolution. *J. Hum. Evol.* 15:313–321.

Templeton, A.R. 1987. Species and Speciation. *Evolution* 41: 235–236.

Templeton, A.R. 1989. The Meaning of Species and Speciation: a Genetic Perspective. Pp. 3–27 in *Speciation and Its Consequences.* D. Otte; J.A. Endler, eds., Mass., Sinauer.

Templeton, A.R. 1994. The Role of Molecular Genetics in Speciation Studies. *Molecular Ecology and Evolution: Approaches and Applications.* B. Schierwater; J.A. Endler, eds. Basel, Birkhauser Verlag.

Terborgh, J. 1986. Community Aspects of Frugivory in Tropical Forests. Pp. 371–384 in *Frugivores and Seed Dispersal.* A. Estrada; T.H. Fleming, eds. Dordrecht, Dr. W. Junk.

Terborgh, J. 1992. *Diversity and the Tropical Rain Forest.* New York, Scientific American Library.

Tiffney, B.H. 1981. Diversity and Major Events in the Evolution of Land Plants. Pp. 193–230 in *Paleobotany, Paleoecology, and Evolution, Vol. II.* K.J. Niklas, ed. New York, Praeger.

Tiffney, B.H. 1984. Seed Size, Dispersal Syndromes, and the Rise of the Angiosperms: Evidence and Hypothesis. *Annals of the Missouri Botanical Garden* 71:551–576.

Tshudy, R.H. 1977. Palynological Evidence for Change in Continental Floras at the Cretaceous-Tertiary Boundary. *J. Paleontol.* 51:29.

Uexkull, J. von 1957. A Stroll Through the World of Animals and Men: A Picture Book of Invisible Worlds. Pp. 5–80 in *Instinctive Behavior: The Development of a Modern Concept.* C.H. Schiller, ed. New York, International Universities Press.

Upchurch Jr., G.R.; Wolfe J.A. 1987. Mid-Cretaceous to Early Tertiary Vegetation and Climate: Evidence from Fossil Leaves and Woods. Pp. 75–105 in *The Origins of Angiosperms and Their Biological Consequences.* E.M. Friis, W.G. Chaloner, P.R. Crane, eds. Cambridge, Cambridge University Press.

Van Valen, L. 1965. Tree Shrews, Primates, and Fossils. *Evol.* 19: 137–151.

Vrba, E.S. ed. 1985. *Species and Speciation.* Trasvaal, Transvaal Museum.

Wible, J.R.; Covert, H.H. 1987. Primates: Cladistic Diagnosis and Relationships. *J. Hum. Evol.* 16:1–22.

Wing, S.L.; Tiffney, B.H. 1987a. The Reciprocal Interaction of Angiosperm Evolution and Tetrapod Herbivory. *Review of Palaeobotany and Palynology* 50:179–210.

Wing, S.L.; Tiffney, B.H. 1987b. Interactions of Angiosperms and Herbivorous Tetrapods Through Time. Pp. 203–224 in *The Origins of Angiosperms and Their Biological Consequences.* E.M. Friis, W.G. Chaloner, P.R. Crane, eds. Cambridge, Cambridge University Press.

PART II

The Ecology of Prosimians

INTRODUCTION

Prosimians are the earliest primates of modern aspect to appear in the fossil record. Found in Eocene deposits in Europe, North America, Asia, and recently in Africa, they lived in areas tropical in climate. Two Eocene families of Prosimii are recognized: Adapidae and Omomyidae. Although both families were very diverse, it can be stated generally that the Adapidae were lemurlike, and the Omomyidae were tarsier-like. Recent fossil finds in China extend modern Tarsiidae back to the Eocene (Beard 1998). The early Tertiary fossil record of Africa is almost entirely unknown. However, a loris-like primate recently has been discovered in Eocene deposits of Egypt (Simons and Rasmussen 1994).

By the Oligocene, prosimians seem to largely disappear, and most fossil discoveries for this time period are of anthropoids, mainly in Africa. There is one record of an African species of Tarsiidae during the Oligocene. By the early Miocene prosimians extremely close to modern lorisoids occurred in Africa while a tarsier is known from Thailand. However, the Miocene lorisoids have not helped to close the morphological gap between the Eocene and recent prosimians because they were so much like modern forms (Simons 1972, Conroy 1990, Fleagle 1999). For a detailed summary of current knowledge on the evolutionary history of the Lorisiformes see Rasmussen and Nekaris 1998.

Because of the present-day distribution of prosimians, their current ecological adaptations, and the rarity of these forms in the post-Oligocene fossil record, it is likely that only a few fairly specialized species survived past the Eocene. The Lemuriformes are found now only on the island of Madagascar and on the Comores (Figure 3-1). They have been isolated on Madagascar at least since the Eocene, and it is probably due to this isolation from competition with the anthropoids and other evolving mammalian groups that they owe their continued existence and diversification. Lorisiformes are found in Africa and Asia, and modern Tarsiiformes are found only in Southeast Asia, on oceanic islands.

71

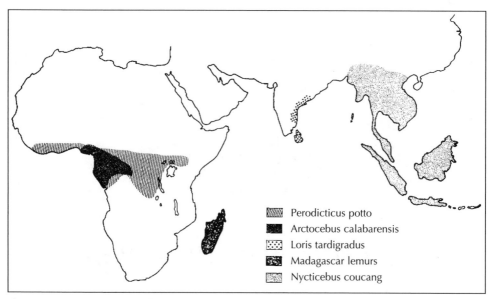

Figure 3-1 Occurrence of the Asiatic lorisines (Nycticebus coucang and Loris tardigradus), African lorisines (Perodicticus potto and Actocebus calabarensis and the Madagascar lemurs (Lemuriformes). [Adapted from Charles-Dominique 1977a]

CHAPTER 3

Lorisiformes

INTRODUCTION

The living Lorisiformes are represented by one family (Lorisidae) that is divided into two subfamilies: Lorisinae and Galaginae (see Table 2-4). These two subfamilies had diverged from each other by the late Miocene (Rasmussen and Nekaris 1998). All Lorisiformes are nocturnal (active at night and asleep during the day). Lorisines are slow moving animals, and as I describe, this specialized locomotion pervades all aspects of their morphology and behavior and is the major determinant of their habitat preferences. There are four genera of lorisines currently recognized, all but one of which are monotypic. They are found in the equatorial forests of Africa *(Arctocebus* and *Perodicticus)* and in the tropical forests of India and Sri Lanka *(Loris)* and Southeast Asia *(Nycticebus)*. As shown in Figure 3-1, *Arctocebus* has a restricted range in west central Africa and is sympatric with *Perodicticus* throughout its range. The two Asian genera *(Loris* and *Nycticebus)* are allopatric.

Currently there is a great deal of disagreement concerning the number of genera existing within the Galaginae (Nash et al. 1989). Furthermore, within the past 15 years, the number of recognized species of galagos has increased from 6 to 12 (Olson 1979, Nash et al. 1989, Bearder 1995). This is because taxonomists in the past placed undue emphasis on visually detectable characters in formulating their classifications of galagos (this is most likely also true of the classification of lorises, nocturnal lemurs and tarsiers). In these nocturnal primates, many morphological differences between species are subtle, rendering separation of species difficult. However, using Paterson's (1985) suggestion that sexually reproducing species often can be distinguished on the basis of specific-mate recognition systems (see previous chapter), researchers have found that galago species can be identified by a combination of species-specific signals, among which certain vocalizations appear to be particularly important (e.g., Masters 1988, 1991, 1993, Zimmermann et al. 1988, Courtenay and Bearder 1989, Zimmermann 1990, 1995, Bearder 1995, Masters 1998).

Table 3-1 Common Names of Lorisiformes

Family LORISIDAE

Species	Common Name
Perodicticus potto	Potto
Arctocebus calabarensis	Angwantibo, Golden potto
Loris tardigradus	Slender loris
Nycticebus coucang	Common slow loris
N. pygmaeus	Pygmy slow loris

Family GALAGINAE

Species	Common Name
Galago crassicaudatus	Large-eared greater bushbaby
G. garnettii	Small-eared greater bushbaby
G. senegalensis	Senegal bushbaby
G. gallarum	Somali bushbaby
G. moholi	Moholi's, South African lesser bushbaby
G. elegantulus	Needle-clawed bushbaby
G. matschiei	Matschie's needle-clawed bushbaby
G. demidoff	Dwarf, Demidoff's bushbaby
G. thomasi	Thomas' bushbaby
G. orinus	Malawi bushbaby
G. zanzibaricus	Zanzibar bushbaby
G. alleni	Allen's bushbaby

Until the taxonomists and galago specialists agree on the genetic classification of this group, I prefer to place them all into one genus, *Galago*. The twelve species recognized are listed in Table 3-1. All galagos are fast moving, leaping animals with powerfully developed hind limbs and long tails (in contrast to the lorises which have greatly reduced tails, Figure 3-2). Galagines are found throughout sub-Saharan Africa (Figure 3-3). The weights and body measurements of the Asian and African lorisids are given in Table 3-2.

Unfortunately, nocturnal primates are difficult to study. Very little is known about the Asian lorises, and there has been little or no field research on the following five galago species: *Galago gallarum, G. orinus, G. matschiei, G. senegalensis,* and *G. thomasi*. There are, however, excellent studies available on the other African forms (see Baldwin and Teleki 1977, Williams 1983, 1991, and Rasmussen 1997 for detailed bibliographies).

THE LORISIDS OF GABON: A CASE STUDY

The most detailed research to date on Lorisiformes in their natural habitat is that of Charles-Dominique (1971, 1974, 1977a). Between 1965 and 1973, Charles-Dominique spent 42 months studying the ecology and social behavior of five sympatric species:

Arctocebus calabarensis, Perodicticus potto, Galago alleni, G. demidoff, and *G. elegantulus.* The study was carried out in a field station supported by the French Centre National de Recherche Scientifique (C.N.R.S.) in the primary equatorial rain forest of Gabon, West Africa (Figure 3-4). This research provides us with a basic understanding of the particular morphological and behavioral adaptations of the Lorisiformes.

Figure 3-2a Photo of Potto. [Photo by Ben Freed]

As stated above, the lorises are slow-moving climbers. They move cautiously through the forest and never leap. Galagos are fast and their customary mode of locomotion is leaping. These behavioral differences can be related to a number of differences in locomotor anatomy, especially in the limbs, hand, and tail (McArdle 1981, Preuschoft et al. 1995, Sellers 1996). By contrast, there are only minor differences between the two subfamilies in the skull, dentition, digestive tract, and reproductive organs (Charles-Dominique 1974). The extreme differences in locomotor capabilities of the two taxa can be directly related to where they choose to locomote, to what they eat, to how they avoid being eaten, to where they sleep, and to their social behavior. In his field

Figure 3-2b Photo of Senegal Bushbaby. [Photo by Simon Bearder]

study, Charles-Dominique focuses on how these five species of prosimians select different habitats within the same forest and thus avoid competing for resources.

SPATIAL LOCALIZATION

One of the first questions that a field investigator asks is where are the animals most likely to be found? The methods used to answer this question vary with habitat, degree to which the animals are habituated to the observer, activity pattern (i.e., are they nocturnal or diurnal), and a number of other factors. To study animals that are active at night, it is necessary to use headlamps to locate the animal and then powerful spotlights to observe detailed behavior (see Charles-Dominique and Bearder 1979). In many cases, it is difficult or impossible to follow the animal for any length of time. One conducts searches along predetermined paths and observes a sighted animal as long as possible, then searches for another one. To study individuals, animals are often trapped and marked for identification and then retrapped over the tenure of the study (referred to as the *mark-recapture technique*). Finding animals can be facilitated by radio-tracking. The location of each sighted animal is mapped (Figure 3-5).

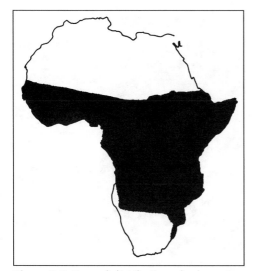

Figure 3-3 General distribution of galagos in Africa.

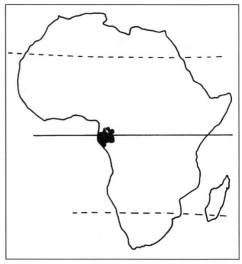

Figure 3-4 Map showing the geographical location of the study zone of Charles-Dominique (1977a).

Charles-Dominique wanted to determine the type of vegetation and substrate that each of the five species of lorisids most commonly used. He spent 2500 hours conducting searches with a battery-operated headlamp during the 42 months in Gabon. He took systematic notes on the characteristics of the supports on which the prosimians were observed during the night. The following characteristics were noted: type (ground, thin trunk, large vine, large trunk, large branch, foliage, foliage mixed with vines), height, diameter, and orientation (horizontal, oblique or vertical). Since the light-beam often had an effect on subsequent movements of an animal, only the support on which it was *first* sighted, at any one time, was recorded. Charles-Dominique made 836 sightings in this manner on the five species in primary and secondary forest. These data were supplemented by detailed observations of a few individuals who were radio-tracked. An animal is trapped and fitted with a miniaturized radio-transmitter; the tagged animal can then be located and followed using a directional antenna.

Perodicticus potto (Figure 3-2) was found to spend most of its time in the highest canopy portion of the forest (10–30 m. in primary forest, 5–15 m. in secondary forest). It uses a variety of supports of a wide range of sizes (1-30 cm.). The potto, a quadrupedal climber, moves from tree to tree by initially using large limbs and gradually moving out onto smaller branches and vines until it reaches the fine branches of the adjacent tree (Figure 3-6). Pottos use branches at all angles though they are seen on horizontal supports most frequently (39%, Table 3-3). In Gabon, the potto rarely descends to the ground. Outside of the tropical forests of West Africa, the potto is found commonly in secondary forests, in clearings along forest margins and in savannah (Jewell and Oates 1969, Kingdon 1971, Sabater Pi 1972, Oates 1984). In these areas it frequently descends to the ground.

Unlike the potto, *Arctocebus calabarensis* (Figure 3-7), commonly called the angwantibo, lives between 0-5 m. in both primary and secondary forest, only going higher in the trees when greatly alarmed. The angwantibo explores the foliage of

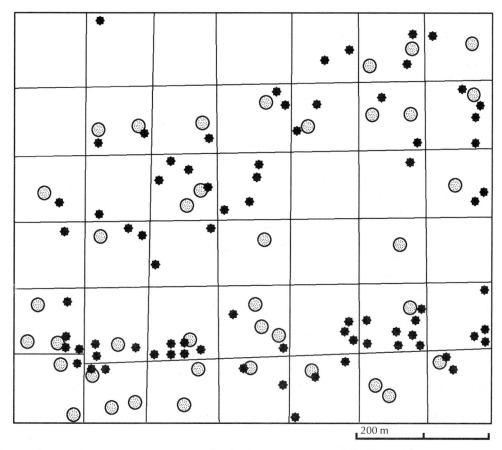

Figure 3-5 Diagram illustrating the distribution of a gum-producing liane *(Entada gigas)* and sightings of needle-clawed bushbabies during the night. (Circles = lianes; stars = bushbaby sightings.) The data were obtained by equally distributed surveys conducted along transects at intervals of 100 m. (77 sightings of *Euoticus elegantulus* in 42 1-hectare squares containing 39 identified bases of *Entada gigas).* In this region, the needle-clawed bushbabies feed primarily on gums from *Entada gigas* and follow the heterogeneous distribution of this liane species. (N.B. The transects, which were not exactly parallel, are indicated by lines.) [From Charles-Dominique 1977a]

bushes in the undergrowth for insects and fruit, using small vines to pass between bushes. In the primary forest, the angwantibo is often found in tree-fall zones and forest edges (i.e. ecotones; see Chapter 1), where dense undergrowth is more abundant. It uses supports having small diameters (40% less than 1 cm.; 52% between 1-10 cm.), and often descends to the ground to eat fallen fruit or to search for insects. Angwantibos use all angles of support but are most frequently found on vertical branches and vines (Table 3-3).

The galagos, commonly referred to as bushbabies, are characterized by their fast speed and leaping locomotion. *Galago demidoff,* Demidoff's or the dwarf galago, is the smallest species of bushbaby (Figure 3-8), weighing only 60 gm (Table 3-2). It inhabits dense vegetation, moving up to the terminal branches in the primary forest (10–30 m.) and to low branches in the dense foliage of the secondary forest (0–10 m.). The vertical distribution of this species is not determined by height but by the presence of suitable habitats of dense vegetation.

The dwarf galago occasionally comes to the ground. The angle of the supports utilized by this galago are similar to those used by the angwantibo (Table 3-3). Outside the tropical forest belt, *G. demidoff* is found in bamboo, swamp, gallery, lowland, and mountain forests (Kingdon 1971).

Galago alleni, or Allen's bushbaby, is located mainly at low levels of the forest, at heights of between 0–2 m. Its locomotion is characterized by what is called vertical clinging and leaping (Napier and Walker 1967). The animal jumps from one vertical support to another, clinging to the vertical support between leaps. Allen's bushbaby mainly leaps between thin trunks and vines and

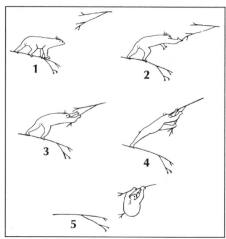

Figure 3-6 Potto moving between two trees separated by a small gap. [From Charles-Dominique 1977a]

in five seconds can cover 12 meters with five or six leaps. Approximately 70% of the supports utilized by this galago are less than 5 cm in diameter and it uses vertical supports over 80% of the time. However, by following radio-tracked individuals, Charles-Dominique found that Allen's bushbabies spend a great deal of time on the ground foraging for prey and fallen fruit. In some areas, they move out of the forest at night to forage in grasslands or bush vegetation (Jewell and Oates 1969).

Galago elegantulus is active exclusively in the forest canopy and moves in large trees whose larger branches are usually oblique or horizontal. It has a special adaptation for climbing on large and smooth branches and trunks. The nails on digits 2, 3, 4, and 5 of the hand and 3, 4 and 5 of the foot are keeled and elongated to form a

Figure 3-7 Photo of angwantibo.
[Photo by Simon Bearder]

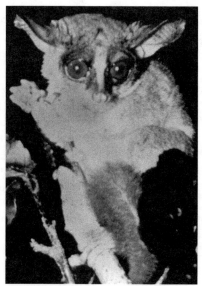

Figure 3-8 Photo of *G. demidoff.*
[Photo by Simon Bearder]

Table 3-2 Field Weights and Measurements of Some Lorisiformes[1]

	Mean Body Weights (gm) (sample size)	Head & Body Length (range in mm)	Tail Length (mm)	Reference
Perodicticus potto	850 – 1600 (33) x = 1100	305 – 370 x = 327	37 – 70 x = 50	Charles-Dominique 1977a
Arctocebus calabarensis	150 – 270 (30) x = 210	230 – 260 x = 244	15	Charles-Dominique 1977a
Loris tardigradus	85 – 348 (23)* 283 (4)**	186 – 264*	—	*Napier & Napier 1967 **Singh & Udhayan unpub. ms.
Nycticebus coucang	M679 (56)* F626 (44)	265 – 380 (9)**	—	*Bearder 1986; **Napier & Napier 1967
N. pygmaeus	190 – 230 (3)	190 (1♂)	—	Bearder 1986
Galago crassicaudatus	M 1510 (8)* F 1258 (9)	M 335 (4)** F 315 (4)	M 458 (4)** F 421 (4)	*Charles-Dominique 1977a **Napier & Napier 1967
G. garnettii	M 822 (8) F 721 (5)	260 – 294 (7) x = 278	330 – 410 (14) x = 360	Harcourt 1984
G. moholi[2]	M 210 (21)* F 193 (14)	150 – 173 (2)**	205 – 250 (2)**	*Olson 1979; **Napier & Napier 1967
G. elegantulus	270 – 360 (39) x = 300	182 – 210 (39) x = 200	280 – 310 (39) x = 290	Charles-Dominique 1977a
G. demidoff	46 – 88 (66) x = 61	105 – 123 x = 123	150 – 205 x = 172	Charles-Dominique 1977a
G. zanzibaricus	M 159 (12) F 136 (14)	149 – 168 (32) x = 154	200 – 230 (24) x = 213	Harcourt 1984
G. alleni	188 – 340 (5) x = 260	185 – 205 (5) x = 200	230 – 280 x = 255	Charles-Dominique 1977a

1 All measurements are from wild caught adults except those from Napier and Napier whose source is unknown.
2 Could include three species: *G. moholi*, *G. senegalensis* and *G. gallarum*. Harcourt (1984) reports no significant differences between *G. senegalensis* and . (See also Smith & Jungers, 1997).

79

Table 3-3 Orientations of supports utilized by Lorisiformes

	Percent Utilized		
	Horizontal	Obliques	Vertical
Perodicticus potto	39	35	26
Arctocebus calabarensis	20	30	50
Galago demidoff	22	30	48
Galago alleni	6	13	81
Galago elegantulus	22	51	27

claw (Figure 3-9). Because of this adaptation, *G. elegantulus* is commonly referred to as the "needle-clawed" bushbaby. Even though it is often seen on large branches, large vines, or tree trunks, the needle-clawed galago does not seem to be at a disadvantage on small supports; about 40% of sightings were on supports 1–5 cm. in diameter (Charles-Dominique 1977a). Oblique and vertical supports are most commonly used (78%, Table 3-3). Like other bushbabies, the needle-clawed galago is frequently seen leaping. The presence of "clawed" digits allows this species to remain on large, vertical trunks and vines, which cannot be utilized by other species of galago. As I discuss later, this adaptation is related to the diet of *G. elegantulus,* and similar adaptations are found in a number of other species of primates.

Thus, within each subfamily, species are separated by particular substrate and habitat preferences. Among the lorises, the potto remains in the canopy whereas the angwantibo lives in the undergrowth. In the galagos, the needle-clawed galago exploits the canopy but moves up and down on large supports. Allen's galago inhabits the undergrowth. The dwarf galago, facilitated by its small size, follows the dense vegetation, moving up and down in the terminal branches of the canopy and undergrowth. Correlated with the differences in spatial localization of these lorisids are particular dietary patterns.

DIET

Because of the difficulty in following these nocturnal animals throughout the night, Charles-Dominique could not determine their diet by direct observation. He thus collected 201 specimens and examined stomach contents. The large number of specimens was needed because animals eat different things throughout the night and during different seasons. Contents of the stomach were separated into different constituents (e.g., fruit, insects, gums, leaves) and fresh weights of each constituent were obtained for

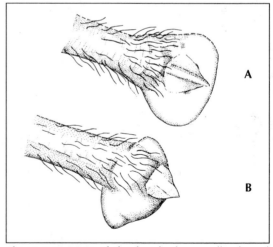

Figure 3-9 Digit of the hand of a needle-clawed bushbaby: (a) usual posture—the tactile pad adheres to the support, and the "claw" is not in contact; and (b) when the terminal phalange is retracted, the tactile pad of the digit is deformed and the "claw" digs into the support. [Drawing prepared from original photographs, from Charles-Dominique 1977a]

each stomach examined. As I describe throughout the book, there are a number of less destructive and equally valid methods of obtaining data on diet. The methods used in Charles-Dominique's study require the sacrifice of an inordinate number of animals and these, of course, cannot be restudied. It is my opinion, and that of many other primatologists, that killing primates for research purposes is rarely justified.

After calculating the percentages of the three principal dietary categories for each species, Charles-Dominique obtained the following results.

On first examination of Table 3-4, one sees a simple division among the species of each subfamily. Among the bushbabies there is one insectivore, one frugivore, and one gumivore. The lorises include one frugivore and one insectivore. Important factors are revealed by a closer examination of these data.

First, although the percent of each dietary component differs radically among the species, the actual weights of animal prey ingested are virtually the same (Table 3-5). This is probably because the foraging patterns and numbers of insects encountered during the night by each of the galago species are similar. The smallest species, *Galago demidoff*, relies almost entirely on insects; the two larger species supplement their diets with fruit *(G. alleni)* and gums *(G. elegantulus)*. The smaller loris, *Arctocebus*, is highly insectivorous, whereas *Perodicticus potto* feeds mainly on fruit and gum (Table 3-4). Similar patterns are found in many other taxa of primates, and it seems to be generally true that small forms can obtain most of their necessary nutritional requirements from animal prey. Large forms must supplement their diets with plant food. I will discuss this further in later chapters.

Table 3-4 Percent of Dietary Components from Stomach Contents

	Animal Prey	Fruits	Gums
Perodicticus potto	10%	65% (+ some leaves and fungi)	21%
Arctocebus calabarensis	85%	14% (+ some wood fibre)	-
Galago demidoff	70%	19% (+ some leaves and buds)	10%
Galago alleni	25%	73% (+ some leaves, buds and wood fibre)	(small amounts only)
Galago elegantulus	20%	5% (+ some buds)	75%

[Adapted from Charles-Dominique, 1974]

Table 3-5 Weights of Dietary Components

	Animal Prey	Fruits	Gums
Perodicticus potto	3.4 gm.	21 gm.	7 gm.
Arctocebus calabarensis	2.0 gm.	0.3 gm.	0 gm.
Galago demidoff	1.16 gm.	0.3 gm.	0.15 gm.
Galago alleni	2.2 gm.	9.2 gm.	(negligible)
Galago elegantulus	1.18 gm.	0.25 gm.	4.8 gm.

[Adapted from Charles-Dominique 1974]

The diets now may be examined more carefully to determine whether the animals are competing for specific dietary components. For example, are any of the species feeding on the same animal prey? Comparing the two subfamilies, 78% of the prey of the galagos are beetles, grasshoppers and nocturnal moths whereas 70% of the prey of lorises are caterpillars and ants. Thus, the fast-moving bushbabies hunt for fast-moving prey and the slow-moving lorises feed on relatively immobile species. This ability to feed on slow-moving prey seems like a simple solution for the lorises. However, it involves a special adaptation. In fact, Rasmussen and Nekaris (1998:276) hypothesize that "this single adaptive shift could be responsible for a cascade of additional, divergent adaptations that followed" (Figure 3-10).

Most slow-moving animals have developed specific adaptations to protect themselves from predation. The insects fed upon by potto and angwantibo are noxious to most predators, because they contain toxins or are unpalatable. This special tolerance of the lorises for noxious prey allows them to capture sufficient quantities easily. The physiological adaptations this requires also have implications for growth and reproduction of lorisines. The prey is usually abundant and quite noticeable, the color and odor signalling their unpalatable qualities to potential predators including the galagos. A predominant part of potto foraging is the olfactory exploration of branch surfaces (Oates 1984). The prey of the potto is ants, which release large quantities of formic acid, centipedes, which release iodine, and malodorous crickets, which also emit repellent chemicals. The angwantibo feeds primarily on caterpillars with stinging hairs. However, in captivity, if given a choice between these noxious prey and immobilized grasshoppers or moths, the two lorises choose grasshoppers and moths. Thus, their choice of prey in the wild appears to be a tolerance, not a preference.

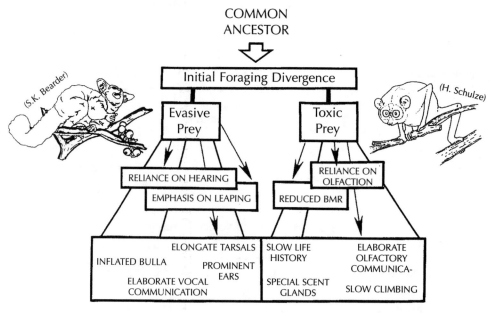

Initial split leads to a cascade of additional divergences in morphology, ecology, physiology, behavior and life history.

Figure 3-10 Evolution of adaptive traits. [Adapted from Rasmussen & Nekaris, 1998]

The three species of bushbabies also choose different prey. The two large species, *G. alleni* and *G. elegantulus,* are active at different heights. *G. demidoff* and *G. elegantulus* both exploit the canopy of the primary forest, but the dwarf bushbaby feeds on small prey while the needle-clawed galago eats larger prey.

Charles-Dominique found that overlap of food items among the frugivores and gumivores was also minimal. *G. alleni* and *P. potto* are mainly frugivores, the other three species eat very little fruit. The potto feeds on fruit in the canopy, Allen's bushbaby on fruit that has fallen to the ground or in the small bushes of the undergrowth. Fruit trees attract predators as well as fruit-eaters, and both the potto and Allen's bushbaby have the ability to distend their stomachs and to eat large quantities of fruit. They then quickly move away from vulnerable locations.

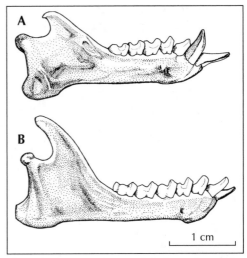

Figure 3-11 Lower jaws of *Galago elegantulus* (A) and *Galago alleni* (B) compared to show the great development of the tooth-scraper ("comb") of *the former,* as an adaptation for gum-eating. [From Charles-Dominique, 1977a]

The potto and the needle-clawed galago both feed on gums. However, in the dry season, when food is generally scarce, these two species limit their dietary choices. The potto eats only fruit and insects, no gums; the needle-clawed bushbaby exists mainly on gums. The claw-like nails of this species allow it to climb and cling to large trunks while feeding on gums. The tongue and tooth-scraper (the lower incisors of *G. elegantulus*) (Figure 3-11) and possibly specializations of the alimentary tract (Hill 1953, Martin et al. 1985) are adaptations for gum feeding. Gums also probably require fermentation for digestion (Nash 1989). A number of studies have illustrated the importance of gums in the diets of several small primates (see Sussman and Kinzey 1984, Nash 1986a). In fact, in seasonal, drier forests, where needle-clawed galagos are not found, the potto and the dwarf galago eat more gum than in Gabon (Kingdon 1971, Oates 1984). Both *G. crassicaudatus* and *G. moholi* also include a large percentage of gum in their diets (see below).

DEFENSE AGAINST PREDATION

Bushbabies are fast-moving animals and can escape from predators by leaping away. To be successful, a predator would probably have to surprise a galago. Galagos give intense alarm calls and often follow potential predators at distances of several meters while constantly calling. They also direct alarm calls at arboreal snakes (Struhsaker 1970). These alarm calls act as a warning system to all other galagos in the vicinity, identifying the location of the predator. Some galagos move more during moonlit nights and, since they use visual detection, mobbing, and escape rather than concealment, Nash (1986b) suggests galagos may be safer from predators in full moonlight.

Risk of predation is greater to young bushbabies. In the day, bushbabies sleep in nests in the dense foliage. At night, the mother carries the infant in her mouth and "parks" it on a thin branch close to where she is foraging. The infant remains completely immobile. However, if the branch shakes, the infant immediately drops from the support and utters a distress call. This call attracts not only the mother but other adult bushbabies, who mob (see Chapter 1) the predator and issue alarm calls. This defense mechanism is quite successful in the tropical forest. During Charles-Dominique's study in Gabon only two young bushbabies disappeared. However, other bushbabies and small nocturnal lemurs are subject to moderate to severe predator pressure (see below and Chapter 4).

As I stated in Chapter 1, because lorises move so slowly, they have specialized mechanisms to protect themselves against predation. The most important defense of the lorises is their cryptic locomotion. They move slowly and silently through the trees, without disturbing vegetation. If there is the slightest disturbance, they freeze and remain immobile. If approached by a predator both the potto and angwantibo move to the dense vegetation and the thinnest branches. These branches offer further concealment and are not stable supports for carnivores.

When there is no possibility of escape, the potto turns and faces the danger, adopting a defense posture (Figure 3-12) and making violent thrusts at the predator. Similar postures and butting have been observed in *Nycticebus* and *Loris* (Rasmussen 1986). Three physical adaptations make this posture a formidable weapon against an attacker. (1) The scapular shield offers protection, defense and acute sensitivity (see Figure 1-13). (2) The muscular hands allow the potto to grip firmly to branches (a potto can support ten times its weight (15 kg.) without falling off a branch). (3) The peculiar circulatory system of the potto's limbs allows continued circulation even under prolonged muscular contraction (Suckling et. al 1969). In extreme danger, the potto simply drops to the ground, moves a few meters [under certain circumstances both the potto and angwantibo are capable of quite rapid movement (Jewell and Oates 1969, McArdle 1981, Oates 1984)], and then remains immobile. This is its likely defense against poisonous snakes. The potto also has a disagreeable odor that warns off predators and may be related to chemical defense mechanisms contained in the saliva (Alterman 1995). *Loris* and *Nycticebus* also emit a strong odor (Petter and Hladik 1970, Erhlich pers. comm.).

The angwantibo does not face a predator, but turns away revealing a conspicuous target-like tail. As with the potto, an *Arctocebus* keeps its head low, between its forelimbs. If the predator gets too close, it lifts its leg and bites the attacker on the nose. Charles-Dominique believes that this causes the predator to violently jerk its head; he thinks the loris then

Figure 3-12 Potto defense posture. [From Charles-Dominique, 1977a]

releases its grip and is thrown off the branch, moving off quickly on the ground and hiding in the dense underbrush.

The ecological studies of these five species of Lorisiformes by Charles-Dominique give us an understanding of the total adaptive complex of each of the species. The strata of the forest, the nature of the supports, the type of locomotor behavior, the variety of food eaten, and the technique of predator avoidance form an integrated system. Charles-Dominique states, they "are all associated with morphological and behavioral adaptations which remain enigmatic without a detailed knowledge of the ecological peculiarities of each species" (1977a: 255).

THE ECOLOGY OF OTHER SPECIES OF LORISIFORMES

Studies of the other species indicate that the behavior of these forms may be similar to one or another of the Gabon lorisids and that they may fill similar ecological roles in the communities in which they are found. However, recent studies reveal some intriguing differences (Nekaris in prep.).

The morphology of the two genera of the Asian lorisines is similar to that of the two African forms. Little is known about these animals. There are two species of *Nycticebus; N. coucang,* extending from eastern India to Java and the Philippine Islands (Fooden 1991), and *N. pygmaeus,* known only from Laos, Vietnam, China and possibly Cambodia (Huynh 1998). *N. coucang* is like *Perodicticus,* being relatively large with short, stout limbs. Field and captive observations indicate that, like the potto, *N. coucang* is primarily frugivorous and mainly inhabits the canopy of the forest (D'Souza 1974, McArdle 1981, Oxnard et al. 1990). However, it may spend more time in the understory and eat more insects than the potto (Eliot and Eliot 1967, Barrett 1984 cited in Bearder 1986). *N. coucang* exploits many types of vegetation including evergreen, bamboo, and scrub forest and it is found in both primary and secondary growth. Little is known of *N. pygmaeus* although studies are currently being conducted in Vietnam (Huynh 1998). It is found in the same types of forest as its cogener but in more isolated areas, with low human population density (Nisbett and Ciochon 1993, Huynh 1998, Ratajszczak 1998). Like many primates, *Nycticebus* is under predation pressure from humans. The fur of *N. coucang* is in demand as a medicine from Burma to Borneo, where it is used to dress wounds (Harrison 1962, see also Huynh 1998). *N. pygmaeus* also is used for medicinal purposes throughout its range (Eudey 1987).

Loris tardigradus, the slender loris of India and Sri Lanka, is small-bodied with thin, elongated arms like *Arctocebus* (Figure 3-13). It is interesting to note that immunological studies indicate that the African lorises may be more closely related to the African galagos than to Asian lorises (Dene et al. 1976, Martin 1990). It has been the subject of short-term studies in Sri Lanka (Petter and Hladik 1970, see Schulze and Meier 1995) and an 11 month study was just completed in India by Nekaris (in prep.).

The slender loris inhabits the forests of Sri Lanka and India south of the Tapti River. Presently, only one species is recognized but there is a great deal of morphological variation among different populations of these wide-ranging animals. Furthermore, it inhabits a wide range of habitats, including dry scrub forests, moist riverine forests, and high altitude forests. As with the galagos, further studies may reveal more species of slender lorises (Nekaris pers. comm.).

The following are preliminary results of the field study
of *Loris tardigradus lyddekeri-
anus* by Nekaris (in prep.) in
Tamil Nadu, southern India.
Nekaris observed that slender
lorises moved quietly in their
dry scrub habitat, skillfully
avoiding thorns. The average
height in which they were seen
was 2-4 m. However, this was
dictated by the prevalence of
short scrubby trees at the field
site. They also were found in
taller trees (10-15 m) when
these trees were available. The
lorises had little hesitation
when they needed to come to

Figure 3-13 Slender loris. [Photo by Kimberly Nekaris]

the ground and even crossed roads. Most of their locomotion was done on small
oblique branches (less than 5 cm).

Slender lorises are fully nocturnal, sleeping throughout the day. They wake
just before dawn and actively groom both themselves and others just after wak-
ing, with sometimes as many as four animals social grooming. They then spread
out from the sleeping site, which is usually located in the core area of the range,
and then traveled separately in a generally shared home range. Slender lorises at
Tamil Nadu did not use nests but slept in dense thorny bushes or in cacti.

Slender lorises were found to be among the most faunivorous primates with
97% of the diet being animal prey, 96% of which were insects. Orders of insects
eaten included: Orthoptera, Hymenoptera, Coleoptera, Lepidoptera, Isoptera,
and Odanata. They also ate slugs, snails, and geckos. The non-animal foods
included *Acacia* gums and, in one instance, a seed pod. Slender lorises fed in all
areas of the trees.

As with the slow loris, the slender lorises are captured for their "health" benefits as well as for the pet trade. Nekaris saw portions of dead lorises being sold
in city markets to be used as medicinal charms and for homeopathic medicines.
The lorises of Asia seem to hint at a new, wide range of primate behavioral variation that has yet to be studied. Future studies of these fascinating creatures are
sorely needed.

Besides the Gabon species, only four other galagos have been studied in any
detail, *Galago crassicaudatus, G. garnettii, G. moholi,* and *G. zanzibaricus.* The former two species are the largest of the galagos and were once considered to be subspecies of *G. crassicaudatus.* However, they are distinct in a number of ways
(Masters 1988). *G. crassicaudatus* is widely distributed, whereas *G. garnettii* is found
only in East Africa (Figure 3-14). *G. crassicaudatus* occurs in a number of forest
types, including woodland/savannah. In East Africa, where it occurs with *G. garnettii,* the two species are ecologically separated, *G. crassicaudatus* preferring woodland and *G. garnettii* preferring riverine and highland forest (Nash et al. 1989).

Medium sized *G. moholi* and *G. zanzibaricus* were once included as subspecies
of *G. senegalensis,* as was *G. gallarum.* Now they are considered to be separate

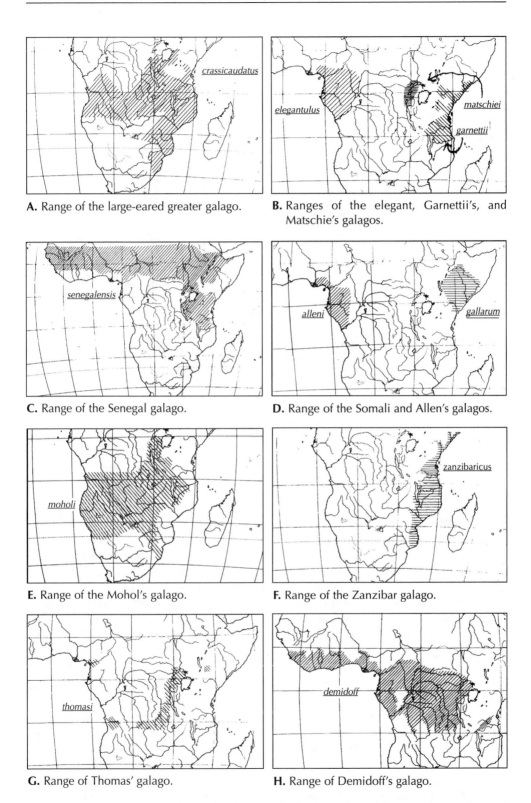

A. Range of the large-eared greater galago.

B. Ranges of the elegant, Garnettii's, and Matschie's galagos.

C. Range of the Senegal galago.

D. Range of the Somali and Allen's galagos.

E. Range of the Mohol's galago.

F. Range of the Zanzibar galago.

G. Range of Thomas' galago.

H. Range of Demidoff's galago.

Figure 3-14 East and West African distribution of galagos. [Adapted from Nash, Bearder, and Olson, 1989]

Table 3-6 Comparison of *G. moholi* in South Africa with *G. zanzibaricus* in Kenya

	G. moholi	*G. zanibaricus*
Habitat	*Acacia* thornveld	Forest
Body color	Gray	Brown
Head shape and color	Rounded; Small white nose stripe	More pointed; Extended white nose stripe
Weight (g)		
Male median	211	149
Female median	188	137
Locomotion	Leaping	Running
imi	51.8	58.4
Advertisement call	Single-, double-, or triple-unit calls; Fundamental frequency, 0.69 kHz	Patterned sequence of 4–18 units; Fundamental frequency, 1.0 kHz
Breeding season	Jan.–Feb. Oct.–Nov.	Feb.–Mar. Sept.–Oct.
Litter size	Twins	Singletons
Diet	Invertebrates and gum	Invertebrates and fruit
Social organization	Male overlapping territories of several females	Male sharing territory with 1 or 2 females
Sleeping groups	Solitary or females and infants; male with them less often	Males regularly with female(s) and infants
Home-range size (ha)		
Male	1.5–22.9	1.9–2.9
Female	4.4–11.7	1.6–2.6
Chromosome No.	38	36

[From Harcourt and Bearder, 1989.]

species, with a number of behavioral, morphological and habitat differences (Table 3-6, see also Anderson 1998, Weisenseel et al. 1998). Both the Senegal and Moholi's bushbabies have a wide distribution in Africa, with the former occurring to the north and the latter to the south (Figure 3-14). These species inhabit open woodland, but also are found in savannah, bush and forest fringes. The Somali bushbaby *(G. gallarum)* and the Zanzibar bushbaby have more restricted ranges in East Africa, with the former occurring north to the Ethiopian Rift Valley, and the latter to the south (Figure 3-14). The Somali galago probably is the most *xerically* (dry forest) adapted of all bushbabies. Zanzibar galagos are found in coastal and evergreen forests. Another medium sized galago, *G. matschiei,* until recently identified as *G. inustus,* is found only in relic forests in Zaire and Uganda. It has keeled nails similar to those of *G. elegantulus* and the two allopatric species may have similar habitat preferences and diets (Kingdon 1971). *G. orinus* and *G. thomasi* (Fig 3-14) are the size of the dwarf galago and until recently were interpreted as subspecies of *G. demidoff.* These species are found in relic primary forest (Nash et al. 1989, Bearder 1995).

Unlike the lorisids of the tropical forests of Gabon, the South and East African bushbabies must tolerate large seasonal and diurnal variations in temperature and rainfall. This variation in weather corresponds to marked seasonal changes in vegetation: leaf growth, fruiting, and flowering are restricted to the warm, humid months of the austral summer. Comparative studies have been conducted on *G. crassicaudatus* and *G. moholi* in South Africa, and on *G. garnettii* and *G. zanzibaricus* in East Africa.

Galago crassicaudatus, the large-eared greater galago, has an uneven distribution in this drier area of Africa, living in subtropical and tropical forests, riverine and coastal forests, and open woodland. It prefers more densely forested habitats in some parts of its range, and woodland/savannah habitats in others parts (Nash et al. 1989). Studies of *G. crassicaudatus* have been conducted in Northeastern Transvaal, South Africa (Bearder 1974, Doyle and Bearder 1977, Harcourt 1980, 1986, Crompton 1984, Clark 1988).

Much like potto, the large-eared greater bushbaby inhabits the thick vegetation of the canopy, moving where vegetation remains unbroken. It will cross open areas on the ground if necessary, but prefers to remain in the trees, usually above 6 meters high (Crompton 1984). Generally, the behavior and ecology of this greater bushbaby parallels that of the potto. It is relatively slow moving, and much of its postural and locomotor behavior has been compared to that of the lorises (Kingdon 1971, McArdle 1981, Oates 1984, Oxnard et al. 1990). The large-eared greater bushbaby is gumivorous and frugivorous, with a small proportion of insects included in its diet. Population density of the species is highest in forests where fleshy fruit are abundant. However, in the dry season, and where fruit is unavailable, *G. crassicaudatus* can subsist almost entirely on gums (Crompton 1984, Harcourt 1986). The diet throughout the year is: 62% gum, 21% fruit, 8% nectar from flowers, 4% seeds, and 5% insects and unidentified items (Charles-Dominique and Bearder 1979).

The most favored habitat of *G. moholi* is the semi-arid thornveld savannah but it also can be found in a wide range of other habitats in temperate Africa. It is replaced by ecologically similar *G. senegalensis* in East and West Africa, and by *G. gallarum* in East Africa. As with *G. crassicaudatus,* field studies have been conducted on *G. moholi* in the Transvaal, South Africa (Doyle and Bearder 1977, Martin and Bearder 1979, Bearder and Martin 1980a,b, Crompton 1984, Harcourt 1986). Moholi's bushbaby moves from the ground to the treetops (6-12 meters, depending on height of the canopy) but is most often found around 3 meters high. Like the Gabon bushbabies, it is a fast-moving, leaping animal. It is a vertical clinger and leaper, landing on its hind feet during jumps and hopping bipedally when on the ground (Crompton 1984). Moholi's bushbabies often descend to the ground where they hunt for insects, but they cross open areas with extreme caution. Being small animals, they are vulnerable to many predators. In a study by Martin and Bearder (1979), six of thirty-seven radio-tracked individuals were killed by large-spotted genets. These and other bushbabies opportunistically are preyed upon by chimpanzees (McGrew et al. 1978, Goodall 1986). During the day, *G. moholi* frequently uses dense thorn trees (especially *Acacia tortilis)* as sleeping sites.

The diet of the moholi's bushbaby is made up mainly of animal prey (mostly arthropods) and gums of *Acacia* trees (especially *A. karroo*). In mild winter months, this medium-sized galago concentrates on insects, and eats comparatively little gum (Harcourt 1986). It also moves lower in the forest, catching much of its insect prey on the ground. In contrast to this, the larger *G. crassicaudatus* reduces its intake and time foraging for insects and feeds mainly on gums higher in the trees (Crompton 1984). However, during severe winters, when insects become scarce, even *G. moholi* subsists almost exclusively on gums (Martin and

Bearder 1979, Bearder and Martin 1980a). At these times, the Moholi's galagos decrease activity and undergo pronounced loss of weight. The strategy of gum feeding and curtailment of activity in adverse conditions apparently has enabled the lesser bushbaby to inhabit areas in Africa having extremely arid and variable climatic conditions. Nocturnal species of lemurs living in seasonal forests in Madagascar show similar adaptations (see Chapter 4).

Although *G. crassicaudatus* and *G. moholi* live in sympatry in part of their range, the studies reported above were done on allopatric populations. It is assumed that differences observed between the two species, such as choice of strata, locomotion, and diet, allow them to coexist in sympatry. These differences also relate to differences in body size, as did those between the lorisids of Gabon. Harcourt and Nash (1986a) studied two East African species with similar size differences, *G. garnettii* and *G. zanzibaricus. G. garnettii* is a large galago, weighing around 800 gm, whereas *G. zanzibaricus* weighs only 140 gm. Research was conducted at two sites in Kenya where the two species coexist.

As in the South African species, differences between *G. garnettii* and *G. zanzibaricus* can be related to differences in body size. The larger Garnetti's galago spends most of its time in trees, higher than 5 meters. The smaller species, the Zanzibar galago, uses the lower levels of the forest, and smaller and more oblique supports more frequently. The Zanzibar galago is not a vertical clinger and leaper like Moholi's galago, however. It runs and leaps quadrupedally both in the trees and on the ground (Harcourt and Bearder 1989). Both species eat insects and fruit. As expected, the smaller *G. zanzibaricus* eats more insects, estimated at 70% of its diet. *G. garnettii* eats an equal proportion of insects and fruit (Harcourt and Nash 1986a). There were no major gum producing trees at either of the study sites in Kenya, and neither of these species was observed eating gums. However, at another site in Kenya, *G. senegalensis* was observed feeding exclusively on gums and insects (Nash and Whitten 1989).

The differences in diet and substrate use in the two species of sympatric bushbabies in coastal forests of Kenya in some ways parallel differences in the two southern African thornveld forms, and in some of the species studied in Gabon, West Africa. Though the relationships are complex, differences in body size between galago species are certainly one of the important factors contributing to differences in behavior and habitat selection among these forms and enabling closely related species to coexist.

THE SOCIAL BEHAVIOR OF THE LORISIFORMES

Gathering data on ranging patterns, population densities and dispersal, and social interactions among nocturnal species requires a variety of specialized techniques. Density and distribution are determined by counts of animals sighted with headlamps along mapped paths. These data are supplemented with systematic trapping and intensive studies of radio-tracked individuals. From 1968 to 1973, Charles-Dominique trapped over 100 individuals of three species in Gabon: *Galago demidoff, G. alleni,* and *Perodicticus potto.* Each animal was marked for individual identification. For more detailed observations on particular animals, 8 *G. alleni* and 2 *P. potto* were fitted with radio collars (Charles-Dominique 1977a and b). In a two year period, Bearder and Martin (1980b) radio-tracked 37 *G. moholi* in South Africa. Similar mark-recapture and radio-tracking techniques

Table 3-7 Home Range Size, Population Density and Group Structure of the Lorisiformes

Species	Home Range (Hectares) ♀	Home Range (Hectares) ♂	Density Per Square Kilometer	Type of Group Structure	Source
Perodicticus potto	7.5 (range: 6–9) N = 5	9–40 N = 5	8–10	solitary but social	Charles-Dominique 1977
Arctocebus calabarensis			2–7*	solitary but social	Charles-Dominique 1977
Loris tardigradus			~35	solitary but social; nesting groups	Nekaris in prep.
Galago demidovff	0.8 (range: 0.6–1.4) N = 6	0.5–2.7 N = 6	50–80	solitary but social; ♀ nesting groups	Charles-Dominique 1977
Galago alleni	10 (range: 8–16) N = 6	30–50 N = 2	15–20	solitary but social; ♀ nesting groups	Charles-Dominique 1977
Galago elegantulus			15–20	solitary but social; ♀ nesting groups	Charles-Dominique 1977
Galago moholi	6.7[1] (4.4–11.7) N = 11	A♂'s 11.0[1] (9.5–15.6) N = 4 B♂'s 15.9 (9.9–22.9) N = 7	95–200[2]	solitary but social ♀ nesting groups	Bearder & Martin 1979[1] Doyle & Bearder 1977[2]
Galago crassicaudatus			72–125	solitary but social ♀ nesting groups	Doyle & Bearder 1977
Galago zanzibaricus	2.3 (1.6–2.8) N = 10		170–180	solitary but social w/ ♂/♀ assoc.	Harcourt & Nash 1986
Galago garnettii	12 N = 4	17 N = 3	31–38	solitary but social ♀ nesting groups	Harcourt & Nash 1986

*2/sq. km. in primary forest as a whole; 7/sq. km. as commonly found in concentrated pockets within the forest.

have been used to study the social behavior of *G. crassicaudatus* (Bearder 1974, Clark 1978, 1985), *G. garnettii* (Nash and Harcourt 1986), and *G. zanzibaricus* (Harcourt and Nash 1986b).

The population densities of the African Lorisiformes that have been studied to date are given in Table 3-7. These density figures, of course, represent averages and certain species may have greater densities in specific habitats within their geographical range. For example, *P. potto* has a density of 28 per square kilometer in areas of the forest that are periodically flooded. The distribution of *G. elegantulus* is directly related to that of *Entada gigas,* a species of vine that provides 80% of its gum resources. At night, *G. moholi* is usually in stands of *Acacia karroo* trees; during the day it sleeps in patches of *A. tortilis* or *A. nilotica,* and density of this galago is dependent upon these tree species. In East Africa, the distribution of *G. senegalensis* also is closely associated with that of *Acacia* (Nash and Whitten 1989).

The galagos studied to date share a similar type of social organization, though there are some variations on the basic theme, especially in the lorises and in *G. zanzibaricus.* The basic theme is as follows (see Charles-Dominique 1995). During the night, when the animals are active, they generally forage on their own (Table 3-8). During the day, most species of galago sleep in nesting groups made up mainly of adult females and their young. Occasionally an adult male and, at least in *G. demidoff,* more than one (Kingdon 1971), can be found sleeping with a nesting group, but males usually sleep alone. Nesting groups of galagos range in size from 2 to 10 individuals. During the night, the females of a nesting group utilize overlapping areas of the forest. Thus, to a great extent, these females share home ranges. Females from different nesting groups have home ranges that only slightly overlap at their borders.

Unlike home ranges of females, those of large, adult males are not usually shared. Adult males avoid each other, and their home ranges overlap only slightly at borders. Females are allowed to stay with their mothers after reaching adulthood, and it is likely that nesting groups consist of individuals of the same matriline (groups of females related to a common mother). In most species, the subadult males are not allowed to stay in the area; resident adult males force

Table 3-8 Comparative Figures for Nocturnal Sightings of the Five Lorisid Species, Showing Percentages of Animals Encountered Singly or in Groups of Various Sizes

	Solitary %	2 together %	3 together %	4 together %	5 together %	Mother & infant %
Perodicticus potto N=105	96%	2%				2%
Arctocebus calabarensis N=99	97%	1%				2%
Galago demidovff N=263	75%	21%	2%	0.5%	0.5%	1%
Galago alleni N=97	86%	8%	4%			2%
Galago eiegantulus N=103	76%	17%	2%	1%		4%

[Adapted from Charles-Dominique, 1977a]

young adult males out. However, the fate of these young males varies considerably in different species and under various ecological conditions. In general, these males become peripheralized to marginal areas, become roaming "vagabond" males, or become residents of small home ranges within the area of the larger adult males, but they do not interact with the resident adult males or females. However, smaller males of *G. moholi* have larger ranges than large resident males and do interact with females (Martin and Bearder 1979, Bearder 1986). Furthermore, tolerance between adult males also may vary.

In some regions, adult male ranges overlap extensively, especially in *G. demidoff, G. senegalensis* (Kingdon 1971), and *G. crassicaudatus* (Clark 1978). Replacement of the resident adult males seems to occur quite often, and many young males become resident males as they reach full adult size and status. Generally, however, there are few data on juvenile sex differences in dispersal patterns (see Nash 1993).

The home ranges of the females are relatively small in all species (Table 3-7); the size of the range seems to depend, to some extent, on availability of resources. Within each species, females living in larger nesting groups have relatively larger home ranges. The home ranges of adult males are much larger then those of females (Table 3-7). Since males generally do not share home range areas, these large ranges contain ample resources. The size of the home range is determined more by the location of females than by food availability. Males attempt to associate with the maximum number of females. Male and female ranges overlap completely, and the two sexes interact throughout the year. Since females have small, overlapping home ranges, the home range of an adult male usually overlaps with that of many females (Figure 3-15). The resident male attempts to maintain the integrity of his range as well as that of his females.

Because of this pattern of home range use, it is often assumed that this establishes a "harem-like" mating system in that one male is associated and probably mates with a number of females. However, at least in the Senegal bushbaby and the Asian loris, estrus females may be followed by up to six males simultaneously (Bearder 1986). Furthermore, there is evidence that females of many species mate with a number of partners and there is no reason to "conclude that sexual contacts are restricted among nocturnal prosimians and that females mate exclusively with a single male whose home range coincides with their own" (Dixson 1995:111).

Though these animals forage alone, social interactions occur throughout the night. For example, Clark (1985) observed over 850 interactions in 350 hours of observations on *G. crassicaudatus*. These involved body contact or overt avoidance of contact, and 69% of the interactions were friendly (Table 3-9). Positive interactions occurred between all age-sex classes, though adult females were the most social (Figure 3-16). The amount of social contact and affiliative behavior has not been documented for most nocturnal species and it is likely that considerable variation exists (Charles-Dominique 1983, Clark 1985). Because these animals are solitary during most of their active period but maintain regular social networks, their social organization has been called solitary but social or nongregarious.

This basic social organization is found in most galago species that have been studied (e.g., *G. alleni, G. demidoff, G. crassicaudatus, G. garnettii,* and *G. moholi*), with some slight variations between species (Bearder 1986). The social organization of *G. zanzibaricus* shows some more serious variations on this theme (Harcourt

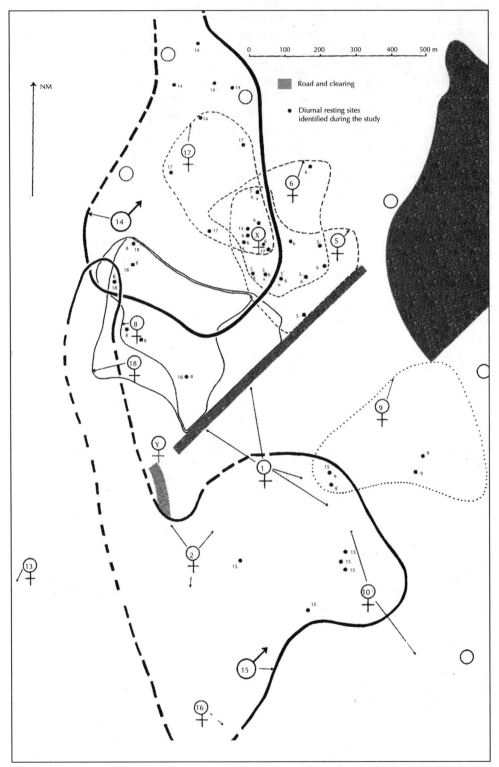

Figure 3-15 Home ranges of Allen's bushbabies followed by radio-tracking in primary forest. [From Charles-Dominique, 1977a]

Table 3-9 Total Interactions Observed, by Type, Galago Crassicaudatus

	Type					
	Positive	Neutral	Ambivalent	Agonistic	Sexual	Total
N	595	59	53	92	66	865
(% total)	(69)	(7)	(6)	(11)	(7)	(100)

[From Clark, 1985]

and Nash 1986b). In this species an adult male will share its home with one or two adult females and their offspring. Male and female home ranges are essentially similar in size, and males and female ranging partners sleep together during the day. The distance traveled per night also is similar between the sexes. Although the animals move independently, they maintain regular vocal contact throughout the night. Thus, for as yet unknown reasons, male-female associations are closer

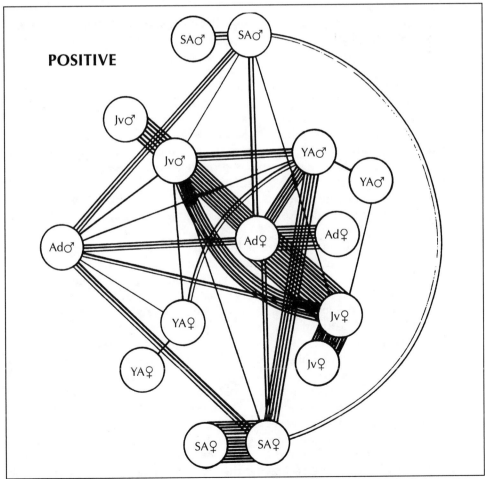

Figure 3-16 Network of positive interactions linking all age–sex classes. Thick lines denote >0.5–1 interaction per potential dyad (one pair of animals of those age–sex classes which were present in overlapping ranges at the same time); thin lines denote ≤0.5 interaction per dyad, e.g., for two potential dyads, only one interaction was observed. [From Clark, 1985]

in *G. zanzibaricus* than in the other species of galago so far studied.

The African lorises, *P. potto* and *A. calabarensis*, because they move so slowly, also have a social organization that differs in some aspects from that of the galagos. The females are too slow to move to a central location each day and will sleep in suitable, dense vegetation wherever they happen to be when the sun begins to rise. Accordingly, the females do not sleep in nests nor are

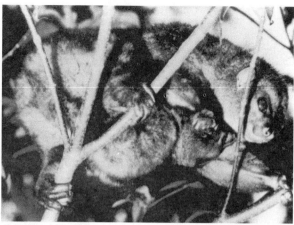

Figure 3-17 Female loris with infant. [From Charles-Dominique, 1977a]

they ever found in nesting groups. Females, like males, do not share their home ranges with individuals of the same sex. However, as in most bushbabies, male and female ranges overlap and the ranges of males are larger and overlap the ranges of more than one female.

The slow locomotion of the lorises can also be related to some differences in infant development and care between the two subfamilies of Lorisiformes. Whereas bushbabies leave young infants in nests at night or park them while they forage, the newly born lorises are able to cling to the fur of their mothers when they are active (Figure 3-17). Later in development, loris babies also are parked (Rasmussen 1986). Galagos have higher rates of growth and development than lorises, another reason that parking behavior differs. Generally, Lorisiformes give birth once a year and normally have one infant per birth (Table 3-10). However, *Galago crassicaudatus, G.moholi,* and at least some *Loris tardigradus* are exceptions and often give birth to twins or triplets.

G. moholi has an extremely high breeding potential; females give birth twice a year and more than half of the births are twins or triplets (Doyle et al 1971). This may be an adaptive response to the harsh environments in which this species is found (Martin and Bearder 1979); however, see Nash (1983) for other possible explanations. Overall, the intrinsic rate of population increase, an important ecological variable, is higher in galagos than in lorises of similar size. This is a good example of how reproduction, metabolic, and dietary ecology are all interrelated.

Unlike most of the lorises of Africa, slender lorises spend a lot of time in spatial proximity to one another (Nekaris in prep.). It is interesting that the social behavior and reproduction of this species is more similar, in some ways, to the galagos than to African lorises, and thus may represent a more primitive characteristic among the lorisiformes. As many and eight animals slept together and interactions between these animals occurred at the sleeping site as well as throughout the night. At dawn, animals converged from different directions to join the sleeping group. Prior to leaving the sleeping tree and upon joining the group in the morning, animals were playful and social grooming bouts were common. Groups can contain more than one adult female, more than one adult male, juveniles and infants.

Table 3-10 Reproductive Parameters for Some Lorisiformes

Species	Birth Seasonality	Gestation Length (days)	Modal Litter Size (maximum)	Interbirth Interval (months)	Age at Sexual Maturity (months)	Use of Nest or Hollow	Method of Infant Carriage
Galago demidoff	None	111–114	1 (2)	12	8–10	Yes	Mouth
Galago zanzibaricus	Feb–Mar Aug–Oct	120	1 (2)	8 and 4	?	Yes	Mouth
Arctocebus calabarensis	None	131–36	1	4–5	9–10	No	Fur
Galago senegalensis moholi	Oct–Nov Jan–Feb	121–25	2	8 and 4	9–12	Yes	Mouth
G. alleni	None	133	1 (2)	12	8–10	Yes	Mouth
Loris tardigradus	None	163–l67	2	12	l2	No	Fur
Nycticebus coucang	None	192–193	1	12	>12	No	Fur
Galago garnettii	Aug–Nov	130–35	1	12	12–18	Yes	Mouth
Perodicticus potto	Aug–Jan	170	1	12	18	No	Fur
Galago crassicaudatus	Oct–Nov	136	2 (3)	12	18–24	Yes	Mouth and fur

[Adapted From: Bearder, 1986, with modifications from Rasmussen and Izard, 1988 and Nekaris in prep.]

Slender lorises ranges overlap a great deal. Males range further from the core area than do females, and each animal appeared to have a core area of their own. Both males and females ranged in and out of an approximately 50X50 m central area throughout the night where they engaged in feeding, play, or social grooming. Animals from other sleeping groups entered one another's central area, however, this could lead to a vocal exchange or, in rare instances, physical fighting. Though females stay with one sleeping group, adult males sometimes switched between groups. Male ranges were much larger than those of females (3X as large) and overlapped with more than one female. However, both males and females were seen to consort with more than one adult of the opposite sex.

Nekaris observed four singletons and five sets of twins and believes that twins are generally more common than single births. She made detailed observations on two sets of twins and one singleton infant during her study, although three of these infants died. Gestation was approximately five months and there was no indication of a birth season. Infants continue to suckle for about 2–4 months but begin to hunt for insects at around 6 weeks. For the first 3–4 weeks, the mother carries the infant on her belly where they constantly suckle. After this time, infants are parked and are left in a tree near the sleeping site. Infants are picked up by their mothers in the morning. Some mothers checked on the infant throughout the night, sometimes grooming and playing with them, whereas others did not. Males from the sleeping group also checked on infants during the night and sometimes played with them. No animal other than the mother was ever seen to carry an infant.

Infants are parked for the first 3–4 months and then begin to venture more and more away from the sleeping area. Eventually they follow the mother throughout the night. Nekaris has evidence that females may leave the sleeping group and migrate to another range when her young one is about to reach sexual maturity.

The social organization of a species is maintained by its particular pattern of social communication. Generally, both diurnal and nocturnal species use vocalizations to facilitate social spacing, the cryptic African lorises are an exception. Visual and olfactory signals also are used for social spacing in primates. As might be expected, however, visual signals (e.g., postures, gestures, and facial expressions) are used more frequently in diurnal forms whereas olfactory and vocal communication is more important in nocturnal species. As discussed in the previous chapter, scent-marks and especially advertising calls (used for spacing between members of the same species) are components of species-mate recognition systems in nocturnal primates.

As in many species of mammals, the Lorisiformes often mark certain areas within their ranges with secretions from specific glands or with urine (Sussman 1992). Urine is deposited on the hands and feet or directly onto a branch. Besides being used in communication, urine washing also may serve to facilitate grip (Harcourt 1981). *Galago crassicaudatus* also has a prominent chest gland from which chemicals are secreted directly on a branch in a specific chest rubbing display (Clark 1975, Katsir and Crewe 1980). The Asian lorises have glands on the upper arm (brachial glands) (Rasmussen 1986). Marking leaves a chemical deposit on the branch that serves as an olfactory signal to a passing conspecific. Unlike vocal or visual communication that is instantaneous, olfactory signals remain after the signaling animal is gone. Olfactory signals can communicate information about the amount of time passed since the signal was given and information about the sex, sexual condition, and identity of the signaler.

In a study of *G. alleni,* Charles-Dominique modified his radio-tracking equipment so that a different signal was given whenever the animal urinated. Thus, he could determine not only where an animal was located but also precisely when and where urine marking was taking place. He found that females patrolled overlapping borders once a week for about two hours. In these border areas they urine marked at a rate 3.7 times per hour; in non-overlap zones they marked an average of once per hour. *G. alleni* has two loud vocalizations. One is the mobbing call described earlier. This call is not answered but calls attention to the presence and location of a predator. The second loud call is a social call that is always answered even if the calls are tape recorded and played back. This call, referred to by Charles-Dominique as the croaking call, allows the male to localize and join the female, i.e., it serves as a distance reducing signal (Marler 1968) between the male and female. However, the same call warns individuals of the same sex of each other's presence, thus also serving as a distance maintaining or distance increasing signal within sexes.

Females gave croak calls eight times more frequently in border areas (1.6/hour vs. 0.2/hour) than in the rest of their home range. The pattern among males was similar. The males urine-marked in border areas at a rate of 3.6 time per hour and in the rest of the home range only 1.2 times per hour. They also gave croak calls more frequently at borders. These olfactory marks and calls allowed animals of the same sex to avoid confrontation at borderline areas and to maintain the integrity of their home ranges. They also allowed opposite sexed individuals to maintain continuous social contact.

In *G. alleni* the male had a home range of 32 hectares, 3 times as large as the home range of the female. He also interacted with up to eight adult females. The large home range was determined by the location of females and not solely by abundance of resources. The male did not spend a great deal of time running over his entire home range. Instead, he spent much of his time in the overlap zones of the females. It is in these zones that most social communication took place and where he had the best chance of encountering a female (Figure 3-15).

SUMMARY OF THE ECOLOGY AND SOCIAL BEHAVIOR OF LORISIFORMES

Intensive studies have been carried out on ten of the nineteen species of extant Lorisiformes. Little if nothing is known about the other forms. In his study of five sympatric species in Gabon, West Africa, Charles-Dominique illustrates how each of these species has different habitat preferences. *Perodicticus potto* and *Galago demidoff* mainly utilize the canopy, while *Arctocebus calabarensis* and *G. alleni* occupy the undergrowth of the forest. *G. elegantulus* uses its clawed digits to move up and down large trunks and branches. The bushbabies are quick moving and prey on fast moving insects whereas the lorisines are exclusively slow climbers and feed on slow moving, noxious prey. The smaller species, *G. demidoff* and *A. calabarensis,* are highly insectivorous; the potto and Allen's bushbaby are highly frugivorous. The needle-clawed galago eats mainly gums. Observations of *Loris tardigradus* in India indicate that some interesting differences exist between these lorises and those of Africa.

The other species of galago which have been studied in some detail, *G. crassicaudatus, G. moholi, G. garnettii,* and *G. zanzibaricus,* occur outside of the tropical forest regions in more arid and seasonal forests of South and East Africa. The smaller forms, *G. moholi* and *G. zanzibaricus,* eat mostly insects whereas *G. crassicaudatus* and *G. garnettii* eat mostly gum and fruit, respectively. The two larger species are usually found high in the trees, whereas the smaller forms remain lower in the forest and often hunt insects on the ground. *G. crassicaudatus* reaches its highest population densities in riverine, subtropical evergreen forests whereas *G. moholi* reaches its highest densities in semi-arid savannah vegetation. It is well adapted to this habitat. The two species occur sympatrically in some portions of their range. *G. garnettii* and *G. zanzibaricus* occupy coastal and riverine forests in East Africa and are sympatric over much of their range. Differences in diet and substrate preferences described above are related to the body size, and enable the species to coexist where they occur together. In India, *Loris tardigradus* uses mainly the small branch niche of the undergrowth and canopy. This species is one of the most faunivorous of the primates.

There is some similarity in the social organization of many of the Lorisiformes studied to date, but with variations on a generally similar theme. These variations can be related to specific differences in morphology, i.e., differences in the locomotor anatomy of the lorises and galagos, and to specific differences in tolerances between adult males. Female ranges are small and overlap to a variable degree, no overlap of ranges occurs among the African lorises. Adult male ranges are larger and, in most cases, scarcely overlap with one another although they overlap with one or more of the female home ranges. Adult male tolerance towards one another, however, is one of the most variable aspects of lorisiform sociality. During nocturnal activity, the animals move about alone; however, during the day, females, except among African lorises, often form nesting groups of 2 to 10 individuals. The social organization of *G. zanzibaricus* differs in that an adult male and one or two adult females share a common home range, often stay in vocal contact throughout the night, and sleep together during the day. The social structure of the Indian slender loris differs from that of African lorises and in some ways is more similar to that of some galagos.

Patterns of mating are variable, to some extent paralleling those of gregarious, group-living primates. From what is known, it appears that the following mating systems can be found among the Lorisiformes; one-male, multi-female systems, age-graded systems (possibly with only larger males mating), multi-male/multi-female promiscuous systems, and monogamous mating systems.

Social interactions occur between individuals, including adult males and females, throughout the year. Olfactory (urine marking, arm rubbing, and chest rubbing) and vocal signals (the African cryptic lorises, again, an exception) are the most important modes of communication, and are utilized to regulate social distance and to maintain the species specific spacing patterns. They are also important components of the species-mate recognition system.

"Nongregarious" primates include the lorisiformes, the nocturnal lemuriformes, tarsiers, and orang utans. These animals usually do not form groups during periods of activity. However, there is a great deal of variability in mating systems and in individual distribution patterns. In some species adult females share home ranges, and in others they are intolerant of one another. The same is true of adult males. The number of males and females sharing home ranges also is variable. Most theories attempting to relate patterns of social organization to ecology have been developed from research on more gregarious, group-living diurnal primates. The species diversity found among these nongregarious species offers opportunities to test these theories. To date there are too few data on dispersal patterns of juveniles and too few long-term studies on identified individuals to make specific predictions about species differences (see Nash 1993). We still know very little about how the ecological differences among these species relate to social differences.

BIBLIOGRAPHY

Alterman, L. 1995. Toxins and Toothcombs: Chemical Defense Mechanisms in *Nycticebus* and *Perodicticus.* Pp. 413–424 in *Creatures of the Night: The Nocturnal Prosimian.* S. L. Alterman; G.A. Doyle, M.K. Izard; eds. New York, Plenum.

Anderson, M.J. 1998. Comparative Morphology and Speciation in Galagos. *Folia Primatol.* 69:325–331.

Beard, K.C.; Tao, Q.; Dawson, M.R.; Wang B.; Li, C. 1994. A Diverse New Primate Fauna from Middle Eocene Fissure-Fillings in Southeastern China. *Nature* 368:604–609.

Baldwin, L.A.; Teleki G. 1977. Field Research on Tree Shrews and Prosimians: An Historical, Geographical, and Bibliographical Listing. *Primates* 18:985–1007.

Barrett, E. 1984. *The Ecology of Some Nocturnal, Arboreal Mammals in the Rainforest of Peninsular Malaysia.* Ph.D. Thesis. Cambridge University, Cambridge.

Beard, K.C. 1998. A New Genus of Tarsiidae (Mammlia: Primates) from the Middle Eocene of Shanxi Province, China, with Notes on the Historical Biogeography of Tarsiers. *Bull. Carnegie Mus. Nat. Hist.* 34:260–277.

Bearder, S.K. 1974. *Aspects of the Ecology and Behaviour of the Thick-Tailed Bushbaby* (Galago crassicuadatus) Ph.D. Thesis, University of the Witwatersrand, Johannesburg.

Bearder, S.K. 1986. Lorises, Bushbabies, and Tarsiers: Diverse Societies in Solitary Foragers. Pp. 11–24 in *Primate Societies.* B.B. Smuts; D.L. Cheney; R.M. Seyfarth; R.W. Wrangham; T.T. Struhsaker, eds., Chicago, University of Chicago Press.

Bearder, S.K. 1995. Species Diversity with Special Reference to Mate Recognition. Pp. 331–352 in *Creatures of the Night: The Nocturnal Prosimians.* L. Alterman; G.A. Doyle, M.K. Izard; eds. New York, Plenum.

Bearder, S.K.; Martin, R.D. 1980a. *Acacia* Gum and Its Use by Bushbabies, *Galago senegalensis.* (Primates: Lorisidae). *Intl. J. Primatol.* 1:103–128.

Bearder, S.K.; Martin, R.D. 1980b. The Social Organization of a Nocturnal Primate Revealed by Radio Tracking. Pp. 633–648 in *A Handbook on Biotelemetry and Radio Tracking.* C.J. Amlaner, Jr.; D.W. Macdonald, eds. London, Pergamon.

Charles-Dominique, P. 1971. Eco-éthologie des Prosimians du Gabon. *Biologica Gabonica:* 7:121–228.

Charles-Dominique, P. 1974. Ecology and Feeding Behavior of Five Sympatric Lorisids in Gabon. Pp. 131–150 in *Prosimian Biology.* R.D. Martin; G.A. Doyle; A.C. Walker, eds. London, Duckworth.

Charles-Dominique, P. 1977a. *Ecology and Behaviour of Nocturnal Primates: Prosimians of Equatorial Africa.* New York, Columbia.

Charles-Dominique, P. 1977b. Urine Marking and Territoriality in *Galago alleni* (Waterhouse, 1837-Loisoidea, Primates)—A Field Study by Radio-Telemetry. *Tierpsychol.* 43:113–138.

Charles-Dominique, P. 1983. Ecology and Social Adaptations in Didelphid Marsupials: Comparison with Eutherians of Similar ecology. Pp. 395–422 in *Advances in the Study of Mammalian Behaviour: ASM Special Publication No. 7.* J.F. Eisenberg; D.G. Kleiman, eds. Shippensburg, Pennsylvania, Am. Soc. Mammal.

Charles-Dominique, P. 1995. Social Organization and Food Constraints. Pp. 425–438 in *Creatures of the Night: The Nocturnal Prosimians.* L. Alterman; G.A. Doyle, M.K. Izard; eds. New York, Plenum.

Charles-Dominique, P.; Bearder, S.K. 1979. Field Studies of Lorisid Behavior: Methodological Aspects. Pp. 567–629 in *The Study of Prosimian Behaviour.* G.A. Doyle; R.D. Martin, eds. New York, Academic.

Clark, A.B. 1975. *Olfactory Communication by Scent Marking in a Prosimian Primate,* Galago crassicuadatus. Ph.D. Thesis. University of Chicago, Chicago.

Clark, A.B. 1978. Sex Ratio and Local Resource Competition in a Prosimian Primate. *Science* 201:163–165.

Clark, A.B. 1985. Sociality in a Nocturnal "Solitary" Prosimian: *Galago crassicaudatus.* *Intl. J. Primatol.* 6:581–600.

Clark, A.B. 1988. Interspecific Differences and Discrimination of Auditory Signals of *Galago crassicaudatus* and *Galago ganettii. Intl. J. Primatol.* 9:557–571.

Conroy, G.C. 1990. *Primate Evolution.* New York, Norton.

Courtenay, D.O.; Bearder, S.K. 1989. The Taxonomic Status and Distribution of Bushbabies in Malawi with Emphasis on the Significance of Vocalizations. *Intl. J. Primatol.* 10:17–34.

Crompton, R.H. 1984. Foraging, Habitat Structure, and Locomotion in Two Species of *Galago.* Pp. 73–111 in *Adaptations for Foraging in Nonhuman Primates: Contributions to an Organismal Biology of Prosimians, Monkeys, and Apes.* P.S. Rodman; J.G. Cant, eds., New York, Columbia University Press.

Dene, H.T.; Goodman; M.; Prychodko, W. 1976. Immunodiffusion Evidence on the Phylogeny of the Primates. Pp. 171–195 in *Molecular Anthropology.* M. Goodman; R.E. Tashian; and J.H. Tashian, eds., New York, Plenum.

Dixson, A.F. 1995. Sexual Selection and the Evolution of Copulatory Behavior in Nocturnal Prosimians. Pp. 93–118 in *Creatures of the Night: The Nocturnal Prosimians.* L. Alterman; G.A. Doyle, M.K. Izard; eds. New York, Plenum.

Doyle, G.A.; Andersson, A.; Bearder, S.K. 1971. Reproduction in the Lesser Bushbaby *(Galago senegalensis moholi)* under semi-natural conditions. *Folia Primatol.* 14:15–22.

Doyle, G.A.; Bearder, S.K. 1977. The Galagines of South Africa. Pp. 1–35 in *Primate Conservation.* H.R.H. Prince Ranier III; G.H. Bourne, eds., New York, Academic.

D'Souza, F. 1974. A Preliminary Field Report on the Lesser Tree Shrew *(Tupaia minor).* Pp. 167–182 in *The Study of Prosimian Behaviour.* G.A. Doyle; R.D. Martin, eds. New York, Academic.

Eliot, O.; Eliot, M. 1967. Field Notes on the Slow Loris in Malaysia. *J. Mammal.* 48:497–498.

Eudey, A.A. 1987. *Action Plan for Asian Primate Conservation: 1987–1991.* Lock Haven, Pennsylvania, Intl. Union for Conservation of Nature.

Fleagle, J.F. 1999. *Primate Adaptation and Evolution.* New York, Academic.

Fooden, J. 1991. Eastern Limit of Distribution of the Slow Loris, *Nycticebus coucang. Intl. J. Primatol.* 12:287–290.

Goodall, J. 1986. *The Chimpanzees of Gombe: Patterns of Behaviour.* Cambridge, Belknap.

Harcourt, C.S. 1980. *Behavioural Adaptations in South African Galagos.* M.SC. Thesis. University of Witwatersrand, Johannesburg.

Harcourt, C.S. 1981. An Examinations of Urine Washing in *Galago crassicaudatus. Z. Tierpsychol.* 55:119–128.

Harcourt, C.S. 1984. *Behaviour and Ecology of Galagos in Kenyan Coastal Forest.* Ph.D. Thesis. University of Cambridge, Cambridge.

Harcourt, C.S. 1986. Seasonal Variation in the Diet of South African Galagos. *Intl. J. Primatol.* 7: 491–506.

Harcourt, C.S.; Bearder, S.K. (1989) A Comparison of *Galago moholi* in South Africa with *Galago zanzibaricus* in Kenya. *Intl. J. Primatol.* 10:35–45.

Harcourt, C.S.; Nash, L.T. (1986a) Species Differences in Substrate Use and Diet Between Sympatric Species Galagos in Two Kenyan Coastal Forests. *Primates* 27:39–50.

Harcourt, C.S.; Nash, L.T. (1986b) Social Organization of Galagos in Kenyan Coastal Forests: 1. *Galago zanzibaricus. Am. J. Primatol.* 10:339–355.

Harrison, J.L. 1962. *The Apes and Monkeys of Malaya (including the Slow Loris).* Malayan Museum Pamphlets. *No. 9.* Singapore.

Hill, W.C.O. 1953. *Primates. Comparative Anatomy and Taxonomy. Vol. I, Strepsirhini.* Edinburgh, Edinburgh University Press.

Huynh, D.H. 1998. Ecology, Biology and Conservation Status of Prosimian Species in Vietnam. *Folia Primatol.* 69:101–108.

Jewell, P.A.; Oates, J.F. 1969. Ecological Observations on the Lorisoid Primates of African Lowland Forest. *Zool. Afr.* 4:231-248.

Katsir, Z; Crewe, R.M. 1980 Chemical Communications in *Galago crassicaudatus:* Investigation of the Chest Gland Secretion. *South African J. Zool.* 15:249–254.

Kingdon, J. 1971. *East African Mammals: An Atlas of Evolution in Africa. Vol. 1.* London, Academic.

Marler, P. 1968. Aggregation and Dispersal: Two Functions in Primate Communication. Pp. 420–438 in *Primates: Studies in Adaptation and Variability.* Jay, P.C., ed. New York, Holt, Rinehart, Winston.

Martin, R.D. 1990. *Primate Origins and Evolution: A Phylogenetic Econstruction.* Princeton, Princeton University Press.

Martin, R.D.; Bearder, S.K. 1979. Radio Bushbaby. *Nat. Hist.* 88:77–81.

Martin, R.D.; Chivers, D.J.; Maclarnon, A.M.; Hladik, C.M. 1985. Gastrointestinal Allometry in Primates and Other Mammals. Pp. 61–89 in *Size and Scaling in Primate Biology.* W.L. Jungers, ed. New York, Plenum.

Masters, J.C. 1988. Speciation in the Greater Galagos (Prosimii: Galaginae): A Review and Synthesis. *Biol. J. Linnean Soc.* 34:149–174.

Masters, J.C. 1991. Loud Calls of *Galago crassicaudatus* and *Galago garnettii* in Relation to their Habitat Structure. *Primates* 32:153–167.

Masters, J.C. 1993. Primates and Paradigms: Problems with the Identification of Genetic Species. Pp. 43–64 in *Species, Species Concepts, and Primate Evolution.* W.H. Kimbel; L.B. Martin, eds. New York, Plenum Press.

Masters, J.C. 1998. Speciation in the Lesser Galagos. *Folia Primatol.* 69:357–370.

McArdle, J.E. 1981. *Functional Morphology of the Hip and Thigh of the Lorisiformes. Contributions to Primatology No 17.* Basil, Karger.

McGrew, W.C.;Tutin, C.E.G.;Baldwin, P.J; Sharman, M.J.;Whiten, A. (1978) Primates Preying Upon Vertebrates: New Records from West Africa. *Carnivore* 1:41–45.

Napier, J.R.; Napier, P.H. 1967. *A Handbook of Living Primates.* New York, Academic Press.

Napier, J.R.; Walker, A.C. 1967. Vertical Clinging and Leaping—A Newly Recognized Category of Locomotor Behaviour of Primates. *Folia Primatol.* 6:204–219.

Nash, L.T. 1983. Reproductive Patterns in *Galago zanzibaricus* and *Galago garnettii* in Relation to Climatic Variability. *Am. J. Primatol.* 5:181–196.

Nash, L.T. 1986a. Dietary, Behavioral, and Morphological Aspects of Gummivory in Primates. *Yrbk. Phys. Anthropol.* 29:113–137.

Nash, L.T. 1986b. Influence of Moonlight Level on Travelling and Calling Patterns in Two Species of Galago in Kenya. Pp. 357–367 in *Current Perspectives in Primate Social Dynamics.* D.M. Taub; F.A. Frederick, ed., New York: Holt Reinhold.

Nash, L.T. 1989. Galagos and Gummivory. *Hum. Evol.* 4:199–206.

Nash, L.T. 1993. Juveniles in Nongregarious Primates. Pp 119–206 in *Juvenile Primates: Life History, Development and Behaviour.* Pereira, M.E.; Fairbanks, L.A. eds., Oxford: Oxford University Press.

Nash, L.T.; Bearder, S.K.; Olson, T.R. 1989. Synopsis of Galago Species Characteristics. *Intl. J. Primatol.* 10:57–80.

Nash, L.T.; Harcourt, C.S. (1986) Social Organization of Galagos in Kenyan Coastal Forests: II. *Galago garnetii. Am. J. Primatol.* 10:357–369.

Nash, L.T.; Whitten, P.L. 1989. Preliminary Observations on the Role of Acacia Gum Chemistry in *Acacia* Utilization by *Galago senegalensis* in Kenya. *Am. J. Primatol.* 10:27–39.

Nekaris, K. In prep. *Socioecology of the Slender Loris* (Loris tardigradus lydykkerianus) *in Dindigul (DT), Tamil Nadu, South India.* Ph.D. Thesis. Washington University, St. Louis.

Nisbett, R.A.; Ciochon, R.L. 1993. Primates in Northern Viet Nam: A Review of the Ecology and Conservation Status of Extant Species, with Notes on Pleistocene Localities. *Intl. J. Primatol.* 14:765–795.

Oates, J.F. 1984. The niche of the potto, *Perodicticus potto. Intl. J. Primatol.* 5: 51–61.

Olson, T.R. 1979. *Studies on Aspects of the Morphology of the Genus* Otolemur Coquerel, *1859.* Unpublished Ph.D. Thesis. University of London, London.

Oxnard, C.E.; Crompton, R.H.; Lieberman, S.S. 1990. *Animal Lifestyles and Anatomies: The Case of the Prosimian Primates.* Seattle, University of Washington Press.

Paterson, H.E.H. 1985. The Recognition Concept of Species. Pp. 21–29 in *Species and Speciation.* Transvaal, Transvaal Museum. (Reprinted in Paterson, H.E.H. 1993. *Evolution and the Recognition Concept of Species,* E.S. Vrba, ed. Baltimore, Johns Hopkins University Press.)

Petter, J.-J.; Hladik, M.C. 1970. Observations sur le Domaine Vital et la Densité de population de *Loris tardigradus* dans les forêts de Ceylon. *Mammalia* 3:394–409.

Preuschoft, H.; Witte, H.; Fischer, M. 1995. Locomotion in Nocturnal Prosimians. Pp. 453–472 in *Creatures of the Night: The Nocturnal Prosimians.* L. Alterman; G.A. Doyle, M.K. Izard; New York, Plenum.

Rasmussen, D.T. 1986. *Life History and Behaviour of Slow Lorises and Slender Lorises: Implications for the Lorisine-Galagine Divergence.* Unpublished Ph.D. Thesis. Duke University, Durham.

Rasmussen, D.T. 1997. Asian and African Prosimian Field Studies. Pp. 22–24, in *History of Physical Anthropology.* F. Spencer, ed. New York: Garland

Rasmussen, D.T.; Izard, M.K. 1988. Scaling of Growth and Life History Traits Relative to Body Size Brain Size and Metabolic Rate in Lorises Galagos (Lorisidae, Primates). *Am. J. Phys. Anthropol.* 75:357–367.

Rasmussen, D.T.; Nekaris, K.A. 1998. Evolutionary History of Lorisiform Primates. *Folia Primatol.* 69 Suppl.1:250–285.

Ratajszczak, R. 1998 Taxonomy, Distribution and Status of the Lesser Slow Loris *Nycticebus pygmaeus* and Their Implications for Captive Management. *Folia Primatol.* 69:171–174.

Sabater Pi, J. 1972. Notes on the Ecology of Five Lorisiformes of Rio Muni. *Folia Primatol.* 18:140–151.

Schulze, H.; Meier, B. 1995. The Subspecies of *Loris tardigradus* and Their Conservation Status: A Review. Pp. 193–210 in *Creatures of the Night: The Nocturnal Prosimians.* S. L. Alterman; G.A. Doyle, M.K. Izard; eds. New York, Plenum.

Sellers, W.I. 1996. A Biomechanical Investigation into the Absence of Leaping in the Locomotor Repertoire of the Slender Loris *(Loris tardigradus).* *Folia Primatol.* 67:1–14.

Simons, E.L. 1972. *Primate Evolution: An Introduction to Man's Place in Nature.* New York, MacMillan.

Simons, E.L.; Rasmussen, D.T. 1994. A Remarkable Cranium of *Plesiopithecus teras* (Primates, Prosimii) from the Eocene of Egypt. *Proc. Natl. Acad. Sci.* 91:9946–9950

Struhsaker, T.T. 1970. Notes on *Glagoides demidovii* in Cameroon. *Mammalia* 34:207–211.

Suckling, J.A.; Suckling, E.E.; Walker, A.C. 1969. Suggested Function of the Vascular Bundles in the Limbs of *Perodicticus potto*. *Nature* 221:379–380.

Sussman, R.W. 1992. Smell as a Signal. Pp. 157–160 in Jones, S.; Martin, R.D.; Pilbeam D. (eds.) *The Cambridge Encyclopedia of Human Evolution.* Cambridge, Cambridge University Press.

Sussman, R.W.; Kinzey, W.G. 1984. The Ecological Role of the Callitrichidae: A Review. *Am. J. Phys. Anthropol.* 64:419–449.

Weisenseel, K.A.; Izard, M.K.; Nash, L.T.; Ange, R.L.; Poorman-Allen, P. 1998. A Comparison of Reproduction in Two Species of *Nycticebus*. *Folia Primatologica* 69:321–324.

Williams, J.B. 1983. *Behavioral Observations of Feral Prosimians: A Bibliography.* Primate Information Center, Seattle, Washington.

Williams, J.B. 1991. *Behavioral Observations of Feral and Free-Ranging Prosimians: a Bibliography, 1983-1991.* Primate Information Center, Seattle, Washington.

Zimmerman, E. 1990. Differentiation of Vocalizations in Bushbabies (Galaginae) and the Significance for Assessing Phylogenetic Relationships. *J. Zool. Syst. Evol.* 28:217–319.

Zimmerman, E. 1995. Acoustic Communication in Nocturnal Prosimians. Pp. 311–330 in *Creatures of the Night: The Nocturnal Prosimians.* L. Alterman; G.A. Doyle, M.K. Izard; New York, Plenum.

Zimmerman, E.; Bearder, S.K.; Andersson, A.B.; Doyle, G.A. 1988. Variations in Vocal Patterns of Senegal and South African lesser Bushbabies and their Implications for Taxonomic Relationships. *Folia Primatol.* 51:87–105.

CHAPTER 4

The Nocturnal Lemuriformes

INTRODUCTION

The living Lemuriformes today are found in the coastal regions of Madagascar (Figure 4-1). There are five families of Lemuriformes: Cheirogaleidae, Lemuridae, Lepilemuridae, Indriidae, and Daubentoniidae (Table 2-4). There are five genera of Cheirogaleidae and eight currently recognized species (Table 4-1). These are all small animals that are active during the night. All of the genera, except *Allocebus*, are quite widely distributed (Figure 4-2). Because of the problems often inherent in identifying species in nocturnal forms discussed in earlier chapters, it is possible that more species of cheirogaleids will be discovered in the future. In fact, Groves and Tattersall (1991) have recognized four varieties of *Phaner*, and Schmid and Kappeler (1994, Atsalis et al. 1996) have identified a new species of *Microcebus (M. myoxinus)* in Western Madagascar and there may be yet

Table 4-1 Common Names of Nocturnal Lemuriformes

Species	Common Name
Family CHEIROGALEIDAE	
Cheirogaleus medius	Fat-tailed dwarf lemur
C. major	Greater dwarf lemur
Microcebus murinus	Gray mouse lemur
M. rufus	Brown mouse lemur
M. myoxinus	Pygmy mouse lemur
Mirza coquereli	Coquerel's dwarf lemur
Phaner furcifer	Fork-marked lemur
Allocebus trichotis	Hairy-eared dwarf lemur
Family LEPILEMURIDAE	
Lepilemur mustelinus	Sportive lemur
Family DAUBENTONIIDAE	
Daubentonia madagascariensis	Aye-aye

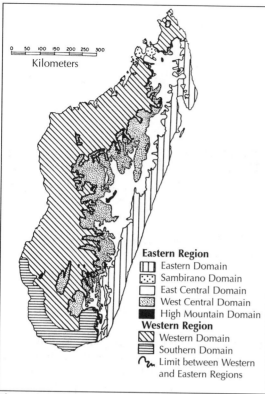

Eastern Region

⊞ Eastern Domain
⊡ Sambirano Domain
☐ East Central Domain
▩ West Central Domain
■ High Mountain Domain

Western Region

▨ Western Domain
☰ Southern Domain
〰 Limit between Western and Eastern Regions

Figure 4-1 Climax vegetation types, as established by H. Humbert (1955), after Tattersall (1982).

unidentified species of this genus (Zimmerman et al. 1998). The systematics of these forms is yet to be determined.

There are four genera and ten species of Lemuridae (Table 5-1). Certain species of this family (i.e., *Lemur catta, Eulemur fulvus, Eulemur macaco,* and *Varecia variegata*) are the most commonly found Malagasy prosimians in zoological parks. In general, these animals are diurnal but some also are *cathemeral*, i.e., active both during the day and at night (Tattersall 1988). Furthermore, *Eulemur mongoz* is nocturnal at least seasonally and in part of its range (see next Chapter).

Madagascar is divided into two major vegetation zones (Figure 4-1), and the distribution of some genera and species is restricted to certain types of vegetation whereas others are more widely distributed. For example, *Lemur catta* is found only in the semi-arid south and in the southwest of the island; *Eulemur fulvus* in all forest regions except the south; *Eulemur coronatus, Eulemur mongoz,* and *Eulemur macaco* in the deciduous forests of the north; and *Eulemur rubriventer* and *Varecia variegata* only in the evergreen rain forests of the east (Chapter 5). The taxonomy of Lepilemuridae is under debate. The one genus *Lepilemur* is divided into seven species by some authors (Rumpler and Albignac 1975, Petter and Petter-Rousseaux 1979, Tattersall 1986, Mittermeier et al. 1994). It is clear that a number of distinct populations of *Lepilemur* exists but exactly how many or at what taxonomic level is not clear (Tattersall 1982, Ganzhorn 1993, Mittermeier et al. 1994). I prefer to consider these types as subspecies of *Lepilemur mustelinus* until we have better data on the distribution and variation of natural populations. *Lepilemur* is nocturnal and is found throughout the coastal regions of the island (Figure 4-2).

The family Indriidae contains three genera and five species. These animals are quite large, especially *Indri* and *Propithecus* weighing about 6000 and 4000 gm, respectively. All of the Indriidae are specialized vertical clingers and leapers (see next chapter). Indriids also have adaptations that allow them to subsist on highly folivorous diets. One species, *Avahi laniger*, is nocturnal. *Propithecus verreauxi* is the only species widely distributed outside of the eastern evergreen forests although a subspecies of *A. laniger* has a localized distribution in the northwest (Chapter 5).

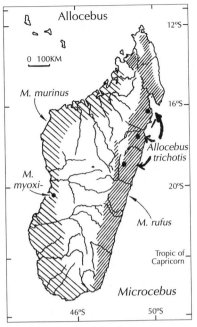

A. Map of *Microcebus* and *Allocebus* distribution.

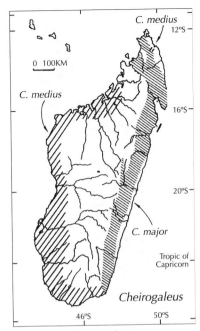

B. Map of *Cheirogaleus* distribution.

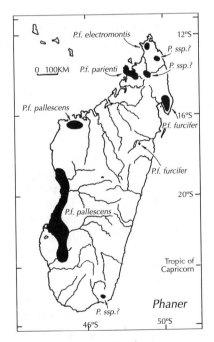

C. Map of *Phaner* distribution.

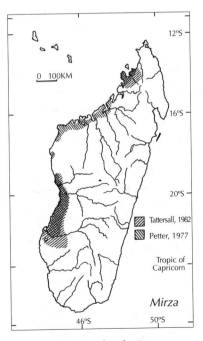

D. Map of *Mirza* distribution.

Figure 4-2 Distribution Maps of *Nocturnal Lemuriformes*. [Adapted from Mittermeier et al. 1994]

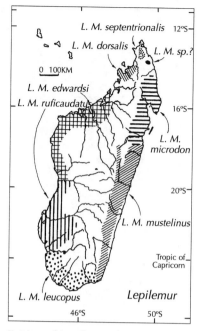

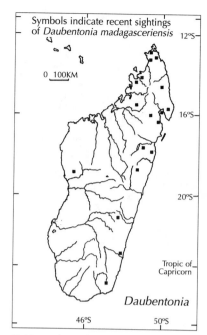

E. Map of *Lepilemur* distribution.

F. Map of *Daubentonia* distribution.

Figure 4-2 *(continued)*

Daubentonia madagascariensis, the aye-aye, is the only living member of the family Daubentoniidae. This strange animal is nocturnal and is found in various localities in Northern and Eastern Madagascar, and possibly in the west (Figure 4-2).

There are, thus, at least twenty-five extant species of lemurs, slightly over half of which are diurnal (13 species), the only diurnal prosimians. The lemurs represent an amazing diversification of primitive primates. Isolated on Madagascar for approximately 60 million years, they were never forced to compete with anthropoids or other major groups of arboreal mammals; thus they have radiated into a number of niches occupied by monkeys and apes in other parts of the world. The first anthropoids to reach Madagascar were human, 1500–2000 years ago (Kent 1970, Richard and Dewar 1991, Verin 1990, Burney 1995). Since that time, because of habitat destruction and hunting, the diversity and numbers of lemurs have greatly diminished. Presently, all of the Lemuriformes are threatened with extinction because habitat loss is progressing so quickly (Harcourt 1990, Green and Sussman 1990, Mittermeier et al. 1994, Patterson et al. 1995).

At least 17 species of lemurs have already become extinct since the arrival of humans (Tattersall 1982, Simons et al. 1995, Simons 1997, Godfrey et al. in press, Table 4-2). These subfossil species must be considered part of the amazing adaptive radiation of the modern Malagasy lemurs. They are referred to as subfossils because in reality they are modern forms. No fossils of great antiquity have been found in Madagascar. Thus, the evolutionary record of the lemurs, since their isolation in the Eocene, is unknown. Many of the subfossil species were quite unlike surviving lemurs (Figure 4-3) and filled ecological niches that have now disappeared or are left unfilled.

Table 4-2 Characteristics and Distribution of Extinct Lemuriformes[1]

Species	Size[2]	Probable Locomotion	Location of Finds	Comments
Varecia insignis	small	arboreal quadraped (cf. modern forms of *Varecia*)	Southwest	moderately abundant
V. jullyi	small	same as *V. insignis*	Central Plateau	moderately abundant
Hapalemur gallieni	small (size of *H. simus*)	unknown	Central Plateau	rare
Megaladapis madagascariensis	large (240 mm cranium)	short limbed, arboreal; possibly modifed Koala-like locomotion (Walker 1974)	South, Southwest	abundant—The animals in this genus probably were arboreal browers, and highly folivorous.
M. grandidiri	large (290 mm cranium)		Central Plateau	
M. edwardsi	largest of three (300 mm cranium) donkey size		South, Southwest	
Mesopropithecus pithecoides	small – medium	uncertain	Central Plateau	moderately abundant—slightly more robust *Propithecus*-type skull.
M. globiceps	small – medium			
M. Sp. (dolicobrachion)	medium	slothlike, climbing suspensory	Extreme Northwest	rare
Paleopropithecus[3] insignis	large	arboreal hanger and climber (cf. orang utan)	Central Plateau, South, Southwest	abundant—long low skull with upturned nasal bones, generally orang-like but dentition like other indrids.
Paleopropithecus maximum	large	arboreal hanger and climber (cf. orang utan)	Central Plateau	
Babakotia radofilai	medium	slothlike, climbing suspensory	Extreme Northwest	rare

111

Table 4-2 (Continued)

Species	Size[2]	Probable Locomotion	Location of Finds	Comments
Archeoindris fontoynonti	large	unknown	Central Plateau	rare—1 specimen. Dentition of modern indriid but very large, high and robust skull.
Archeolemur majori	medium	arboreal and terrestrial quadraped (cf. *Papio* baboon)	South, Southwest and Central Plateau	abundant–Skull-like indriid but highly derived teeth (cf. cercopithecoid monkey) premolars continuous shearing blade unlike any other primate.
A. edwardsi	medium			
Hadropithecus stenognathus	medium	terrestrial quadruped (cf. gelada baboon)	Central Plateau, South, Southwest	rare—Molar dentition highly adapted to grinding probably like gelada baboon—manual grazers. Lost tooth comb.
Daubentonia robustus	small – medium	as living species, but larger	Southwest and possibly Central Plateau	rare—Only represented by long bones and teeth.

[1] All species are probably diurnal except *D. robustus* (for which there is no skull).

[2] small = equivalent to largest living Malagasy primates or somewhat larger.
medium = size of small baboons.
large = approximately chimpanzee size or larger.

[3] A newly discovered species of *Paleopropithecus* from Northern Madagascar is currently being described.
(*Source:* Tattersall 1982, Simons et al. 1992, Jungers pers. comm.)

Figure 4-3 Artistic reconstruction of the fossil site of Ampazambazimba, Madagascar (ca. 8000–1000 B.C.), showing a variety of subfossil prosimians from that locality. At the upper left, a *Megaladapis* feeds on leaves while clinging to a trunk. Below are two individuals of *Pachylemur insignis* and to the right a family of slothlike *Paleopropithecus*. On the ground, an *Archaeoindris* feeds in the background while another individual of *Megaladapis* ambles along to another tree. In the foreground, a group of *Archaeolemur* feed on tamarind pods while a group of *Hadropithecus* wander in from the right. [Adapted from *Primate Adaptation and Evolution* by John G. Fleagle. Academic Press, San Diego, 1999]

As recently as 1987 (Sussman 1988), even year-long field research had been carried out on only four species: *Lemur catta, Eulemur fulvus, Indri indri,* and *Propithecus verreauxi.* Many species had not been studied at all. Since that time there has been a tremendous resurgence of interest in field research on these animals. Furthermore, three new species of lemur have been discovered, *Propithecus tattersalli* (Simons 1988) and *Hapalemur aureus* (Meier et al. 1987), and *Microcebus myoxinus* (Schmid and Kappaler 1994). The first serious attempt in decades to find subfossils has also begun under the leadership of Elwyn Simons of Duke University, and new subfossil species have been uncovered (Godfrey et al. 1990, in press, Simons 1997). In the coming years, we will know a great deal more about these fascinating animals. In this chapter, I will discuss the families containing mainly small, nocturnal animals; Cheirogaleidae, Lepilemuridae, and Daubentoniidae. In Chapter 5, I will focus on the larger, mainly diurnal Lemuridae and Indriidae.

THE ECOLOGY OF THE NOCTURNAL LEMURIFORMES— A CASE STUDY IN WESTERN MADAGASCAR

A preliminary study of five nocturnal Malagasy prosimians has been carried out in a seasonal forest, Marosalaza, in Western Madagascar (Figure 4-4): *Cheirogaleus medius, Microcebus murinus, Mirza coquereli, Phaner furcifer* and *Lepilemur mustelinus* (Petter 1978, Hladik et al. 1980). The study was conducted between October

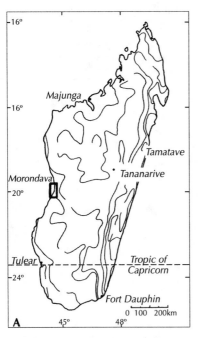

Figure 4-4 Location of the study area according to climate and plant formation. A, The rectangle near Morondava corresponds to the enlarged map in (B). B, Plant formations in the Morondava area. The study area is located 50 km north of Morondava (black square) in Marosalaza forest. [Adapted from Hladik, 1980]

1973 and June 1974, but none of the investigators remained in the field for more than two months at a time. Simultaneous studies were conducted on the same species under simulated climatic conditions by a team from the French National Center of Scientific Research (C.N.R.S.) in Brunoy, France (Charles-Dominique *et al* 1980). The field study of five sympatric nocturnal lemurs provides an excellent comparison to Charles-Dominique's research on lorisiformes in Gabon. Research at the same and other sites in Western Madagascar has focused on the same species: *C. medius* (Petter 1988, Müller 1998, Feitz 1998), *M. murinus* (Martin 1972, 1973, Barre et al. 1988, Pagès-Feuillade 1988, Goodman et al. 1993a, Schmid and Kappeler 1994, 1998, Fietz 1997, Zimmermann et al. 1998); *M. coquereli* (Petter et al. 1971, Andrianarivo 1981, Kappeler 1997); *P. furcifer* (Petter et al. 1975); and *L. mustelinus* (Charles-Dominique and Hladik 1971, Russell 1977, Ganzhorn 1988, 1989, 1993, Nash 1994, 1998, Crompton 1995). There also are some comparative studies of eastern rain forest species. These will be discussed later in the chapter.

In a detailed study of the climate and vegetation, A. Hladik (1980) found that the Marosalaza forest is seasonal in production of flowers, fruit and leaves. There are few of these items available during the dry season which occurs in the austral winter, April/May to September (Figure 4-5). Insect availability is also seasonal, following the pattern of leaf production. The specific dietary and physiological adaptations of the five small nocturnal prosimians are related to this seasonality, which is common throughout the west and southwest of Madagascar.

The most extreme adaptation to the seasonal production of the forest is seen in *Cheirogaleus medius* (the fat-tailed dwarf lemur). Unlike any other primate, *C. medius* actually hibernates for 6–8 months during the year (Hladik et al. 1980, Foerg and Hoffman 1982, Wright and Martin 1995). In Western Madagascar, it emerges in November before the rainy season when food production is at a maximum. During hibernation 3–5 individuals may sleep together in deep hollows of tree trunks (Figure 4-6). During the active period, the animals accumulate large fat

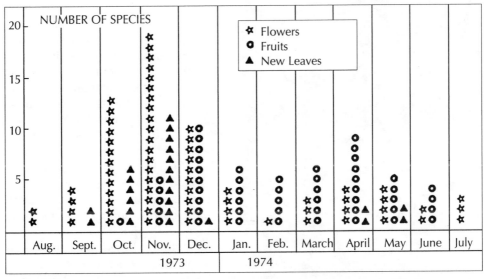

Figure 4-5 Food availability throughout the year at Marosalaza forest. Symbols indicate number of species producing potential foods in a 2,200 m² sample. [Adapted from Hladik et al., 1980]

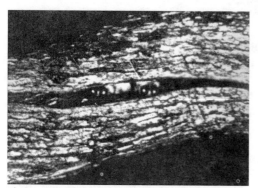

Figure 4-6 *Cheirogaleus* eyes in nest. [Photo by C. M. Hladik]

reserves in the tail (Figure 4-7). Body weights of individuals captured at Marosalaza varied from 142 gm in November to 217 gm in March (Table 4-3). Tail volumes, measured by immersing tails into a graduated cylinder filled with water, ranged from a maximum of 56 cm^3 (mean=42 cm^3) in May before hibernation to a minimum of 9 cm^3 in November (mean=15 cm^3).

C. medius is quadrupedal and not as agile or quick-moving as *Microcebus* or *Phaner* (Petter et al. 1977). The tail, when thickened, is not conducive to acrobatic locomotion! The fat-tailed lemur is active at a height of 2 to 9 meters and selects substrates approximately 5 cm in diameter.

In Marosalaza, the diet of *C. medius* was studied by direct observation and examination of fecal material. The proportions of different dietary items changed throughout the season of activity. Some fruit was eaten in November and December. A number of tree species were in flower from October to January (Figure 4-5), and nectar from flowers made up the major portion of the diet when the animals first emerged from hibernation. Flowers of two plant species were observed being licked by *C. medius*. In fact, one radio-tracked individual fed exclusively upon nectar for six consecutive days in December. The importance of nectar for some lemurs and the role of lemurs as pollinators has only recently been realized (Sussman and Raven 1978, Kress 1993, Nillson et al. 1993, Kress et al. 1994, Birkinshaw and Colquhoun 1998). More on this subject will be found in Chapter 5.

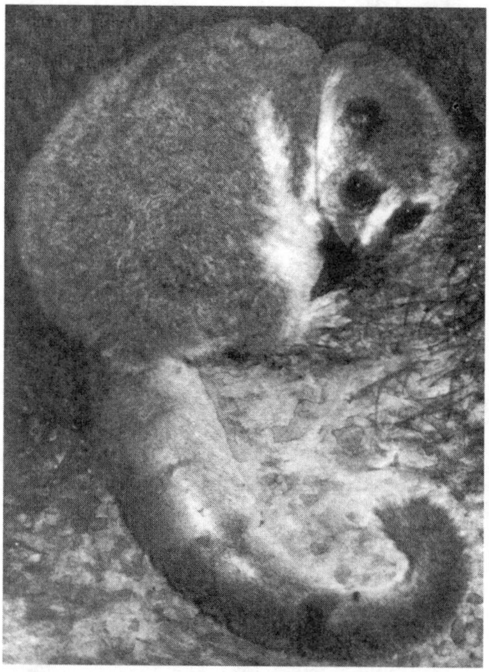

Small and large fruits accounted for the main portion of the diet from February through March. Some gums and other plant and insect secretions made up a small part of the diet. Invertebrates were present in 50% of the feces, but in small amounts. The volume of insect chitin (the material making up the exoskeleton) never exceeded 10% of the food remains. Coleoptera were the major insect prey. The skin of a chameleon also was found in one sample. The proportion of insects in the diet increased in January, corresponding to an increase in the insect population.

In general, *C. medius* is considered to be an opportunistic frugivore. The seasonal variation in diet follows food availability: nectar and fruit in November/December; fruits and an increasing proportion of insects from

Figure 4-7 *Cheirogaleus* tail. [Photo by C. M. Hladik]

Table 4-3 Field (Except Where Specified) Adult Body Weights of Some Nocturnal Lemuriformes

Species	Body Weight in Grams (Range)	Sample Size	Sex	Source
Cheirogaleus medius	142	12 (Nov–Dec)	male and female	Hladik et al. 1980
	217	13 (March)		
C. major	356 (235–470)	7 (October)	male and female	Wright & Martin 1993
Microcebus murinus	57.9 (39–84)	52	male and female	Schmid and Kappeler in press
	62 (54–75)	6	male and female	Pages-Feiullade 1988
Microcebus myoxinus	30.6 (24–38)	32	male and female	Schmid and Kappeler 1994
M. rufus	43 (34–54)	20	male and nonpregnant females	Wright and Martin 1995
	62 (50–75)	6	pregnant females	
Mirza coquereli	~280–335	14	male and female	Kappeler 1997
Phaner furcifer	460	3	1 male, 2 female (captive)	Petter-Rousseaux 1980
Allocebus trichotis	86.5 (75–98)	6	3 male, 3 female	Meier and Albignac 1991
Lepilemur mustelinus	~600–1000	33	males and females	Smith and Jungers 1997
Daubentonia madagascar	2700 (2350–3000)	8	6 males, 2 females	Sterling 1993a

December to January; and a higher proportion of fruits at the time of fattening, preceding hibernation (Hladik *et al* 1980). These patterns of food intake seem to be mirrored in captive *C. medius* given a choice of foods (Petter-Rousseaux and Hladik 1980), and the Brunoy researchers believe that *C. medius* is pre-adapted physiologically to changes in its natural environment. *Microcebus murinus* (the gray mouse lemur) does not hibernate, but it does have a period of lethargy

Figure 4-8 Grey mouse lemur. [Photo by J. Buettner-Janusch]

during the dry season in Marosalaza. At this time, the animals may stay in hollow trunks for several consecutive days, and overall activity is considerably reduced. As with *C. medius,* fat is stored in the tail, but tail fattening is not as spectacular in *M. murinus* (seasonal variations in tail volume being 2 to 9.5 cm³). Russell (1975) found that both *M. murinus* and *C. medius,* housed under constant conditions of ambient temperature, humidity, diet and day length, showed daily patterns of body temperature that varied at different times of the year. During the time equivalent to austral winter, all animals captured while resting had remarkably low temperatures, approaching ambient temperatures, and were lethargic. He also found that some individuals were lethargic and had depressed body temperatures, while others were active. This may be true in the Marosalaza forest, since some gray mouse lemurs were observable throughout the year. Martin (1972) did not observe a period of lethargy in a population living in a warmer, less seasonal littoral rain forest in Southeastern Madagascar, but Atsalis (1998) found some individuals to enter torpor seasonally in a rain forest in the east.

The gray mouse lemur is the size of a mouse, weighing around 60 gm (Table 4-3). Like most other primates, it has hands with five fingers (Figure 4-8). It scurries like a mouse but it also can grip the fine terminal branches of bushes and trees, and can make leaps of over 3 meters (Martin 1972, Pages-Feuillade 1988). It is most commonly found on branches less than 5 cm in diameter. The gray mouse lemur usually lives in areas with dense foliage and fine terminal supports and is active at all heights in the forest where suitable vegetation occurs. However, in the forest reserve of Ampijoroa, Northern Madagascar, Pages-Feuillade (1988) found these little creatures spending most of their time below 10 meters, and 40% of the time below 3 meters. In fact, at the end of the dry season when little plant food was available, some animals spent 70% of their time searching for insects below 3 meters.

While foraging, gray mouse lemurs move slowly, and change height and direction continuously. *M. murinus* descends to the ground to hunt insects and to cross short open areas, but these visits are rare and brief. Insects captured on the ground are transported by mouth up into the branches above. The gray mouse lemur is most commonly found in degraded vegetation or in ecotones, and is not common in areas of primary forest vegetation (Martin 1973, Pages-Feuillade 1988).

The diet of *Microcebus* consists mainly of insects and fruit. Although mouse lemurs are the smallest primates, less than half of their diet is made up of insects (Martin 1972, Hladik et al. 1980, Atsalis 1998). Dietary data on gray mouse lemurs were collected by direct observation, by examination of fecal material (Martin 1972, 1973; Hladik et al. 1980), and by analysis of stomach contents (Martin 1972). In each locality studied, the gray mouse lemur seems to specialize on one or two locally available fruit bearing species at a time (Martin 1972, Petter 1978). This was also found to be the case in the eastern form (Atsalis 1998, see below). Besides fruit, *M. murinus* was observed feeding on small amounts of gums and leaves. It also licked floral nectar (Hladik 1980, Martin 1972, 1973), and very likely serves as a pollinator to some plant species (Hladik 1980, Nilsson et al. 1993).

Beetles were the major source of prey at both Marosalaza and Mandena; however, moths, praying mantids, fulgorid bugs, crickets, cockroaches, spiders and chameleons also were eaten. Most cases of predation on insects occurred on the ground. Martin (1972:58) describes these observations as follows:

> *The animal would descend to a height of about 1 m on a tree trunk or bush and remain motionless for a while, exhibiting alternating movements of the ear pinnae. It would then make a rapid dash across the leaf litter on the forest floor, seize an insect with its hands, bite the prey and then dash back to its former vantage point to eat its prey. . . . It may well be that the large ear pinnae in these prosimians, along with the alternating movement of the pinnae prior to predation, represent an adaptation for precise location of insects moving in leaf litter and other matter.*

Just as with *C. medius*, captive *M. murinus* showed a variation in propensity for different dietary components at different periods of the year. Seasonal variations in body weight, body temperature, basal metabolism, choice of foods, and water intake (Perret 1980) seem to be related to physiologically based endogenous rhythms that enable these two diminutive species to live in forests with seasonal climates and productivity.

As stated above, a second species of mouse lemur may occur in Western Madagascar. Schmid and Kappeler (1994) and Atsalis et al. (1996) have described a population of *M. myoxinus* (the pygmy mouse lemur) that is even smaller (avg. = 31 gm) than either the western grey mouse lemur or the eastern brown mouse lemur. This species differs from both of these forms in a number of other morphological characteristics. From preliminary observations (Zimmermann et al 1998), it appears to prefer parts of the forest with a higher canopy where it leaps more frequently than the sympatric grey mouse lemur. It is more gracile and may not have the same capability to store fat in its tail as the sympatric form.

Mirza coquereli (Figure 4-9) is found in the densest part of the forest (Pages 1980). It locomotes primarily on small branches, under 10 cm in diameter and ranges from the ground, on which it occasionally forages for insects, to about 9 meters in height (Hladik et al. 1980). However, most of its activity and feeding take place low in the trees and bushes (Petter et al. 1971, Pages 1980). *M. coquereli*, unlike *C. medius* and *M. murinus*, shows little seasonal variation in food consumption and body weight (captive adults weigh 280–335 gm with males being heavier than females, Kappeler 1997). Hladik et al. (1980) claim that its peculiar ability to exploit the secretions of insects, which are available during both wet and dry months, enables this species to be active throughout the year. Since this food source is constant but of low density, the population density and biomass of *M. coquereli* at Marosalaza is low compared to that of other species. In a secondary forest dominated by cashew trees *(Anacardium occidentale)* in northern Madagascar, Andrianarivo (1981) estimated the population density to be much higher, 385 per km^2. At this forest, *M. coquereli* feeds mainly on the ripe fruit and gums of the cashew tree during the dry season.

Coquerel's lemur seems to be as insectivorous as *M. murinus* in the forest of Marosalaza, and also eats large amounts of fruit. A few chameleons, frogs, and a

Figure 4-9 Coquerel's lemur. [Photo by Ben Freed]

small amount of floral nectar and gums are found in its diet. However, in the beginning of austral winter, 50% of the observed feeding of *M. coquereli* was on the secretions of homopteran insects, mainly colonial aggregates of *Flatida coccinea* (Pages 1980). At the larval stage, the colonial insects exude a sweet secretion when disturbed. *M. coquereli* regularly visits several colonies each night, licking the liquid deposited. The secretions of *Flatida* have low protein content; insects and small vertebrates provide the major source of protein for Coquerel's lemur at this western site.

Phaner furcifer (Figure 4-10) is about the same size as *Mirza coquereli* (460 gm) and it also shows no seasonal changes in body weight or in basal metabolism (Petter 1978, Hladik et al. 1980). Like the African lorisid *Galago elegantulus, P. furcifer* has specialized to feed on gums and it exists mainly on this resource, which is available throughout the year. *Phaner* is a quick-moving, agile, quadrupedal animal that spends most of its time on the horizontal branches of the canopy, 3 to 4 meters in height. It also ranges through the various vertical strata of the forest, from the ground to above 10 meters (Petter et al. 1975). Like *G. elegantulus, Phaner* has keeled, pointed nails that allow it to climb up large vertical tree trunks and facilitates gum feeding on large branches (Figure 4-11).

The greatest proportion of the diet is made up of gums, mainly from *Terminalia* species. Gums are produced to seal wounds when coleopteran larvae burrow tunnels in the bark of the tree. In the daytime, *Terminalia* are visited for gums by an endemic species of Malagasy bird, *Coua cristata* (Charles-Dominique 1976). Other species of tree are visited less frequently for gums, and especially during the wet season, for floral nectar and fruit. *Phaner* has been observed for long periods clinging to fine terminal branches and licking clusters of flowers (Petter *et al* 1975). *Phaner* also feeds on insect secretions, primarily those from Homopteran larvae. Although some gums of *Terminalia* are proteinaceous, *Phaner* complements its diet with animal food throughout the year. Approximately 10% of its diet is made up of insects although it seems to be much more selective of prey than *Mirza* and searches for insects higher in the canopy (8–10 meters) (Hladik *et al* 1980).

In addition to the presence of claw-like nails, *Phaner* has a number of other morphological characteristics that seem to be related to gum feeding. These

Figure 4-10 *Phaner.* [Photo by Russell A. Mittermeier]

include: special features of the dentition, expecially of the upper incisors, canines and premolars (Figure 4-12); specialized features of the digestive tract, especially the large size of the caecum, and the long, pointed tongue (Petter et al. 1975). Similar anatomical features are found in the African gum-feeder *Galago elegantulus.*

Lepilemur mustelinus is the fifth species of nocturnal primate at Marosalaza. It is unusual for a small nocturnal primate (avg. adult=600–900 gm) in that it is a folivore. Its diet is made up almost exclusively of leaves. Related to

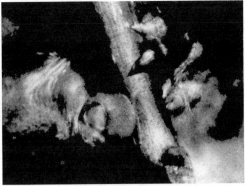

Figure 4-11 *Phaner* nails. [From Petter, Schilling, and Pariente, 1971]

this folivorous diet, it has a specialized alimentary tract (a large caecum and wide colon, Le Gros Clark 1963, Hladik 1979, Chivers and Hladik 1980); and dentition, especially the shearing crests of the molars (Kay et al. 1978, Seligson and Szalay 1978).

Lepilemur is a vertical clinger and leaper (Figure 4-13). It moves horizontally between trees by leaping from one vertical trunk to another, usually maintaining the same height and uses small diameter, vertical substrates most often (Nash 1998, Warren and Crompton 1998). It ascends and descends by limb-over-limb slow climbing or upward hops (Walker 1967, Russell 1977, Crompton 1995, Warren and Crompton 1998). In the Marosalaza forest, *Lepilemur* is typically found in the larger branches of the canopy, between 4 and 11 meters high (Hladik *et al* 1980). In another deciduous forest in Southwestern Madagascar, radio-tracked animals also were observed to be vertical clingers and leapers, and stayed between 5 and 15 meters high (Nash 1994, 1998). In the desert-like vegetation in the south of Madagascar, *Lepilemur* remained on the vertical trunks of cactus-like

Figure 4-12 *Phaner* dentition. [From Petter, Schilling, and Pariente, 1971]

Figure 4-13 *Lepilemeur.* [Photo by Robert W. Sussman]

plants (mainly of the family Didiercheaceae) always above 1.5 meters in height. It rarely, if ever, descends to the ground. *Lepilemur* is one of the few lemurs which lives in the desert-like *Didierea* forests of Southern Madagascar. Vertical clinging and leaping is a necessary mode of locomotion in this type of vegetation for an arboreal species. Because of the necessity of using vertical supports for travelling, one would expect that increasing tree distances by logging would directly effect *Lepilemur* population densities. Ganzhorn (1993) found a significant decline in population density between 1990 and 1992 in an intensively logged portion of a deciduous forest in Western Madagascar.

Lepilemur diet has not been studied in detail at Marosalaza. Charles-Dominique and Hladik (1971) and Russell (1977), however, have done short term but intensive studies of this species in a *Didierea* forest. Ganzhorn (1988, 1993) has studied the diet of *Lepilemur* in two western deciduous forests and in two eastern rain forests, and Nash (1994, 1998) recently completed a 11 month study in which she radio-tracked five individuals. In all localities, the diet of *Lepilemur* is restricted. Russell found that, in a four-month period, it fed on a total of 24 species of plant; an individual ate only 2 to 14 species per month. Two species accounted for 64% of the diet: *Alluaudia procera* (52.8%) and *Grewia* species (11.1%). Nash's (1998) findings were similar; two tree species and one species of vine accounted for 75–85% of the diet of the 5 radio-tracked animals. Of its feeding time, Russell found that *Lepilemur* spent 91% ingesting leaves, 5.9% ingesting fruit and 3.2% ingesting latex. Nash also observed these animals to be almost exclusively folivorous, though some flowers were eaten. They were not observed feeding on animals.

Nash found that 79–85% of *Lepilemur* time was devoted to either feeding (31–35%) or resting (44–54%), depending on season. It rests more and travels less in the cold season but the amount of feeding does not change seasonally, thus probably minimizing expenditure of energy when food is least abundant. In this way *Lepilemur* may be moving little while eating as much as possible at all times (Nash 1998). In any case, the large amount of time spent resting is typical of folivores that need to eat large amounts of low quality foods and then spend long periods digesting it. *Lepilemur* has a much lower resting metabolic rate (RMR) than predicted from its body mass and, in fact, its RMR is among the lowest of folivorous mammals measured so far (Schmid and Ganzhorn 1996). After spending hours watching them rest, Nash (1994) refers to *Lepilemur* as the sloth of the primate world.

Ganzhorn (1988, 1993) compared the chemical composition of the diet of *Lepilemur* with that of the secondarily nocturnal indriid *Avahi* (I will discuss this species further in Chapter 5). He found that the diet of *Lepilemur* was more restricted where the two species are sympatric. Where it occured alone, *Lepilemur* chose higher quality leaves, i.e., leaves with higher protein and lower fiber content. *Lepilemur* also fed on leaves containing alkaloids, chemicals toxic to some animals and not eaten by *Avahi*. However, *Lepilemur* fed on leaves with alkaloids regardless of the presence or absence of *Avahi*. The two nocturnal species fed at the same height and on the same sized trees where together.

Ganzhorn's study suggests that in some forests, *Lepilemur* populations are filling the habitat to carrying capacity set by leaf quality and habitat structure, and that interspecific competition with *Avahi* may be imposing a strong influence in shaping the niche of *Lepilemur*. It is interesting to note, however, that only two of

12 species eaten by the two lemurs were shared (Ganzhorn 1988). Furthermore, Warren and Crompton (1998) found that, even though both *Lepilemur* and *Avahi* are verticle clingers and leapers, they occupied different forest levels, exhibited different frequencies of locomotor behavior, and had different locomotor bout lengths. Certainly, more remains to be learned about these fascinating folivores.

As with the lorisids discussed in Chapter 3, each nocturnal lemur has specific dietary and habitat choices that in most cases appear to minimize competition between species. The dietary pattern of each species can be related to specific morphological and physiological characteristics. In the seasonal forests of the west, three of the species have diets that enable them to have a stable food source even during the severe dry season, when resources are generally limited: *Phaner* a specialized gum feeder; *Lepilemur* a specialized leaf eater; and *Mirza* an insect secretion feeder, at least in some forests (Figure 4-14). During the dry season, *M. murinus* greatly reduces its activity, and *Cheirogaleus medius* hibernates. During the wet season (austral summer, November-April), when all five species are active, there is a maximum amount of food available. Furthermore, *M. coquereli* and *C. medius* are the only species that seem to feed at similar heights and in similar biotopes (Figure 4-15). *Phaner* searches for insects and fruits higher in the canopy. *M. murinus* is found mainly in the lower bushes of secondary vegetation on the forest edge. Its sympatric congener, *M. myoxinus* may live in the higher canopy. *Lepilemur* is the only nocturnal species of

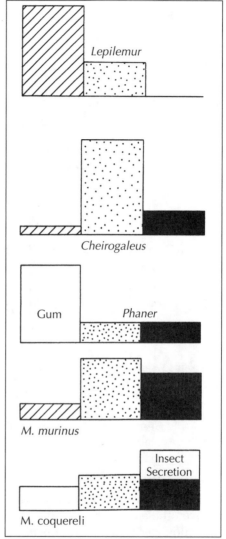

Figure 4-14 Summary of dietary differences between the nocturnal species at Marosalaza forest. Bars represent relative proportions of the diet consisting of leaves and/or gum (left), fruit (center), and animal matter (right). [Adapted from Hladik et al., 1980]

Lemuriformes, except for the secondarily nocturnal *Avahi,* that eats an appreciable quantity of leaves. There is some indication that, where they occur together, *Lepilemur* and *Avahi* may compete for high quality leaves. The feeding behavior of these relatively small, nocturnal primates of Western and Southern Madagascar seems to be well adapted to the long seasonal shortages that occur in the deciduous forests during the dry austral winter.

Each of the six nocturnal species present in the forests of Western Madagascar have a wide geographical distribution and are found in a variety of habitat

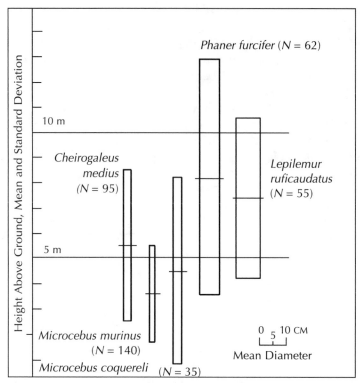

Figure 4-15 Field location of the different nocturnal prosimian species in the Marosalaza forest. The height above ground where the animals were first observed are compared within species, and the thickness of each column is proportional to the diameter of the support used by the animal [after cumulative data of P. Charles-Dominique and C. M. Hladik]

types (Petter et al. 1977, Tattersall 1982, Mittermeier et al. 1994). Even though the above studies illustrate some species-specific patterns of behavior and habitat selection, generalizations concerning these species should be made with caution since they have not been studied throughout their range. Information presently available must be considered to be preliminary and more long-term research is sorely needed.

THE ECOLOGY OF OTHER SPECIES OF NOCTURNAL LEMURIFORMES

Of the twelve species of nocturnal lemurs, *Allocebus trichotis, Cheirogaleus major, Microcebus rufus,* and *Daubentonia madagascariensis* are restricted to the rain forests of Eastern Madagascar. A form of *Lepilemur mustelinus* also is found widely distributed in these rain forests. As far as we know, rain forest-dwelling populations of *Phaner furcifer* are located only in a small area of the northeast (Figure 4-2).

First described in 1875, *A. trichotis* was known only from a few specimens in scattered European museums until 1965. At that time, another individual of this species was captured in the northeast of Madagascar and subsequently the body was sent to Paris for study (Petter-Rousseaux and Petter 1967). *Allocebus* was believed to exist only in a very restricted area of lowland primary forest around Mananara in the northeast, but even after a specific search for the species in 1975, none had been seen since the specimen found in 1965 (Petter et al. 1977). Then in 1989, a German naturalist, David Meier, rediscovered *Allocebus* in the same region (Meier and Albignac

Figure 4-16 *Allocebus.* [Photo by B. Meier]

1991). Observations in the wild only lasted for 2 minutes but a few animals were captured for behavioral observation (Figure 4-16). It was found only in primary lowland rain forest, sleeping in tree holes during the day. More recently, this species has been discovered in highland rain forest in central eastern Madagascar near a very popular tourist area (Rakotoarison et al. 1997). *Allocebus* adults weigh 65–98 gm (n = 9)(Meier and Albignac 1991, Rakotoarison et al. 1997). Its dentition resembles that of *Phaner* in a number of features, and it has an extraordinarily long tongue. Like *Phaner,* it may feed on gums, but almost nothing is known of its behavior or ecology. Meier and Albignac (1991) believe that it has a restricted and patchy distribution and a low population density overall. This species is badly in need of research and conservation programs (see Rakotoarison et al. 1997).

As with *Allocebus,* very little is known about the behavior and ecology of *C. major*. Preliminary data on this species is available from the Ranomafana National Park in Southeastern Madagascar, where it coexists with 12 other species of lemur, including three other nocturnal forms (Wright and Martin 1995). (I will describe this forest in more detail in the next chapter.) *C. major* is larger than *C. medius,* adults weighing 235—470 gms (Table 4-3). Like its western congener, *C. major* is dormant during the Austral winter. It begins to emerge shortly after the spring equinox in late September, and becomes inactive around March/April. When active, animals move alone usually between 7 to 20 meters high. The diet is made up mainly of fruit and floral nectar. In November-December, the nectar of one widespread vine made up the major portion of the diet.

Fruit abundance is low at Ranomafana for nearly 6 months, and unlike many forests in other parts of the world, nectar is available when fruits are abundant, not when fruits are scarce (Overdorff 1991, Wright and Martin 1995). There are few primate frugivores in this forest. Wright and Martin (1995) believe that *C. major* exhibits torpor in the Malagasy rain forest because of the extended period of fruit/nectar scarcity and the lack of other keystone food resources. *C. major* is widely distributed in Eastern Madagascar but has not been the subject of study at any other location.

An excellent, 17 month study on the feeding ecology and yearly activity levels of *Microcebus rufus* recently has been completed at Ranomafana by Atsalis (1998). Atsalis found that these small lemurs ate a mixed diet of fruit and insects. Contrary to predictions based on body size, *M. rufus* relied heavily on fruit, eating possibly as many as 75 species. These mouse lemurs increased the quantity and diversity of fruit eaten during rainy season when fruit productivity was high. This corresponded with the period of seasonal fattening in preparation for the depletion of resources that occurred during the dry season, and with the time when the young began to feed independently. The brown mouse lemur relied heavily on the fruits of several varieties of mistletoe *(Bakerella)* a semi-parasitic epiphyte that was found in small patches but widely distributed in the forest. The small berries of this plant are high in fat content and have twice the energy content of most small fruits which normally are high in carbohydrates. Atsalis hypothesizes that this plant may serve as a keystone species to this population of mouse lemurs. There is a similar close interaction in some forests between certain birds and these high energy mistletoes (Stiles 1993).

In contrast to fruit consumption, insect feeding did not increase during the rainy season when insect abundance was at its highest. In fact, fruit was totally absent from the fecal samples less frequently than was insect matter. It appeared that Coleoptera were a regular part of the mouse lemur diet and other types of insects were added as they became available and as they were nutritionally needed. Atsalis' data indicate that *M. rufus* is a highly frugivorous species. These findings agree with earlier studies by Martin (1972) and Hladik et al. (1980). Atsalis (1998:210) summarizes that "*Microcebus rufus* can broadly be described as a 'frugivore-faunivore' with seasonal patterns of fruit intake and a preference for beetles which along with one high lipid fruit act as dietary staples."

Atsalis also studied the seasonal changes in activity cycle and body size in this population. She found that some members of both sexes increased body weight and tail circumference and displayed winter lethargy. These animals resumed activity after losing body fat. Some of the mouse lemurs, however, remained active throughout the dry season without changes in body fat. It seemed that particular individuals did not adopt the same behavior each year. Finally, although some males and females continue to be active throughout the dry season, more males remain active than females. The sex ratio of trapped animals was approximately 1:1 between January and April, but began to favor males in May and became highly biased in favor of males from May through September (see also Harcourt 1987, and Wright and Martin 1995). Atsalis (1998) suggests some of these males are young adults dispersing from their natal area. Males also emerge from hibernation a month earlier than females. Home ranges of animals overlap to a great degree and as many as 15 mouse lemurs were trapped at the same site (see below).

Atsalis (1998) found little difference in body weights between males and females. Excluding pregnant and lactating females, rufus mouse lemurs weighed between 30–50 g, until the onset of the dry season. At this time, animals with increased body weight (>50 g) were observed, with some animals reaching almost 90 g. At this study site, the forest was relatively short, usually not taller than 20 meters. *M. rufus* was active at all levels of the forest. In areas where the forest was short, they were found at low heights, from 0.5 m. In those areas where the forest was tall, mouse lemurs habitually stayed at 15–20 m. Atsalis (pers. comm.) had the impression that the animals choose their height depending on the availability of suitable supports, much like *Galago demidoff*.

Little is known of *Lepilemur* living in eastern rain forests. As in the western dry forests, it eats mainly leaves of lower quality when in forests with *Avahi*, and it does not discriminate against leaves containing alkaloids. In the rain forest at

Andasibé, *Lepilemur* feeds in a lower strata of the forest, and on trees with smaller crown diameters than does *Avahi* (Ganzhorn 1988, 1989).

Daubentonia madagascariensis (the aye-aye) is one of the strangest of all mammals (Figure 4-17). It has a number of peculiar anatomical characteristics, the most notable being: rodent-like, continuously growing, enlarged incisors; a greatly reduced posterior dentition; an extremely long, thin and mobile third digit on the hand (Figure 4-18); claws rather than nails on all digits except the first toe (Owen 1866); and a peculiarly shaped head with large mobile ears.

A single specimen was first brought to Europe in the mid-1800s and this was the only example available for nearly one hundred years. Because of its peculiar features, it was not known if the animal was a rodent or a primate. With the arrival of a

Figure 4-17 *Daubentonia*. [Photo by Ben Freed]

number of new specimens, Owen (1866) was able to detail its primate-like characteristics. Sterling (1993a, 1994a) gives an excellent review of the history of taxonomic debates surrounding the aye-aye and of its current taxonomic status.

Until 1957, when *D. madagascariensis* was rediscovered by Petter and Petter-Rousseaux (1959), it was thought to be extinct and almost nothing was known of its behavior. Soon after its rediscovery, the Malagasy Government, believing the aye-aye was on the brink of extinction, established Nosy Mangabe, a small island off the east coast of Madagascar, as a Special Reserve and refugium in 1965. Under the direction of Jean-Jacques Petter, a French scientist who pioneered lemur field studies, 4 males and 5 females were captured on the mainland and released onto Nosy Mangabe in 1966–67.

Preliminary field observations were conducted at that time (Petter and Petter-Rousseaux 1967, Petter and Peyrieras 1970, Petter 1977). However, the status of these released animals was unknown until 1975 when Elizabeth Bomford (a photographer who won a photographic contest, the prize being a ticket to any place in the world), chose to visit Nosy Mangabe to search for the aye-aye. She discovered two aye-ayes on the island (Bomford 1981). Since that time, a number of sightings from various localities on the mainland have confirmed the existence of *Daubentonia* in

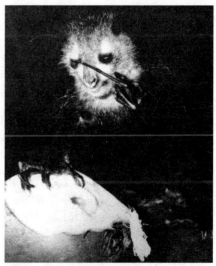

Figure 4-18 *Daubentonia* finger. [Photo by J. J. Petter]

many localities in the north and east coast of Madagascar, and possibly in the west (Figure 4-2)(Constable et al. 1985, Sterling 1993a, 1994a).

In 1989, Sterling (1993a,b, 1994b,c, Sterling et al. 1994) began an excellent two year study of the aye-aye on Nosy Mangabe. She captured, weighed, measured, and radio-collared eight of the animals (six males, two females). Two of each sex were followed for 5 to 11 months. Unfortunately, the transmitters on the other four animals failed in less than a month, a common problem with this technology. The four radio-tracked animals were observed from dusk to dawn on 46 nights, for approximately 500 hours.

On Nosy Mangabe, the aye-aye chose habitats with a mix of features including large trees, as well as small trees and shrubs. They nested in locations with a high concentration of lianas. On the mainland, they have been seen in disturbed and undisturbed rain forest, dry deciduous forest, coastal forest, and mangrove. Sterling found that aye-ayes used all levels of the forest up to 35 m high. Their mean height while travelling was about 7 m, but while feeding they were usually around 11m high. On Nosy Mangabe, the aye-ayes used the ground 25% of the time, more than any other nocturnal prosimian (Sterling 1993a). In a short-term study on the mainland, however, Ancrenaz et al. (1994) observed these animals to spend >80% of the time in the two highest forest levels and rarely come to the ground.

Aye-ayes weigh around 2700 gm and are relatively slow and deliberate quadrupeds. Their claws enable them to move on large vertical and angled branches, but they also are quite adept at moving in the fine terminal branches of the canopy and bush, and at leaping. When the aye-aye is on the ground, it usually moves very slowly with the fingers of the hand raised and not touching the ground. Thus pressure is exerted on the extremity of the palm (Petter 1977). In the daytime, the aye-aye sleeps in well-constructed nests, high in the trees (Sterling 1993a, Ancrenaz et al. 1994). Sterling observed 130 nests, averaging 17.6m in height.

Before Sterling's study, it was known that the *D. madagascariensis* ate both hard fruits and insect larvae (Petter 1977) but the relative importance of each of these foods was unknown. Sterling (1993a,b, 1994c, Sterling et al. 1994) found that aye-ayes ate five types of food, each represented by very few species (Table 4-4). Of these foods, *Canarium* seeds, larvae, a fungus growing on leguminous *Intsia* trees, and nectar from the Madagascar travellers palm made up over 90% of the overall diet. The first three of these resources were available over 10 months of the year. Aye-ayes spent most of their time eating *Canarium* seeds, except during the cool wet season (mid-May through mid-September) when they ate proportionately more insects and fungus. Nectar from the travelers palm was eaten predominantly in the hot wet season (January to mid-May) at Nosy Mangabe. On the mainland, Ancrenaz et al. (1994) found this nectar to be the main food source in the cool dry season.

Although aye-ayes use their long, middle finger and/or their incisors to exploit all of these foods, only hard seeds and wood-boring larvae required the use of both. These foods are structurally defended either by hard outer coats or by wood or bark, and are fairly abundant, not being exploited by many animals. The aye-ayes gnaw the endocarp of hard seeds with their anterior teeth and then extract the cotyledon with their middle finger (Petter 1977, Iwano and Iwakawa 1988, Sterling 1993a). Sterling measured the hardness of the endocarp of the seeds eaten by aye-ayes and found them to be much harder than those eaten by South America primate seed predators. The ability of aye-ayes to eat hard seeds and nuts has prompted Iwano and Iwakawa (1988) to specu- late that these primates are filling the ecological niche of squirrels which are absent in Madagascar.

Table 4-4 Major Food Types Eaten by the Aye-Aye

Food Type	# of Species Eaten
Seeds	3
Larvae	6–9
Fungus	1–2
Nectar	2
Adult Insect	1

Although aye-ayes are not primarily insectivorous, insect larvae are an impor- tant componant of their diet. The amount of time spent eating larvae did not dif- fer across seasons, and aye-ayes extracted larvae from fallen dead wood, dead branches on live trees, living trees, dead and living lianas, and inside parasitized seeds (Sterling 1993b). Many species of insects, especially among the Lepidoptera and Coleoptera, feed as larvae on the cambium and woody tissue of trees. These wood-boring adaptations provide nurture for the larvae and pupae and also pro- tect them from most vertebrate predators. Because all of the unusual morpholog- ical adaptations of the aye-aye are utilized in the exploitation of these wood-boring larvae, it is in this context that these traits can best be understood.

Cartmill (1974) pointed out that the behavior and morphology of *Daubentonia* has an almost exact parallel in a genus of diprodont marsupials of New Guinea, *Dactylopsila*. Both species hunt larvae by tapping branches, trunks or fruit with a long, thin finger, and turning their ears toward the spot being investigated. Petter and Petter-Rousseaux (1967:202) describe the rest of the process as follows.

> *When an Aye-aye discovers a larva, it gnaws the bark ferociously with its incisors. Once it has made a hole, it inserts its third finger, which it moves exactly like a pipe cleaner, sometimes turning it, pulling it out and pushing it in many times; then it puts its finger quickly in its mouth before beginning all over again. It seems to squash the larva, bringing out the bits to put in its mouth rather than extract the larva whole.*

It is the elongated fourth digit of *Dactylopsila*, rather than the third, which is used to probe and snare the burrowing larvae (Figure 4-19). Woodpeckers, mainly of the family Picidae, are the principal vertebrate predators on wood-boring insect larvae in many forest communities. Where woodpeckers are absent, as in Australia and many oceanic islands, other species of bird have developed adaptations anal- ogous to those of the woodpeckers. Cartmill points out that New Guinea and Madagascar are the only extensive regions in the world which have not been col- onized by woodpeckers or by other wood-boring island birds. *Daubentonia mada- gascariensis,* he hypothesizes, fills the woodpecker niche and can be considered a member of a specialized foraging guild (see Chapter 1) shared by *Dactylopsila* in New guinea and by wood-boring birds in the rest of the world. Though, in fact, there are several avian species in Madagascar, the vangas, which probe for insect larvae from the bark of trees (Langrand 1990, Erickson 1991), it is not known to what degree they might compete with aye-ayes.

Some excellent studies have been conducted to determine precisely how aye- ayes locate and extract larvae from woody sources. Erickson (1991, 1994, 1995, 1998, see also Goix 1993), working on four captive animals at the Duke University

Figure 4-19 *Dactylopsila.* [From *Possums and Gliders* (ed. A. Smith and I. Hume). Surrey Beatty and Sons Publishing, Chipping Norton. 1984]

Primate Center, found that aye-ayes can locate prey in cavities in the absence of visual or olfactory cues, and in the absence of movement by the prey. He further found that as the animal taps the surface of an object with its long, thin finger, it receives either auditory, or possibly surface vibration cues about the subsurface hollow spaces, thus forming a mental map of the depth and location of cavities. The tapping also may stimulate the unfortunate larvae to move thus further giving away its location. As early as 1859, Sandwich, an early visitor to Madagascar, provided Owen (1866) with a precise description of this finger-tapping behavior.

Thus, the aye-aye depends heavily on tapping to locate the tunnels of the larvae. It then gnaws the tough bark to get at these tasty, nutritious morsels. In this way it utilizes the morphological adaptations of its teeth, ears, and hands to exploit wood-boring larvae. Its strong jaws and continuously growing incisors, however, also allow the aye-aye to exploit extremely hard seeds and nuts. In this way, the aye-aye appears to fill a unique ecological role that combines some aspects of the niches of both the squirrel and of wood-boring woodpeckers. Because the structurally protected resources used by the aye-aye are difficult to harvest, they are relatively abundant and not utilized by many vertebrate competitors.

Although the aye-aye is now known to be more widespread than was recently believed its densities may be quite low. Furthermore, the habitat that aye-ayes and other Malagasy lemurs need for continued survival is rapidly disappearing (Green and Sussman 1990, Ganzhorn et al. 1996, 1997). Let us hope that rational means of preserving habitat and yet improving the quality of life of the rural Malagasy can be accomplished before all species on the island overreach the capacity of the island to sustain them.

PREDATION

Most nocturnal species of Lemuriformes are small and thus can serve as prey for many Madagascan predators. Potential predators include a number of species of viverrids, feral cats (Felidae), domestic dogs, snakes, and birds of prey. There are few extant large predators in Madagascar. However, recently an extinct large eagle, *Aquila* sp., has been identified (Goodman 1994). Since predation occurs infrequently and acts of predation are of brief duration, they are observed relatively rarely by primatologists. Indeed, it is argued by some that predation has not played a primary role in the evolution of social behavior or morphology of primates (e.g., Wrangham 1983, Cheney and Wrangham 1986). Others, however, believe that predation has been a major selective force in molding the behavior and morphology of primates and other mammals (e.g., Alexander 1974, van Schaik 1983, Terborgh and Janson 1986, Goodman et al. 1993a).

Table 4-5 Predators of Nocturnal Lemuriformes.
Data Derived from Goodman et al. 1993a.

Prey Species	Predator	Remains	Direct Observation
Microcelous murinus	Goshawk *(Accipiter hentsii)*	x	
	Barn owl *(Tyto alba)*	x	
	Long-eared owl *(Asio madagascariensis)*	x	
	Colubrid snake *(Ithycyphy miniatus)*		x
	Boa *(Sanzania madagascariensis)*		x
	Fossa *(Crytoprocta ferox)*		x
	Bokiboki *(Mungotictus decemlineata)*		x
M. rufus	Barn owl	x	
	Long-eared owl	x	
	Hook-billed vanga *(Vanga curvirostris)*		x
	Harrier hawk *(Polyboroides radiatus)*		x
	Galidia *(Galidia elegans)*		x
	Domestic dog *(Canis domesticus)*		x
Mirza coquereli	Buzzard *(Buteo brachyterus)*		x
Cheirogaleus major	Galidia		x
	Buzzard		x
Phaner furcifer	Buzzard		x
	Cuckoo falcon *(Aviceda madagascariensis)*		x
Lepilemur mustelinus	Barn owl	x	
	Boas		x
	Harrier hawk		x

A detailed search of the primate literature, in fact, reveals that there are a large number of observations of attempted and successful predation on primates (Hart in prep.). Furthermore, primates display many of the anti-predator behaviors generally used by mammals to reduce their vulnerability to predators (Endler 1991). Research on predators gives further evidence of the importance of primates as prey. By studying carnivore scats, or raptor prey remains, primates have been found to be important prey items for many predators.

Goodman et al. (1993a) review information available on the predation of lemurs, and estimate predation rates for a population of *Microcebus murinus* in light of heavy predation pressure from two species of owl. Carnivores, including domestic dogs, have been observed preying on western and eastern species of mouse lemurs and on *Cheirogaleus major* (Table 4-5). Small cat-like viverrids, *Mungotictis decemlineata* in the west and *Galidia elegans* in the east, search in the dense vegetation for prey. They often discover the tree holes serving as nests of small nocturnal lemurs and eat the occupants after enlarging the opening (Petter et al. 1977, Rasoloarison et al. 1995). The larger viverrid, *Cryptoprocta ferox,* would be capable of hunting most nocturnal lemurs since it hunts at night using auditory, visual and olfactory cues (Albignac 1973). Russell (1977), however, observed a *C. ferox* (commonly called the fossa) using the same techniques described above in searching mouse lemur nests: "the fossa had unsuccessfully attempted to claw a large opening in a tree hole that housed a mouse lemur (p. 103)." Thus, the diameter of the nest hole opening and the deepness of the nest are under substantial selection pressure (see Martin 1972). *C. medius* has been found in the scat of fossa (Rasoloarison et al. 1995). *Mirza coquereli* nests are not in tree holes but are close to the top of the trees, associated with thick lianas, and located in particular species of trees to avoid terrestrial predators (Sarikaya and Kappeler 1997).

Larger species of snakes are potential predators to nocturnal lemurs. A colubrid snake, *Ithycyphys miniatus,* and a boa, *Sanzinia madagascariensis,* have been observed preying on western gray mouse lemurs (Richard 1978, Randrianarivo 1979). The larger *Lepilemur* has been

Figure 4-20 Barn owl. [Photo by Robert W. Sussman]

observed to give a specific alarm vocalization in response to large boas (McGeorge cited in Russell 1977), and these snakes may hunt *Lepilemur* in its nests (Ratsirarson 1986).

Most birds of prey in Madagascar are diurnal and hunt by visually spotting vulnerable, active prey. Nevertheless, Russell (1977) observed *Polyboroides radiatus,* a harrier hawk, clinging to trees and tearing out sections of bark with their talons in pursuit of reptiles and perhaps mouse lemur prey, and *Lepilemur* has been observed being hunted in the same manner (reported in Goodman et al. 1993a). Goodman et al. review cases of observed predation on nocturnal lemurs by the following diurnal raptors: Henst's Goshawk, Madagascar harrier hawk, Madagascar buzzard, cuckoo falcon, and hook-billed vanga (Table 4-5). Large raptors elicit alarm calls from diurnal lemurs (see next chapter). Even with all of these potential predators, it is the nocturnal barn owl, *Tyto alba,* and long-eared owl, *Asio madagascariensis,* that are the most threatening predators of the nocturnal lemurs, especially the diminutive mouse lemurs (Figure 4-20).

Goodman and colleagues studied the predation pressure on *M. murinus* by these two species of owl at the Beza Mahafaly Special Reserve in Southwestern Madagascar and at Kirindy in the west (Goodman et al. 1993a,b,c, Rasoloarison et al. 1995). At Beza Mahafaly, they collected regurgitated pellets from a pair of barn owls twice a month for one year (11/90–11/91). Pellets also were collected from a nearby roost of a long-eared owl on three occassions. A minimum of 58 mouse lemurs (mainly adult animals) were eaten by the pair of barn owls, and at least 16 were represented in the long-eared owl samples. Goodman estimates that there are 9-10 pairs of barn owls in the Beza Mahafaly Reserve and that approximately 520–580 mouse lemurs are taken each year by this owl alone. This would represent a predation rate (percent population taken by predator per year) of approximately 25%, the highest known for any primate species. Although it is one of the major prey items of these owls, mouse lemurs also are vulnerable to many other predators. However, Goodman et al. (1993a) believe that this level of predation is not drastic given the high reproductive potential of the mouse lemur (see below).

Mouse lemurs were also a prominent prey item at Kirindy, though *Lepilemur* were also found in pellets of the larger *Asio madagascariensis* (Rasoloarison et al. 1995).

All nocturnal prosimians are relatively small and non-gregarious. They cannot protect themselves from predators by direct defense nor by social facilitation. These species appear to use three major strategies to counteract predation (Goodman et al. 1993a). These include choosing well-protected nest sites during the day, spending most of their active periods in dense vegetation, and having high reproductive rates. In any case, it seems obvious that predation is an important factor in the life of small nocturnal prosimians. I will discuss the potential effects of predation on diurnal lemurs in Chapter 5.

THE SOCIAL ORGANIZATION OF THE NOCTURNAL LEMURIFORMES

Preliminary studies of social organization have been done on *Microcebus* (Martin 1972, 1973, Pages-Feuillade 1988, Fietz 1997, Atsalis 1998), *Mirza coquereli* (Pages 1980; Andrianarivo 1981, Kappeler 1997), *Cheirogaleus medius* (Müller 1998), *Lepilemur mustelinus* (Charles-Dominique and Hladik 1971; Russell 1977, 1985, Nash 1998), *Phaner furcifer* (Charles-Dominique and Petter 1980), and *Daubentonia madagascariensis* (Petter et al. 1977, Sterling 1993a,b, 1994b). Except for *Phaner* these species normally are found alone while active at night, but form groups to sleep during the day. However, there is substantial variation in social behavior and organization among them. Table 4-6 gives the sizes of home ranges and population densities of these nocturnal species.

Nesting groups of *Microcebus*, *Mirza* and *Cheirogaleus* appear to occur in clusters ("localized population nuclei," Martin 1973; "villages," Andrianarivo 1981) and the ranges of animals living in these nuclei overlap, whereas those living in separate nesting groups have ranges with little or no overlap. *Mirza* and *Cheirogaleus* were found to sleep alone or in small groups of two to three individuals; males were not found to nest together (Andrianarivo 1981, Müller 1998) and in the coquerel's dwarf lemur adult males were not found in female nests (Keppeler 1997). Müller (1998) hypothesizes that *C. medius* may associate in stable family groups, but there is little evidence for this.

During a 4-year study of *Mirza coquereli* in Western Madagascar, Kappeler (1997) captured and followed 88 marked individuals. He found that females had quite stable home ranges of between 2.5 and 4.5 ha. Male home ranges, however, increased to more than 4 times as large during the mating season (from about 4+ to 17+ ha). Female home ranges overlapped with as many as 8 other females. Male home ranges did not overlap with other males except during the mating season, when up to at least 11 of the 12 radio-collared male's ranges overlapped.

Radio-collared males ranges overlapped with at least 4–15 of the 19 known females during the mating season, whereas these numbers ranged between at least 2–6 after the mating season. Kappeler (1997) also found that the males were significantly larger than females and showed pronounced seasonal variation in testis size. Males were observed with significantly more external injuries than females during the mating season. His observations suggest that coquerel's dwarf lemur is polygynous, and that male-male competition occurs during the mating season.

Martin (1972, 1973) studied the gray mouse lemur for six months in Southern Madagascar. In this population, nesting groups consisted mainly of females and ranged in size from 2 to 15 individuals. Males usually slept alone in rudimentary nests, but sometimes, especially when females were in estrus, up to 3 males were found nesting with females. Martin observed more overlap between female gray mouse lemur ranges than between those of males. He also found that some males were peripheralized. The male to female ratio was 1:4. Immature and smaller adult males were often in habitats less favorable to the species. This social organization is similar to that of *Galago demidoff,* and to some other galagos (Bearder 1986, Chapter 3).

Pages-Feuillade (1988) radio-tracked five male and four female gray mouse lemurs for 30 nights over a six week period at Ampijoroa. She found that these 9 individuals, and possibly more, made up a nesting group which shared a total area of 7 hectares (a hectare is 100X100 meters). Male home ranges were larger (> 3 ha) and overlapped more (65%) than those of females (< 2 ha, with 44% overlap). In the center of the 7 ha study area, the nine animals shared a common area where interactions between individuals took place. Males usually slept alone (n = 9), with another male (n = 4), and rarely with a female (n = 1). They chose nest sites opportunistically, either where they ended their night's activity or near an estrus female. Female nest sites were more stable. During the 30 day study period, three females repeatedly used three nests located within a 20 meter radius. Two or all three of these females slept together over 90% of the time.

Animals normally foraged alone, but 11% of the time mouse lemurs were less than 10 meters apart. This usually occurred towards the end of the night as females followed one another into their mutual nests, during mating season when males often grouped together and followed estrus females, and sometimes when animals fed in adjacent fruit trees. Animals were more separated when they hunted for insects.

Although actual interactions occurred infrequently, Pages-Feuillade observed 195 of these. Most took place in the area where the home ranges of all 9 animals overlapped, and were of short duration. Almost 70% lasted less than 3 minutes. In most cases, animals simply paused as they passed one another. Interactions were often accompanied by vocalizations and olfactory marking. However, as in *Galago crassicaudatus* (Clark 1985, see Chapter 3), there was a definite pattern of social interactions, with some individuals interacting more often and for longer periods (Figure 4-21). Only 23 interactions were agonistic, and in 18 of these in which sex could be identified, 3 were between females, and none between males. The remainder occurred when estrus females chased persistent males.

During breeding season, males and females exchanged calls, and males exhibited elaborate olfactory marking behavior. Females also marked but more discreetly and sporadically. Although many males, some of which were not in the nesting group, remained near the female and in close proximity to one another for long periods of time. In a short-term study in Kirindy, Fietz (1997), observed that male gray mouse lemurs home ranges overlapped with one another and with smaller home ranges of females. The large overlapping home ranges of males, lack of sexual dimorphism, and relatively large testis, led Fietz to predict a promiscuous, multi-male mating system (rather than a polygynous system).

In Fietz's study males seemed to establish a hierarchy before females entered estrus. This conclusion was reached because it seemed that heavier males

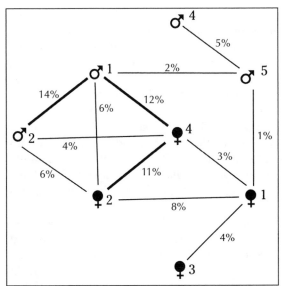

Figure 4-21 Diagram of the most frequent dyadic relations among *Microcebus murinus*. [Adapted from Pages-Feuillade, 1988]

occurred closer to females, which may have facilitated "access to breeding opportunities" (Fietz 1997:95). Pages- Feuillade, however, observed no aggression or visible competition for estrus females among males. By contrast, in captivity male mouse lemurs become very aggressive during breeding season, and develop a strict dominance hierarchy. These captive males may show the highest plasma testosterone levels found in mammals, and even the odor of a dominant male can lower testosterone levels and sexually inhibit a subordinate animal (Petter-Rousseaux and Picon 1981, Schilling et al. 1984). In the Pages-Feuillade's population, there was no indication of a dominance hierarchy or of sexual inhibition among males. These patterns of behavior and the differences between the behavior of some wild and captive mouse lemurs are fascinating subjects that await further study.

The findings of Atsalis (1998, pers. comm.) on the eastern rufus mouse lemur were similar to those of Pages-Feuillade. She observed as many as 15 animals to share a home range. Individuals often congregated in large trees when food sources were sufficient. In one case, several individuals, including one identified, collared animal, returned to the same tree each night for two weeks to feed on insects attracted to the large quantity of flowers available. Atsalis observed that, although the mouse lemurs were usually solitary during their nocturnal activities, they do not simply ignore one another. Rather, through a system of indirect modes of communication (e.g., olfactory), they are participating in continuous social interaction. Thus, radio-tracked individuals of both sexes had a central area of overlap in their shared home ranges (Atsalis pers. comm.). Atsalis (1998) reports an eight-fold increase in testicular size during the mating season, however, male ranges overlap and she did not report any incidence of aggressive interactions between males.

Charles-Dominique and Hladik (1971) studied *Lepilemur* in southern Madagascar. They believed that individuals occupied areas ("territories") that were defended against intrusion by animals of the same sex. The home ranges of males and females, however, overlapped. Russell (1977) found that although males were not observed to have overlapping home ranges, several pairs of females, as well as males and females did have mutual ranges. Like other nocturnal species, *Lepilemur* usually fed and moved individually but Russell observed that from one to three times per night "range-mates" would come together in the same tree. They then often fed, played, groomed or rested together. These rendezvous would last from 5 min. to an hour. Range-mates often slept together during the day. They included at least two adult animals—female-female, or male-female often with young, but never male-male. Animals occupying adjacent ranges usually avoided one another.

The social organization of *Phaner* is similar to that of *Lepilemur*. Of the five adult males followed by Charles-Dominique and Petter (1980), one was solitary, three shared ranges with one female, and one was the range-mate of two females. No females were seen without male range-mates. Rather than meeting a few times each night like *Lepilemur*, *Phaner* range-mates maintained continuous vocal contact throughout the night and were in close proximity at least half the night. They usually slept together during the day. Vocalizations were used both to maintain continuous contact between range-mates and to maintain the integrity of range boundaries (Charles-Dominique and Petter 1980). *Phaner* is very noisy; individuals exchanged as many as 30 loud calls (or series of loud calls) per hour during certain periods of the night.

The social organization of *Daubentonia* is unique in a number of ways. The two aye-aye males followed by Sterling (1993a,b) had large home ranges, 126 and 215 ha; 3–6 times as large as those of females (Table 4-6). Furthermore, the home ranges of males overlapped considerably, whereas those of females did not. In fact, no female was seen within another's range (see also Ancrenaz et al. 1994). During Sterling's study on Nosy Mangabe, individuals slept alone, but males (Sterling 1993a,b) and mothers and their young (Petter et al. 1977) have been observed to sleep together.

Occasionally two or three individuals foraged as a unit, exchanging calls as they moved. These units included two adult males, young and adult males, and adult males and females. No mutual grooming was observed. For less than 14% of the time, animals were within 20 meters of each other, but, as with *Galago crassicaudatus* and *Microcebus murinus*, patterns of interindividual interactions emerged. For example, some males avoided others, and an adult male and female pair was observed foraging and traveling together several times when the female was not in estrus.

Sterling (1993a, 1994b) found aye-ayes to be only one of two Malagasy prosimians for which mating occurs throughout the year, though the number of times an individual female cycles per year is unknown. In fact, the aye-aye mating system is unique among primates. Approximately 10 days prior to estrus, females increased the frequency of scent-marking and visited male nests. Estrus lasted for about three days and, during this time, females called repeatedly. These calls attracted males and up to six would surround an estrus female. Aye-aye males often chased and bit one another during these gatherings. Eventually a male would succeed in copulating with the female, and then continue to hold her for the next hour or so. The female then quickly moved 500–600 meters and started calling again. Females mated with more than one male on the first day of estrus, and only one male on the next two days. Because of their elongated baculum and other specializations of the genitalia, Dixson (1987) had predicted that male aye-aye would have prolonged intromissions. In this way, males guard females, preventing immediate insemination by other males. Mate guarding also has been observed in *G. crassicaudatus* and *G. demidoff* (Clark, cited in Bearder 1986).

Thus, adult females are solitary. Males have large overlapping ranges and travel over a wide area in search for estrus females. During mating, many males surround the female and compete with one-another. Sterling refers to this type of mating system as "scamble competition polygyny." Similar mating systems are

Table 4-6 Home Range Size, Population Density and General Group Structure of Some Nocturnal Lemuriformes

Species	Home Range in Hectares	Density per sq. km.	Type of Group Structure	Sources
Cheirogaleus medius	4	A. 350 B. 200 C. 20	solitary but social(?)	Hladik et al. (1980)
Cheirogaleus major		75–110	solitary but social(?)	Petter and Petter-Rousseau (1964)
Microcebus murinus	2 (1) ♂3.2 (n = 4) ♀1.8 (n = 4) (2)	A. 400 B. 60 (1) C. 130 143 (2)	solitary but social plus nesting groups. Many males follow estrus females. Variability in male-male tolerance.	(1) Hladik et al. (1980) (2) Pagès-Feuillade (1989)
Mirza coquereli	♀4.5 ♂4.0 (1) ♀2.5 (4) ♂4.5–17+ (4)	A. 30 B. 30 (1) C. 0 250 (2) 385 (3)	solitary but social plus nesting groups	(1) Hladik et al. (1980) (2) Petter et al. (1971) (3) Andrianarivo (1981) (4) Keppeler (1997)
Phaner furcifer	♀4.0 ♂3.8 (1)	A. 50–60 B. 100 (1) C. 40 550–870 (2) 400 (3)	loose pairs or 1 male with 1 or 2 females (range-mates interact throughout the night)	(1) Hladik et al. (1980) (2) Patter et al. (1971) (3) Sussman (1972)
Lepilemur mustelinus	♀0.9–3.2 (1)(2) ♂.20–.46	A. 250 B. 330 (1) C. 40 220–81 (3) 200 (4)	male-female or female-female range-mates. (Intermittent interaction throughout the night)	(1) Hladik et al. (1980) (2) Russell (1997) (3) Charles-Dominique and Hladik (1971) (4) Sussman (1977)
Daubentonia madagascariensis	♀35.6 (n = 2) ♂170.3 (n = 2)	3–6	male—highly overlapping ranges; compete for estrus females. female—non-overlapping home ranges. Males guard females after mating.	Sterling 1993a

A. Marosalaza study area
B. Transact in closed forest on west coast
C. Transact in dry brush forest on west coast

seen in red kangaroos, North American moose, and polar bears (Peterson 1955, Croft 1981, Ramsay and Stirling 1986). It is not shared by any other primate, though mating systems in which several males mate with a single estrus female are very common throughout the order (and as we have seen in many of the nocturnal prosimians), and there are some interesting similarities between the aye-aye mating system and that of orang utans.

In all species of Lemuriformes, except *Daubentonia* and *Hapalemur griseus,* births are seasonal and usually synchronized so that females are lactating and infants are becoming independent during periods of maximum food availability (Martin 1990, Meyers and Wright 1993). In many primates, including humans, females increase calorie and protein intake during lactation (see Sauther and Nash 1987). The time of the mating and birth seasons, thus, vary from species to species according to the gestation and lactation period, to the rate of development of the young, and to the food preferences of the species.

REPRODUCTION

Species of *Microcebus* and *Cheirogaleus* usually have more than one infant per birth (Table 4-7); for example, *M. murinus* and *C. medius* have 1–4 infants and most commonly 2–3 infants (Martin 1972, Petter et al. 1977, Russell 1977, Foerg 1982). Furthermore, because of postpartum estrus cycles (PPE), females have the potential of having 2 litters during each breeding season, though impregnation during PPE is rare (Andriantsiferana et al. 1974, Foerg 1982). Sexual maturity is reached within the first year in these species. Given these life history traits, the potential for rapid population growth is considerable. For example, in relation to predation pressure at Beza Mahafaly, Goodman et al. (1993a) estimate that, with a base population of 2100–2300 mouse lemurs in the reserve and two young weaned successfully per year, the population would increase to 4200–4600. Thus, the estimated 520–580 mouse lemurs taken by owls each year would not represent an intolerable predation rate.

Phaner, Lepilemur and *Daubentonia* commonly have only one infant per birth season and, at least in the latter two species, the maturation rate is relatively slow (Table 4-7). *Microcebus, Cheirogaleus,* and *Lepilemur* infants are left in the nest or

Table 4-7 Gestation Periods, Litter Sizes and Sexual Maturation in Various Nocturnal Lemuriformes

Species	Avg. Gestation Period (days)	Modal Litter Size	App. Age at Sexual Maturity (months)
Microcebus murinus	63	2–3	8–12
Mirza coquereli	86	2	—
Cheirogaleus medius	62	2–3	12–24
Cheirogaleus major	70	2-3	—
Phaner furcifer	167	1	—
Lepilemur mustelinus	135	1	18
Daubentonia madagascariensis	167	1	36–48 ?

Adapted from Tattersall 1982; Data for *Daubentonia* from Glander (in press, pers. comm.).

carried in the mouth by the mother and "parked" on a branch, as in many species of Lorisiformes. From observations of captive animals and confirmed by observations of one infant in the wild (Andriamamisanana 1994), it appears that *Daubentonia* infants remain in the nest for approximately 3 months. They then slowly and carefully begin to venture out and to follow their mother (Glander pers. comm.). By 5–6 months, wild aye aye infants easily follow their mothers during nighttime activities (Andriamamisanana 1994). *Phaner* infants are first left in the nest, but later they grip the fur on the mother's belly and eventually ride on her back, a pattern similar to that found in most diurnal lemurs. As in the Lorisiformes, detailed studies of the behavior of juveniles have not been conducted and patterns of dispersal of individuals as they reach adulthood are unknown.

SUMMARY OF THE ECOLOGY AND SOCIAL BEHAVIOR OF NOCTURNAL LEMURIFORMES

Although studies on the nocturnal lemurs of Madagascar must be considered preliminary, they give us a background on the basic habitat preferences of these forms. The six nocturnal species of Western Madagascar are sympatric in many forests. Four of these species obtain protein from insects and carbohydrates and sugar (energy) from fruits, gums and nectar. However, *Cheirogaleus medius* and *Microcebus murinus* are less active or are inactive during the dry, winter months when food is more scarce. When active in the austral summer, *C. medius* is found mainly in the center of the forest whereas *M. murinus* is more common in forest edges. *Mirza coquereli* and *Phaner furcifer* are active all year round but eat relatively few insects. *M. coquereli* feeds mainly on insect secretions, and *Phaner* mainly on gums. *Lepilemur* is unusual for a small, nocturnal primate in that it is a folivore. Little is known about the newly discovered *Microcebus myoxinus*.

When the dietary composition and the biomass of the nocturnal lemurs of Western Madagascar are compared to those of the Gabon lorisiforms, a predictable pattern emerges (Figure 4-22). The characteristic pattern of food choices reflects the Eltonian pyramid of energy transfer (see Chapter 1). Highly insectivorous species are less densely populated, and frugivorous and gumivorous species are more densely populated and have a higher biomass. Because there is a greater production and density of leaves than fruit and insects, the density and biomass of *Lepilemur* is correspondingly greater than the other nocturnal prosimian species. Also illustrated in Figure 4-22 is the fact that the African and Madagascan nocturnal prosimians eat mainly plant foods. The African subfamily Galaginae and the Madagascan subfamily Cheirogaleidae are basically nocturnal, arboreal, fast-moving, quadrupedal and omnivorous, and these characteristics are most likely primitive primate evolutionary characters (see Charles-Dominique 1978, Martin 1990).

Little has been published on eastern rain forest populations of *Phaner furcifer*, *Cheirogaleus major* or *Lepilemur mustelinus*. From what is known, it appears that these species may be ecological equivalents to their western counterparts, although some distinctive adaptations will very likely be discovered. *Allocebus trichotis* has not been studied. However, it has morphological features that resemble those of *Phaner*, and may fill a similar gum-feeding niche. *Daubentonia madagascariensis* is a specialist at obtaining defended resources that are difficult to access, such as hard seeds and nuts, and wood-boring insect larvae. These resources are

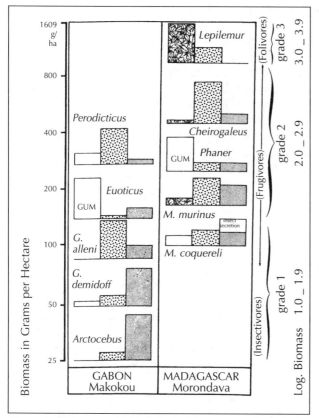

Figure 4-22 A comparison of diets of the nocturnal prosimian species of Madagascar (Marosalaza forest) to those of the Gabon rain forest. [after Charles-Dominique, 1977] Each species' diet is illustrated by the relative amount of leaves and/or gums (left rectangle), fruits (central rectangle), and animal matter (right rectangle) which have been ingested.

normally exploited by squirrels and woodpeckers, animals that are not found in Madagascar.

Although the nocturnal lemurs normally travel and forage alone during the night, there is a great deal of variation in the social organization among species, and possibly among populations of the same species. Some populations of *M. murinus* and *M. coquereli* have an organization similar to that found in some of the galagos, with shared female home ranges, but little or no overlap in adult male ranges. Other gray mouse lemurs live in neighborhoods of interacting males and females with little apparent competition between males. Finally, some populations of coquerel's dwarf lemur appear to be polygynous with male-male competition during the mating season. *Lepilemur* range mates (pairs of females or a male and a female) meet 2–3 times a night and often sleep together. *Phaner* range mates (one male with one or two females) nest together and remain in vocal contact throughout the night. *Daubentonia* males have large ranges which overlap with females and other males; adult females do not have overlapping ranges nor do they interact. Many males follow and compete for estrus females. Aye-ayes are one of only two lemurs known in which breeding is not seasonal.

Given this amazing variability it is difficult to make any generalizations about the social organization of the nocturnal Malagasy lemurs. Current socioecological theory assumes that the distribution of females is related to the distribution of the major resources utilized by the species. The distribution of males, on the other hand, is related to that of females. Nocturnal prosimians are ideal subjects to test hypotheses concerning relationships among resource distribution and individual spacing patterns since (1) individuals do not travel in permanant groups throughout the night, and especially when searching for resources, and (2) since there is such a great variety of patterns of distribution of both males and females. However, to date it has not been possible to find consistent, meaningful patterns among the nocturnal primates, and the relationships between ecology and social structure remain a mystery.

BIBLIOGRAPHY

Albignac, R. 1973. *Faune de Madagascar. No. 36. Mammiferes Carnivores.* Paris, Orstom/Cnrs.

Alexander, R.D. 1974. The Evolution of Social Behavior. *Ann. Rev. Ecol. Syst.* 5:325–383.

Ancrenaz, M.; Lackman-Ancrenaz, I.; Mundy, N. 1994. Field Observations of Aye-Ayes *(Daubentonia madagascariensis)* in Madagascar. *Folia Primatol.* 62:22–36.

Andriamasimanana, M. 1994. Ecoethological Study of Free-Ranging Aye Ayes *(Daubentonia madagascariensis)* in Madagascar. *Folia Primatol.* 62:37–45.

Andrianarivo, A.J. 1981. *Etude Comparee de L'organisation Sociale Chez Microcebus murinus. Memoire.* Tananarive, Université de Madagascar.

Atsalis, S. 1998. *Feeding Ecology and Aspects of Life History in Microcebus rufus.* Ph.D. Thesis. CUNY, New York.

Atsalis, S.; Schmid, J.; Kappeler, P.M. 1996. Metrical comparisons of three species of mouse lemur. *J. Hum. Evol.* 31:61–68.

Barre, V.; Levec, A; Petter, J.J.; Albignac, R. 1988. Etude du Microcèbe par Radio-tracking dans la Forêt de l'Akarafantsika. Pp. 61–71 in *L'equlibre des Ecosystemes Forestiers a Madagascar: Actes D'un Seminaire International.* L. Rakotovao; V. Barre; J. Sayer, eds., Gland, I.U.C.N.

Bearder, S.K. 1986. Lorises, Bushbabies, and Tarsiers: Diverse Societies in Solitary Foragers. Pp. 11–24 in *Primate Societies.* B.B. Smuts; D.L. Cheney; R.M. Seyfarth; R.W. Wrangham; T.T. Struhsaker, eds., Chicago, University of Chicago Press.

Birkinshaw, C.R.; Colquhoun, I.C. 1998. Pollination of *Ravenala madagascariensis* and *Parkia madagascariensis* by *Eulemur macaco* in Madagascar. *Folia Primatol.* 69:252–259.

Bomford, E. 1981. On the Road to Nosy Mangabe. *International Wildlife* 11:20–24.

Burney, D.A. 1995. Theories and Facts on Holocene Environmental Change Before and After Colonization. Pp. 41–42. In *Environmental Change in Madagascar.* B.D. Patterson; S.M. Goodman; J.L. Sedlock, eds., Chicago, Field Museum.

Cartmill, M. 1974. *Daubentonia, Dactylopsila,* Woodpeckers and Klinorhyncy. Pp. 655–670 in *Prosimian Biology.* R.D. Martin; G.A. Doyle; A.C. Walker, eds., London, Duckworth.

Charles-Dominique, P. 1976. Les Gommes dans le Regéme Alimentaire de *Cuoa cristata* à Madagascar. *Oiseau* 46:174–178.

Charles-Dominique, P. 1978. Solitary and Gregarious Prosimians: Evolution of Social Behavior in Primates. Pp. 139–149 in *Recent Advances in Primatology.* D.J. Chivers; K.A. Joysey, eds., New York, Academic Press.

Charles-Dominique, P.; Cooper, H.M.; Hladik, A.; Hladik, C.M; Pages, E.; Pariente, G.F.; Petter-Rousseaux, A.; Petter, J.J.; Schilling, A., eds. 1980. *Nocturnal Malagasy Primates: Ecology, Physiology, and Behavior.* New York, Academic Press.

Charles-Dominique, P.; Hladik, M. 1971. Le *Lépilemur* du Sud de Madagascar: Écologie, Alimentation, et Vle Sociale. *Terre Vie* 25:3–66.

Charles-Dominique, P; Petter, J.J. 1980. Ecology and Social Life of *Phaner furcifer*. Pp. 75–95 in *Nocturnal Malagasy Primates: Ecology, Physiology, and Behavior.* P. Charles-Dominique; H.M. Cooper; A. Hladik; C.M. Hladik; E. Pages; G.F. Pariente; A. Petter-Rousseaux; J.J. Petter; A. Schilling, eds. New York, Academic Press.

Cheney, D.L.; Wrangham, R.W. 1986. Predation. Pp. 227–239 in *Primate Societies.* B.B. Smuts; D.L. Cheney; R.M. Seyfarth; R.W. Wrangham; T.T. Struhsaker, eds., Chicago, University of Chicago Press.

Chivers, D; Hladik, M. 1980. Morphology of the Gastrointestinal Tract in Primates. Comparisons with Other Mammals in Relation to Diet. *J. Morphol.* 166:337–386.

Clark, A.B. 1985. Sociality in a Nocturnal "Solitary" Prosimian: *Galago crassicadatus. Intl. J. Primatol.* 6:581–600.

Clark, W.E. LeGros (1963) *The Antecedents of Man,* New York: Harper and Row.

Constable, I.D.; Mittermeier, R.A.; Pollock, J.I.; Ratsirarson, J; Simons, H. 1985. Sightings of Aye-Ayes and Red Ruffed Lemurs on Nosy Mangabe and the Masoala Peninsula. *Primate Conserv.* 5:59–62.

Croft, D. B. 1981. Behaviour of Red Kangaroos (*Macropus rufus*) in Northwestern New South Wales. *Australian Mammol.* 4:5–58.

Crompton, R. 1995. Ecology and Locomotion in *Tarsius, Avahi,* and *Lepilemur:* Biomechanics. Unpublished manuscript.

Dixson, A.F. 1987. Baculum Length and Copulatory Behavior in Primates. *Am. J. Primatol.* 13:51–60.

Endler, J.A. 1991. Interactions Between Predators and Prey. Pp. 169–196 in *Behavioural Ecology.* J.R. Krebs; N.B. Davies, eds. London, Blackwell

Erickson, C.J. 1991. Percussive Foraging in the Aye-Aye, *Daubentonia Madagascariensis. Anim. Behav.* 41:793-801.

Erickson, C.J. 1994. Tap-sacanning and Extractive Foraging in Aye-Ayes, *Daubentonia madagascariensis. Folia Primatol.* 62:125–135.

Erickson, C.J. 1995. Perspectives on Percussive Foraging in the Aye-Aye, *Daubentonia madagascariensis.* Pp. 251–259 in *Creatures of the Dark: The Nocturnal Prosimians.* L. Anderson; G.A. Doyle; K. Izard eds., New York, Plenum.

Erickson, C.J. 1998. Cues for Prey Location by Aye-Ayes *(Daubentonia madagascariensis). Folia Primatol.* 69, Suppl. 1:35–40.

Fietz, J. 1997. Comparison of the Sexes in *Microcebus murinus* (J.F. Miller). *Folia Primatol.*

Fietz, J. 1998. Parental Care and Monogamy in a Nocturnal Lemur *(Cheirogaleus medius). Folia Primatol.* 69:210.

Foerg, R. 1982. Reproduction in *Cheirogaleus medius. Folia Primatol.* 39:49–62

Foerg, R.; Hoffmann, R. 1982. Seasonal and Daily Activity Changes in Captive *Cheirogaleus Medius. Folia Primatol.* 38:259–268.

Ganzhorn, J.U. 1988. Food Partitioning Among Malagasy Primates. *Oecologia* 75:436–450.

Ganzhorn, J.U. 1989. Niche Separation of Seven Lemur Species in the Eastern Rainforest of Madagascar. *Oecologia* 79:279–286.

Ganzhorn, J.U. 1993. Flexibility and Constraints of *Lepilemur* Ecology. Pp. 153–165 in *Lemur Social Systems and Their Ecological Basis.* P.M. Kappeler; J.U. Ganzhorn, eds. New York, Plenum.

Ganzhorn, J.U.; Langrand, O.; Wright, P.C.; O'Conner, S.; Rakotosamimanana, B.; Feistner, A.T.C.; Rumpler, Y. 1996/97. The State of Lemur Conservation in Madagascar. *Primate Conserv.* 17:70–86.

Glander, K. in press. Morphology and Growth in Captive Aye-Aye. *Folia Primatol.*

Godfrey, L.R.; Simons, E.L.; Chatrath, P.J.; Rakotosamimanana, B. 1990. A New Fossil Lemur (*Babakotia,* Primates) from Northern Madagascar. *C.R. Acad. Sci. Paris:* 310, Série II: 81–87.

Godfrey, L.R.; Jungers, W.L.; Simons, E.L.; Chatrath, P.J.; Rakotosamimanana, B. In press. Past and Present Distributions of Lemurs in Madagascar. In *New Directions in Lemur Studies.* H. Rasamimanana; B. Rakotosamimanana; J. Ganzhorn; S. Goodman, eds. New York, Plenum.

Goix, E. 1993. L'Utilisation de la Main Chez le Aye-Aye en Captivité *(Daubentonia madagascariensis)*(Prosimiens, Daubentoniidés). *Mammalia* 57:171–188.

Goodman, S.M. 1994. The Enigma of Antipredator Behavior in Lemurs: Evidence of a Large Extinct Eagle on Madagascar. *Intl. J. Primatol.* 15:129–134.

Goodman, S.M.; O'Conner, S.; Langrand, O. 1993a. A Review of Predation on Lemurs: Implications for the Evolution of Social Behavior in Small, Nocturnal Primates. Pp. 51–66 in *Lemur Social Systems and Their Ecological Basis.* P.M. Kappeler; J.U. Ganzhorn, eds. New York, Plenum.

Goodman, S.M.; Langrand, O.; Raxworthy, C.J. 1993b. Food Habits of the Madagascar Long-Eared Owl, *Asio madagascariensis,* in Two Habitats in Southern Madagascar. *Ostrich* 64:79–85.

Goodman, S.M.; Langrand, O.; Raxworthy, C.J. 1993c. Food Habits of the Barn Owl *Tyto alba* at Three Sites on Madagascar. *Ostrich* 64:160–171.

Green, G.M.; Sussman, R.W. 1990. Deforestation History of the Eastern Rain Forests of Madagascar from Satellite Images. *Science* 248:212-215.

Groves, C.P.; Tattersall, I. 1991. Geographical Variation in the Fork-Marked Lemur, *Phaner furcifer* (Primates, Cheirogaleidae). *Am. J. Primatol.* 56:39–49.

Harcourt, C. 1987. Brief Trap/Retrap Study of the Brown Mouse Lemur *(Microcebus rufus). Folia Primatol.* 49:209–211.

Harcourt, C. 1990. *Lemurs of Madagascar and the Comoros: The IUCN Red Data Book.* Gland, The World Conservation Union.

Hart, D.L. In prep. *Primates as Prey: Ecological, Morphological, and Behavioral Interrelations Between Primates and Their Predators.* Ph.D. Thesis. Washington University, St. Louis.

Hladik, A. 1980. The Dry Forest of the West Coast of Madagascar: Climate, phenology, and Food Availability. Pp. 3–40 in *Nocturnal Malagasy Primates: Ecology, Physiology & Behavior.*

Hladik, P. 1979. Diet and Ecology of Prosimians. Pp. 307–339 In *The Study of Prosimian Behavior.* G.A. Doyle; R.D. Martin, eds. New York, Academic.

Hladik, P.; Charles-Dominique, P.; Petter, J.J. 1980. Feeding Strategies of Five Nocturnal Prosimians in the Dry Forest of the West Coast of Madagascar. Pp. 41–73 in *Nocturnal Malagasy Primates: Ecology, Physiology, and Behavior.* P. Charles-Dominique; H.M. Cooper; A. Hladik; C.M. Hladik; E. Pages; G.F. Pariente; A. Petter-Rousseaux; J.J. Petter; A. Schilling, eds. New York, Academic Press.

Iwano, T.; Iwakawa, C. 1988. Feeding Behaviour of the Aye-Aye *(Daubentonia madagascariensis)* on Nuts of Ramy *(Canarium madagascariensis). Folia Primatol.* 50:136–142.

Kappeler, P.M. 1997. Intrasexual Selection in *Mirza coquereli:* Evidence for Scramble Competition Polygyny in a Solitary Primate. *Behav. Ecol. Sociobio.* 45:115–127.

Kay, R.F.; Sussman, R.W.; Tattersall, I. 1978. Dietary and Dental Variations in the Genus *Lemur,* with Comments Concerning Dietary-Dental Correlations Among Malagasy Primates. *Am. J. Phys. Anthropol.* 49:119–128.

Kent, R.K. 1970. *Early Kingdoms in Madagascar, 1500–1700.* New York, Holt, Rinehart and Winston.

Kress, W.J. 1993. Coevolution of Plants and Animals: Pollination of Flowers by Primates in Madagascar. *Current Science* 65:253–257.

Kress, W.J.; Schatz, G.E.; Andrianifahanana, M.; Morland, H.S. 1994. Pollination of *Ravenala madagascariensis* (Strelitziaceae) by Lemurs in Madagascar: Evidence for an Archaic Coevolutionary system? *Am. J. Botany* 81:542–551.

Langrand, O. 1990. *Guide to the Birds of Madagascar.* New Haven, Yale University Press.

Martin, R.D. 1972. A Preliminary Field-Study of the Lesser Mouse Lemur *(Microcebus Murinus* J.F. Miller 1777). *J. Comp. Ethol. Suppl.* 9:43–89.

Martin, R.D. 1973. A Review of the Behavior and Ecology of the Lesser Mouse Lemur *(Microcebus murinus* J.F. Miller 1777). Pp. 1–68 in *Comparative Ecology and Behavior of Primates.* R.P. Michael; J.H. Crook, eds. New York, Academic.

Martin, R.D. 1990. *Primate Origins and Evolution: A Phylogenetic Reconstruction.* Princeton, Princeton University Press.

Meier, B.; Albignac, R. 1991. Rediscovery of *Allocebus trichotis,* Gunther 1875 (Primates) in Northeast Madagascar. *Folia Primatol.* 56:57–63.

Meier, B.; Albignac, R.; Peyrieras, A.; Rumpler, Y; Wright, P. 1987. A new species of *Hapalemur* (Primates) from South East Madagascar. *Folia Primatol.* 48:211-215.

Meyers, D.M.; Wright, P.C. 1993. Resource Tracking: Food Availability and *Propithecus.* Pp. 179–192 in *Lemur Social Systems and Their Ecological Basis.* P.M. Kappeler; J.U. Ganzhorn, eds. New York, Plenum.

Mittermeier, R.A.; Tattersall, I.; Konstant, W.R.; Meyers, D.M.; Mast, R.B. 1994. *Lemurs of Madagascar.* Washington, D.C., Conservation International.

Müller, A.E. 1998. A Preliminary Report on the Social Organization of *Cheirogaleus medius* (Cheirogaleidae; Primates) in North-West Madagascar. *Folia Primatol.* 69, Suppl. 1:160–166.

Nash, L.T. 1994. *Behavior and Habitat Use of Lepilemur at Beza Mahafaly Special Reserve, Madagascar.* Paper delivered at Am. Soc. Primatol. Annual Meetings. Seattle, WA.

Nash, L.T. 1998. Vertical Clingers and Sleepers: Seasonal Influences on the Activities and Substrate Use of *Lepilemur leucopus* at Beza Mahafaly Special Reserve, Madagascar. *Folia Primatol.* 69, Suppl. 1:204–217.

Nillson, L.A.; Rabakonandrianina, E.; Pettersson, B. 1993. Lemur Pollination in the Malagasy Rainforest Liana *Strongylodon craveniae* (Leguminosae). *Evol. Trends in Plants* 7:49–56.

Overdorff, D.J. 1991. *Ecological Correlates of Social Structure in Two Prosimian Primates: Eulemur fulvus and Eulemur rubriventer* in Madagascar. Ph.D. Thesis, Duke University, Durham, NC.

Owen, R. 1866. On the aye-aye (*Chiromys, Cuvier; Chiromys madagascariensis,* Desm.; *Sciurus madagascariensis,* Gmel., Sonnerat; *Lemur psilodactylus,* Schreber, Shaw). *Trans. Zool. Soc. Lon.* 5:33–101.

Pages, E. 1980. Ethoecology of *Microcebus coquereli* During the Dry Season. Pp. 97–116 in *Nocturnal Malagasy Primates: Ecology, Physiology, and Behavior.* P. Charles-Dominique; H.M. Cooper; A. Hladik; C.M. Hladik; E. Pages; G.F. Pariente; A. Petter-Rousseaux; J.J. Petter; A. Schilling, eds. New York, Academic Press.

Pages-Feuillade, E. 1988. Modalités de l'Occupation de l'Espace et Relations Interindividuelles Chez un Prosimien Nocturne Malgache *(Microcebus murinus).* *Folia Primatol.* 50:204–220.

Patterson, B.D.; Goodman, S.M.; Sedlock, J.L., eds. 1995. *Environmental Change in Madagascar,* Chicago, Field Museum.

Perret, M. 1980. *Influence De La Captivite et du Groupement Social sur la Physiologie du Microcebe (Microcebus murinus, Cheirogaleus,* Primates). Thèse de Doctorat d'Etat. University of Paris, Paris.

Peterson, R.L. 1955. *North American Moose.* Toronto, University of Toronto Press.

Petter, J.J. 1977. The aye-aye. Pp. 38–59 in *Primate Conservation.* H.S.H. Prince Ranier III; G.H. Bourne, eds. New York, Academic.

Petter, J.J. 1978. Ecological and Physiological Adaptations of Five Sympatric Nocturnal Lemurs to Seasonal Variations in Food Production. Pp. 211–223 In *Recent Advances in Primatology.* D.J. Chivers; K.A. Joysey, eds., New York, Academic Press.

Petter, J.J. 1988. Contributions à l'Étude du *Cheirogaleus medius* dans la Fôret de Morondava. Pp. 57-60 in *L'equilbre des Ecosystemes Forestiers a Madagascar: Actes d'un Seminaire International.* L. Rakotovao; V. Barre; J. Sayer, eds., Gland, I.U.C.N.

Petter, J.J.; Peyrieras, A. 1970. Nouvelle Contribution à l'Étude d'un Lemurien Malgache, le Aye-Aye (*Daubentonia madagascariensis,* E. Geoffroy). *Mammalia* 34:167–193.

Petter, J.J.; Petter-Rousseaux, A. 1959. Contributions à l'Étude du Aye-Aye. *Natur. Malgache* 11:165–173.

Petter J.J.; Petter-Rousseaux, A. 1967. The Aye-Aye of Madagascar. Pp. 195–205 in *Social Communication Among Primates.* S.A. Altmann, ed. Chicago, University of Chicago Press.

Petter, J.J.; Petter-Rousseaux, A. 1979. Classification of the Prosimians. Pp. 1–44 In *The Study Of Prosimian Behavior.* G.A. Doyle; R.D. Martin, eds. New York, Academic.

Petter, J.J.; Schilling, A.; Pariente, G. 1971. Observations Éco-Éthologique sur Deux Lemuriens Malgaches Nocturnes: *Phaner furcifer* et *Microcebus coquereli. Terre Vie* 25:287–327.

Petter, J.J.; Schilling, A; Pariente, G. 1975. Observations on the Behavior and Ecology of *Phaner furcifer.* Pp. 209–218 in *Lemur Biology.* I. Tattersall; R.W. Sussman, eds. New York, Plenum.

Petter, J.J.; Albignac, R.; Rumpler, Y. 1977. *Faune de Madagascar 44: Mammiferes Lemuriens (Primates, Prosimiens).* Paris, Orstom/Cnrs.

Petter-Rousseaux, A.; Hladik, C.M. 1980. A Comparative Study of Food Intake in Five Nocturnal Prosimians in Simulated Climatic Conditions. Pp. 169–179 in *Nocturnal Malagasy Primates: Ecology, Physiology, and Behavior.* P. Charles-Dominique; H.M. Cooper; A. Hladik; C.M. Hladik; E. Pages; G.F. Pariente; A. Petter-Rousseaux; J.J. Petter; A. Schilling, eds. New York, Academic Press.

Petter-Rousseaux, A; Picon, R. 1981. Annual Variation in the Plasma Testosterone in *Microcebus murinus. Folia Primatol.* 36:183–190.

Rakotoarison, N.; Zimmermann, H.; Zimmermann, E. 1997. First Discovery of the Hairy-Eared Dwarf Lemur *(Allocebus trichotis)* in a highland rain forest of Eastern Madagascar. *Folia Primatol.* 68:86–94.

Ramsay, M.A.; Stirling, I. 1986. On the Mating System of Polar Bears. *Can. J. Zool.* 64:2142-2151.

Randrianarivo, R. 1979. *Essai d'Inventaire Des Lemuriens de la Future Reserve de Beza Mahafaly.* Memoire de Fin d'Étude. University of Madagascar, Tananarive.

Rasoloarison, R.; Rasolonandrasana, B.; Ganzhorn, J.; Goodman, S. 1995. Predation on Vertebrates in the Kirindy Forest, Western Madagascar. *Ecotropica* 1:59–65.

Ratsirarson, J. 1986. *Contribution a L'etude Comparee du l'Eco-Ethologie de Deux Expeces de Lemuriens:* Lepilemur mustilenus (I. Geoffroy 1850) et Lepilemur septentrionalis (Rumpler et Albignac 1975). Thèse Doct. Etat. Université Louis Pasteur, Strasbourg.

Richard, A.F. 1978. *Behavioral Variation: Case Study of a Malagasy Prosimian.* Lewisburg, Bucknell University Press.

Richard A.F.; Dewar, R.E. 1991. Lemur Ecology. *Ann. Rev. Ecol. System.* 22:145–175.

Rumpler, A.; Albignac, R. 1975. Intraspecific Chromosome Variation in a Lemur from the North of Madagascar: *Lepilemur septentrionalis,* Species Nova. *Am. J. Phys. Anthropol.* 38:425–429.

Russell, R.J. 1975. Body Temperatures and Behavior of Captive Cheirogaleids. Pp. 193–206 in *Lemur Biology.* I. Tattersall; R.W. Sussman, eds. New York, Plenum.

Russell, R.J. 1977. *The Behavior, Ecology, and Environmental Physiology of a Nocturnal Primate.* Ph.D. Thesis. Duke University, Durham.

Russell, R.J. 1985. Lepilemur: *A Study of Natural History and Evolution.* Unpublished manuscript.

Sarikaya, Z.; Kappeler, P.M. 1997. Nest Building Behavior of Coquerel's Dwarf Lemur *(Mirza coquereli). Primate Report* 47:3–9.

Sauther, M.L.; Nash, L.T. 1987. Effect of Reproductive State on Food Consumption in Captive *Galago senegalensis braccatus. Am. J. Phys. Anthropol.* 73:81–88.

Seligson, D.; Szalay, F.S. 1978. Relationship Between Natural Selection and Dental Morphology: Tooth Function and Diet in *Lepilemur* and *Hapalemur.* Pp. 289–307 in *Studies in the Development, Function and Evolution of Teeth.* P.M. Butler; K.A. Joysey, eds., London, Academic.

Schilling, A.; Perret, M.; Predine, J. 1984. Sexual Inhibition in a Prosimian: A Pheromone-like Effect. *J. Endocrinol.* 102:143–151.

Schmid, J.; Ganzhorn, J.U. 1996. Resting Metabolic Rates of *Lepilemur ruficaudatus. Am. J. Primatol.* 38:169–174.

Schmid, J.; Kappeler, P.M. 1994. Sympatric Mouse Lemurs (*Microcebus* spp.) in Western Madagascar. *Folia Primatol.* 63:162–170.

Schmid, J.; Kappeler, P.M. 1998. Intrasexual Selection in *Microcebus murinus. Folia Primatol.* 69:211.

Simons, E.L. 1988. A New Species of *Propithecus* from Northeast Madagascar *Folia Primatol.* 50:143–151.

Simons, E.L. 1997. Lemurs: Old and New. Pp. 142–166. in *Natural Change and Human Impact in Madagascar.* S.M. Goodman; B.D. Patterson; eds., Washington, D.C., Smithsonian Institution.

Simons, E.L.; Godfrey, L.R.; Jungers, W.L.; Chatrath, P.S.; Rakotosamimanana, B. 1992. A New Giant Subfossil Lemur, *Babakotia,* and the Evolution of the Sloth Lemur. *Folia Primatol.* 58:197–203.

Simons, E.L.; Godfrey, L.R.; Jungers, W.L.; Chatrath, P.S.; Ravaoarisoa; J. 1995. A New Species of *Mesopithecus* (Primates, Palaeopropithecidae) from Northern Madagascar. *Int. J. Primatol.* 16:653–682.

Smith, R.J.; Jungers, W.L. 1997. Body Mass in Comparitive Primatology. J. Hum. Evol. 32:523–559.

Sterling, E.J. 1993a. *Behavioral Ecology of the Aye-Aye (Daubentonia madagascariensis) On Nosy Mangabe, Madagascar.* Ph.D. Thesis. Yale University, New Haven.

Sterling, E.J. 1993b. Patterns of Range Use and Social Organization in Aye-Ayes (*Daubentonia madagascariensis*) on Nosy Mangabe. Pp. 1–10 in *Lemur Social Systems and Their Ecological Basis.* P.M. Kappeler; J.U. Ganzhorn, eds. New York, Plenum.

Sterling, E.J. 1994a. Taxonomy and Distribution of Daubentonia: A Historical Perspective. *Folia Primatol.* 62:8–13.

Sterling, E.J. 1994b. Evidence for Nonseasonal Reproduction in Wild Aye-Ayes (*Daubentonia madagascariensis*). *Folia Primatol* 62:45–53.

Sterling, E.J. 1994c. Aye-Ayes: Specialists on Structurally Defended Resources. *Folia Primatol.* 62:142–154.

Sterling, E.J.; Dierenfeld, E.S.; Ashbourne, C.J.; Feistner, A.T.C. 1994. Dietary Intake, Food Composition and Nutrient Intake in Wild and Captive Populations of *Daubentonia madagascariensis. Folia Primatol.* 62:115–124.

Stiles, E.W. 1993. The Influence of Pulp Lipids on Fruit Preference by Birds. *Vegatatio* 107/108:227–235.

Sussman, R.W. 1988. The Adaptive Array of the Lemurs of Madagascar. *Monogr. Syst. Bot. Missouri Bot. Gard.* 25:215–226.

Sussman, R.W.; Raven, P.H. 1978. Pollination by Lemurs and Marsupials: An Archaic Coevolutionary System. *Science* 200:731–736.

Tattersall, I. 1982. *The Primates of Madagascar.* New York, Columbia University Press.

Tattersall, I. 1986. Systematics of the Malagasy Strepsirhine Primates. Pp. 43–72 in *Comparative Primate Biology, Vol. 1: Systematics.* D.R. Swindler; J. Erwin, eds. New York, Liss.

Tattersall, I. 1988. Cathemeral Activity in Primates: A Definition. *Folia Primatol.* 49:200–202.

Terborgh, J.; Janson, C.H. 1986. The Socioecology of Primate Groups. *Ann. Rev. Ecol. Syst.* 17:111–135.

van Schaik, C.P. 1983. Why Are Diurnal Primates Living in Groups? *Behaviour* 87:120–144.

Verin, P. 1990. *Madagascar.* Paris, Karthala.

Warren, R.D.; Crompton. R.H. 1998. Diet, Body Size and the Energy Costs of Locomotion in Saltatory Primates. *Folia Primatol.* 69, Suppl. 1:86–100.

Wrangham, R.W. 1983. Social Relationships in Comparative Perspective. Pp. 325–334 in *Primate Social Relationships: An Integrated Approach.* R.A. Hinde, ed., Oxford, Blackwell.

Wright, P.C.; Martin, L.B. 1995. Predation, Pollination and Torpor in Two Nocturnal Prosimians: *Cheirogaleus major* and *Microcebus rufus* in the Rain Forest of Madagascar Pp. 45–60 in *Creatures of the Dark: The Nocturnal Prosimians.* K. Izard; L. Anderson; G.A. Doyle, eds., New York, Plenum.

Zimmermann, E.; Cepok, S.; Rakotoarison, N. Zietemann, V.; Radespiel, U. 1998. Sympatric Mouse Lemurs in North-West Madagascar: A New Rufous Mouse Lemur Species *(Microcebus ravelobensis). Folia Primatol.* 69:106–114.

Chapter 5

The Ecology of the Diurnal Lemuriformes

INTRODUCTION TO NOCTURNAL VS. DIURNAL PRIMATE ADAPTATIONS

The earliest primates were small nocturnal forms (Chapter 2). These primates probably were in competition with other nocturnal mammals, especially bats and rodents, for the resources available on the recently evolved flowering trees. In North America, Europe, Africa and Asia the small nocturnal prosimians did not fare well, leaving only the few species in Africa and Asia discussed in Chapter 3. The larger diurnal anthropoids evolved from a prosimian ancestor in the Oligocene or earlier. In Madagascar, a number of taxa of large, diurnal Lemuriformes evolved along parallel lines, and it is likely that they did not compete with the nocturnal prosimians but filled new niches not previously exploited in forest communities.

It often is stated that the plight of the Asian and African prosimians was brought about by direct competition with the subsequently evolving anthropoids. However, two lines of evidence argue against this interpretation. First, prosimians and anthropoids of Africa and Asia generally exploit different resources and probably always have. Second, in Madagascar where diurnal and nocturnal lemurs co-exist, but other nocturnal mammals are relatively absent, approximately half of the extant primate species are nocturnal.

A comparison of communities in two tropical forest ecosystems further emphasizes some differences between nocturnal and diurnal mammals. The two forests are Makokou, Gabon and Barro Colorado, Panama (Charles-Dominique 1975). In Gabon, there are 121 species of mammals and 216 species of birds living sympatrically. Seventy percent of the mammalian species are nocturnal and 96% of the bird species are diurnal. In Barro Colorado, 86% of the 86 mammalian species are nocturnal. Generally speaking, birds occupy diurnal canopy niches and small mammals, including flying or gliding species, occupy nocturnal ones.

149

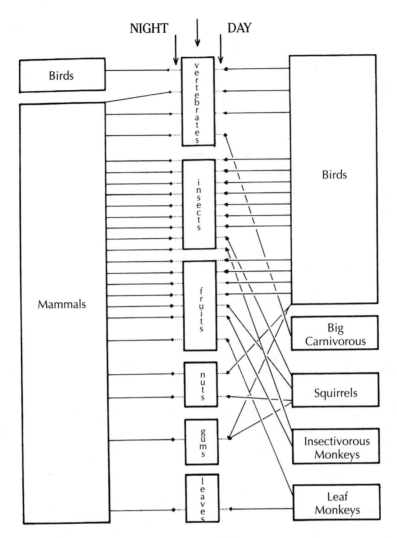

Figure 5-1a Utilization of principal food sources by birds and mammals during night and day. Community structure. [Adapted from *Nocturnality and Diurnality: An Analysis of These Two Models of Life by an Analysis of the Higher Vertebrate Fauna in Tropical Forest Ecosystems,* by P. Charles-Dominique]

As a general rule, diurnal mammals are larger than nocturnal species (Figure 5-1) and only a few diurnal forest mammals are adapted to arboreal life. In Gabon there are 9 species of squirrels, 11 of monkeys and apes, and one anteater. In Barro Colorado, there are 4 monkey, 2 squirrel, one anteater and one carnivore species.

Thus, the only major taxonomic groups of mammals which have developed mainly diurnal, arboreal forms are squirrels and primates. These diurnal mammals have developed specializations which permit them to exploit food sources generally inaccessible to birds. Adaptations in diurnal canopy-living mammals include: (1) the continuously growing incisors and claws of squirrels

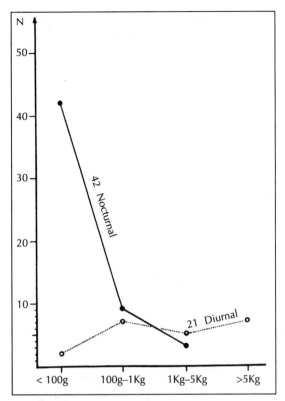

Figure 5-1b Weight categories of nocturnal (42 species) and diurnal (21 species) arboreal mammals in the equatorial primary forest of Gabon (region of Makokou). Community structure. [Adapted from *Nocturnality and Diurnality: An Analysis of These Two Models of Life by an Analysis of the Higher Vertebrate Fauna in Tropical Forest Ecosystems,* by P. Charles-Dominique]

enabling them to open tough fruits and nuts, scrape bark, and move and posture on large vertical trunks; and (2) the large body size, and digestive tract and masticatory specializations which have enabled primates to feed on large fruits and pods with tough rinds, and on leaves. Increase in body size seems to have been a major factor enabling anthropoids and some lemur species to successfully adopt diurnal activity patterns and to coexist with birds in the forest canopy: a weight of one kilogram appears to be the upper limit for nocturnal primates and the lower limit for diurnal primates (Figure 5-1) (Charles-Dominique 1975). The only major exception to this rule is the South American family Callitrichidae, which may have developed small body size from a large-bodied ancestor to fill a particular ecological niche (Garber 1980, Sussman and Kinzey 1984, Martin 1992, Kay 1994).

Along with the larger body size, there are several other characteristics that generally distinguish diurnal from nocturnal primates. Some of these are listed in Table 5-1. It should be noted that these are general traits with exceptions found in each category. Some of these exceptions have already been discussed (for example, folivory in *Lepilemur* and clinging to the fur of the mother by infant *Perodicticus*) and others will be discussed in the next volume (e.g., small body size in callitrichids and nest building by some pongids). The exceptions appear to be specialized, derived characteristics in some cases, and retained generalized characters in others.

Many of the characteristics associated with a diurnal activity cycle are related to large body size. However, one of these, the propensity to live in groups, is unique for a major taxon of mammals. In all other mammalian orders, most species are essentially solitary and only a few taxon are group-living. All diurnal species of primate, with the exception of orang utans, live in relatively stable social groups. The adaptive significance of this as a general, pervasive trait of diurnal primates remains an intriguing problem.

Table 5-1 Some Characteristics Distinguishing Nocturnal from Diurnal Primates

Nocturnal	Diurnal
1. Small Body Size (<1 Kg.)	1. Large Body Size
2. Small day and home ranges (related to body size)	2. Large day and home ranges
3. Generally higher % of insects in diet (related to body size)	3. Lower % of insects, more fruit and leaves
4. Nests	4. No nests
5. "Park" infants; carry infants by mouth	5. Infants grip fur and are carried at all times
6. Higher % of twinning or multiple births	6. Usually single birth
7. Oral grooming	7. Groom with hands (except diurnal Lemuriformes)
8. Mainly olfactory and vocal communication	8. Mainly vocal and visual communication
9. Solitary when active	9. Group living

INTRODUCTION TO THE ECOLOGY OF THE DIURNAL LEMURIFORMES

As stated in Chapter 4, Madagascar is divided into two major climatic zones: the wet, evergreen zone of the east (oriental zone) and the dry west and south (occidental zone) (Figure 4-1). The primary vegetation of the occidental zone is characterized mainly by deciduous forests in the north and west, and by dry brush, and desert-like forests in the south (Koechlin 1972). Until recently, most long-term studies of diurnal lemurs had been conducted in the occidental zone.

As I discussed earlier, modern Lemuriformes probably evolved from the Eocene Adapidae. This fossil taxon was quite diverse but generally showed strong resemblance postcranially to modern lemurs (Simons 1972, Conroy 1990, Martin 1990). Fossil evidence of the Eocene adapids is mainly from North America and Europe. However, it is likely that an African form (or forms) reached Madagascar early in the Eocene and was isolated there as the Mozambique channel gradually increased in width (Rabinowitz et al. 1983). No fossils of great antiquity have been found on Madagascar itself. The oldest "subfossil" lemurs are dated at 26,000 years old or less (Simons 1997, Godfrey et al. in press). At least 17 species of lemur have become extinct since the arrival of humans, 2000 years ago. Most were large, diurnal forms (Table 4-2) and should be considered together with extant species as an integrated lemur fauna (Godfrey et al. 1997, in press).

Of the fifteen species of diurnal or cathemeral Lemuriformes, *Lemur catta, Eulemur fulvus* and *Propithecus verreauxi* are the only species distributed over a wide area in the Occidental Zone (see, Table 5-2). *Hapalemur griseus* and *Avahi laniger* have very localized distributions in this region. Seven species are found in the northwest and north of the island. *E. mongoz, E. macaco,* and *E. coronatus* only occur here and may replace one another from northwest to further north, respectively. Subspecies of *P. verreauxi* are found in the northwest and one of *E. fulvus* occurs throughout the north. Again, *H. griseus* and *A. laniger* are found in localized areas in the north. Ten diurnal species occur in the eastern forests. Seven are

Table 5-2 Common Names, General Distribution, and Research Status of Diurnal Lemuriformes

Family Lemuridae

Species	Common Name	Geographical Distribution	Research Status
Lemur catta	ringtailed lemur	S, W	LT
Eulemur fulvus	brown lemur	—	—
E. f. rufus	red-fronted or rufus lemur	W, E	LT
E. f. fulvus	brown lemur	N.W., E	ST
E. f. sanfordi	Sanford's lemur	N	LT
E. f. mayottensis	Mayotte lemur	Co	ST
E. f. albifrons	white-fronted lemur	E	LT
E. f. albocollaris	white-collared lemur	E	O
E. f. collaris	collared lemur	E	O
Eulemur mongoz	mongoose lemur	N, Co	LT
Eulemur macaco	black lemur	—	—
E. m. macaco	black lemur	N	LT
E. m. flavifrons	Sclater's lemur	N	O
Eulemur coronatus	crowned lemur	N	LT
Eulemur rubriventer	red-bellied lemur	E	LT
Varecia variegata	ruffed or variegated lemur	—	—
V. v. variegata	black & white ruffed lemur	E	LT
V. v. rubra	red ruffed lemur	E	LT
Hapalemur griseus	gentle lemur	—	—
H. g. griseus	grey gentle lemur	E	ST
H. g. alaotrensis	Alaotran gentle lemur	E	LT
H. g. occidentalis	western gentle lemur	W	ST
Hapalemur aureus	golden bamboo lemur	E	ST
Hapalemur simus	greater bamboo lemur	E	ST
Propithecus verreauxi	Verreauxi's sifaka	—	—
P. v. verreauxi	Verreauxi's sifaka	S, W	LT
P. v. deckeni	Decken's sifaka	W	O
P. v. coquereli	Coquerel's sifaka	NW	ST
Propithecus tattersalli	Tattersall's sifaka	N	LT
Propithecus diadema	diademed sifaka	—	—
P. d. diadema	diademed sifaka	E	LT
P. d. candidus	silky sifaka	E	O
P. d. edwardsi	Milne-Edward's sifaka	E	LT
P. d. perrieri	Perrier's sifaka	N	O
Indri indri	indri	E	LT
Avahi laniger	woolly lemur	—	—
A. l. laniger	eastern woolly lemur	E	ST
A. l. occidentalis	western woolly lemur	W	ST

Geographical Distribution: S=south, W=west, NW=northwest, N=north, E=east, Co.=Comoros.
Research Status: LT=long term (>I years), ST=short term (<I yr.), O=no intensive studies.

found here exclusively: *E. rubriventer, Varecia variegata, H. simus, H. aureus, P. diadema, P. tattersalli,* and *Indri indri.* Three species, *E. fulvus, H. griseus,* and *A. laniger,* are the only ones found in all three of these regions. In table 5-2, I summarize the present status of studies on each of these species. I will review the literature from each of these three regions separately.

THE ECOLOGY OF THE DIURNAL LEMURS OF WESTERN MADAGASCAR

Modern field research on lemurs began in 1956, when Jean-Jacques Petter (1962, 1965) carried out surveys, noting the ecology and behavior of a number of species throughout the island (see Tattersall 1982, 1997 for earlier history of lemur studies). In 1963/64, Jolly (1966) conducted the first intensive field study on *Lemur catta* and *Propithecus verreauxi* at Berenty, a protected private reserve in Southern Madagascar (Figure 5-2). A number of studies followed at this site and information from 30 years of research is available on population dynamics of *L. catta* and *P. verreauxi* in this forest (see partial bibliography in Jolly et al. 1993).

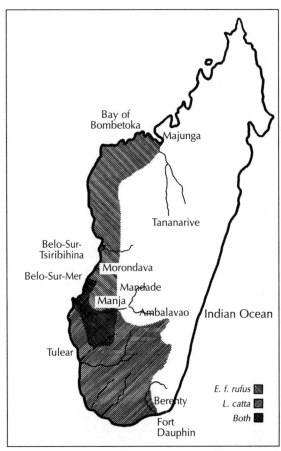

Outside of the Berenty Reserve, comparative studies of *L. catta* and *E. fulvus* were conducted in forests in Western Madagascar (Sussman 1972, 1974) and *P. verreauxi* was studied in two forests with very different vegetation in the north and west (Richard 1973, 1978a). In the mid-1980's, long-term research on *L. catta* and *P. verreauxi* was initiated at the Beza Mahafaly Reserve in the southwest (e.g., Richard et al. 1987, Sussman 1991, Richard et al. 1991, Sauther 1992, Brockman 1994, Gould 1994, Kubzdela 1998). The ranges of these three species overlap, in a region of approximately 250 km. of the southwest. Each fills a particular niche (Sussman 1988), and detailed studies of their ecology and social behavior have revealed very different adaptive complexes.

Figure 5-2 Distribution of *Eulemur fulvus rufus* and *Lemur catta*. Populations are not continuous within these areas but are only found where suitable primary vegetation exists.

A STUDY OF TWO SYMPATRIC SPECIES OF LEMUR

METHODOLOGY

The methods used to study diurnal and nocturnal species of primates differ. Ideally diurnal animals can be seen and followed throughout much of the day if the density of vegetation permits. After choosing a site in which the animals are visible most of the time, the researcher attempts to habituate them without unduly influencing their behavior. The time it takes to accomplish this is quite variable, depending on a number of factors e.g., visibility of the group, whether hunting has been commonly practiced, species temperament, etc.

After habituation is accomplished, the observer collects data on the behavior and ecology of the study population. Because primate ecologists are usually interested in relationships among habitat selection, morphology and behavior, some of the most informative studies can be conducted on a number of different species living in the same forest or on populations of the same species living in forests with very different habitat structure. It is desirable for different investigators to compare studies on the same and different species. For such comparisons to be made, systematic methods must be employed and quantitative data collected.

One of the most commonly used methods of collecting data on free-ranging primates is *scan-sampling*: several individuals of a group are "scanned" at predetermined times and their behavior is scored (Altmann 1974, Lehner 1996). This method was first applied to free-ranging baboon populations by Crook and Aldrich-Blake (1968) who were influenced greatly by their colleague, the late K.R.L. Hall. Hall (1963, 1965) emphasized the need for quantitative methods in the study of free-ranging primates.

A modification of this method was applied to my study of *Lemur catta* (the ringtailed lemur) and *Eulemur fulvus* (the brown lemur) (Figure 5-3). At five minute intervals, on prepared data sheets (Figure 5-4), I recorded the number of individuals engaged in each of six mutually exclusive activities—feeding, grooming, resting, moving (individual displacement), travel (group displacement) and other (e.g., fighting, sunning, play, etc.)—and the levels of the forest at which activities were performed (Figure 5-5). This provided time samples from which the sum of activity records observed could be calculated. Thus, the percentage of the total number of records that are accounted for by feeding, resting, etc. can be determined. Field notes also were kept and cross referenced with data sheets. These included written comments on rare events, descriptions of behavior, travel patterns, census data, etc. This type of data gathering (i.e., normal field notes) is referred to as *ad libitum* sampling (Altmann 1974).

Figure 5-3a *L. catta.* [Photo by Robert W. Sussman]

Figure 5-3b *E. fulvus.* [Photo by Robert W. Sussman]

Figure 5-4 Data Sheet

Time	Hthr	Horiz	Fdg	Grmg	Rest	Mvmt	Trvl	Other

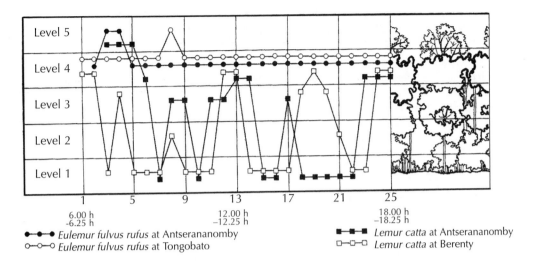

●—●—● *Eulemur fulvus rufus* at Antserananomby   ■—■—■ *Lemur catta* at Antserananomby
○—○—○ *Eulemur fulvus rufus* at Tongobato   □—□—□ *Lemur catta* at Berenty

Figure 5-5 The forest level at which the highest percentage of animals was observed during each of the twenty-five half hour time periods for the species studied at each of the forests.

Similar data collection techniques can now be entered directly into hand-held computers (see, Sauther 1992, Freed 1996).

Quantitative data were collected on populations of ringtailed and brown lemurs in three separate forests. One in which the two species coexisted (Antserananomby), one in which *E. fulvus* was found alone (Tongobato) (both near Manja) and one in which *L. catta* was found alone (Berenty) (Figure 5-2). Both of these species are cat-sized (@ 2000 gm.) (Sussman 1991, Smith and Jungers 1997), quadrupedal animals (Table 5-3), and very similar in general morphology. In the following, I will refer mainly to my studies of these two species but the results from other studies on these species will be cited where appropriate.

VERTICAL AND HORIZONTAL RANGING PATTERNS AND INTERGROUP SPACING

In Western Madagascar, *L. catta* and *E. fulvus* use the forest strata in very different ways and do not seem to influence one another (Table 5-4). Groups of *E. fulvus* spent over 70% of their day in the continuous canopy (level 4) and less than 2% of the time on the ground. Approximately 85% of group travel took place in the canopy. Brown lemurs were rarely seen in areas which necessitated terrestrial locomotion. *L. catta* utilized all of the strata of the forest. Generally, they moved and travelled on the ground, rested during the day in low trees, slept at night in the closed canopy, and fed in all forest strata.

At both forests, the ringtailed lemur spent over 30% of the time, and @ 70% of group travel, on the ground. In Antserananomby, they spent 58% of the day in areas without a canopy, although this represented only 30% of the total area (Figure 5-6). The brown lemur was rarely seen outside the canopy. There are a number of differences in the morphology of these two species which correlate with differences in locomotion and habitat preferences (Ward and Sussman 1979).

The home ranges of brown lemur groups were very small—approximately 1.0 hectare—and overlapped extensively (Figure 5-6). Borders of ranges were not

Table 5-3 Field (Except Where Specified)
Adult Body Weights of Some Diurnal Lemuriformes

Species	Body Weight in Grams (Range)	Sample Size	Location	Source
Lemur catta	Avg. 2,211	65	Beza Mahafaly (southwest)	Sussman 1991, nd
	M = 2,213 (1850–2775)	41		
	F = 2,207 (1850–2750)	24		
Eulemur fulvus rufus	Avg. 2,207	33	Ranomafana (southeast)	Glander et al. 1992
	M = 2,178 (1750 – 2650)	20		
	F = 2,251 (1810–2650)	13		
Eulemur mongoz	Avg. 1,670	27	Captive (Duke Univ.)	Kappeler 1990
	M = 1,682	14		
	F = 1,658	13		
Eulemur macaco	Avg. 2,456	39	Captive (Duke Univ.)	Kappeler 1990
	M = 2,397	22		
	F = 2,532	17		
Eulemur rubriventer	Avg. 2,004	22	Ranomafana (southeast)	Glander et al. 1992
	M = 2,067 (1800 – 2400)	9		
	F = 1,960 (1650–2200)	13		
Varecia variegata	Avg. 2,958	12	Nosy Mangabe (east)	Morland 1991
	M = 2,875 (2400–3700)	4		
	F = 3,000 (2450–3400)	8		
Hapalemur griseus	Avg. 931	15	Captive (Duke Univ.)	Kappeler 1990
	M = 943	9		
	F = 912	6		
	770	1	Ranomafana	Glander et al. 1992

Table 5-3 *(continued)*

Species	Body Weight in Grams (Range)	Sample Size	Location	Source
Hapalemur aureus	Avg. 1,585	4	Ranomafana (east)	Glander et al 1992
	M = 1,613 (1540–1660)	3		
	F = 1,500	1		
Hapalemur simus	M = 2,365	1	East	Meier et al 1987
Propithecus verreauxi	Avg. 3,627	20	Captive (Duke Univ.)	Kappeler 1990
	M = 3,543	9		
	F = 3,696	11		
Propithecus tattersalli	Avg. 3,069	8	Captive (Duke Univ.)	Kappeler 1990
	M = 3,039	4		
	F = 3,098	4		
Propithecus diadema edwardsi	Avg. 5,721	14	Ranomafana (east)	Glander et al. 1992
	M = 5,590 (5200–6100)	8		
	F = 5,895 (5200–6500)	6		
Propithecus diadema diadema	Avg. 6428		East	Glander & Powzyk pers. comm.
	M = 6,496 (6000–7380)	5		
	F = 6360 (5050–7250)	5		
Indri indri	Avg. 6480		East	Glander & Powzyk pers. comm.
	M = 5,825 (5750–5900)	2		
	F = 7,316 (6750–7520)	2		
Avahi Laniger	Avg. 1,178		Ranomafana (east)	Glander et al. 1990
	M = 1,033 (900–1200)	4		
	F = 1,136 (1200–1600)	4		

Table 5-4a. Total percentage of time in different forest strata by diurnal species inhabiting the southwest. 1=ground level; 2=small bushes; 3=small trees and vertical trunks; 4=continuous canopy; 5=emergent layer

Species	Stratum				
	1	2	3	4	5
Eulemur fulvus	1	3	9	71	16
Lemur catta	33	13	23	25	6
Propithecus verreauxi	0	7	27	32	35

Table 5-4b. Percentage of travel in different forest strata by diurnal species inhabiting the southwest.

Species	Stratum				
	1	2	3	4	5
Eulemur fulvus	<1	2	4	85	29
Lemur catta	68	4	9	17	1
Propithecus verreauxi	0	9	50	22	19

Table 5-4c. Percentage of feeding in different forest strata by diurnal species inhabiting the southwest.

Species	Stratum				
	1	2	3	4	5
Eulemur fulvus	1	7	10	52	29
Lemur catta	29	14	33	16	8
Propithecus verreauxi	0	5	10	28	57

defended but groups maintained spatial separation by means of frequent vocalizations. Groups moved little within a day, only 125–150 meters. The population density in both forests was very high, averaging about 1000 per sq. km. Home and day range size, and population density are quite variable in brown lemurs living in different regions of the island (Table 5-5).

Encounters between groups were frequent. They occurred most often when groups were moving, or when one group entered a tree occupied by another. Animals from each group faced each other in adjacent trees or branches, vocalized, wagged their tails and jockeyed for position. Within a few minutes one group would move to another portion of its range. Encounters were important in maintaining group cohesion. This pattern of intergroup spacing allows many groups of *E. fulvus* to utilize overlapping resources in close temporal and spatial proximity while the integrity of the group is maintained.

Home ranges of ringtailed lemur groups have highly overlapping boundaries, with little or no areas of exclusive use (Figure 5-7). The size of ranges varies with habitat and location, with averages at different localities ranging from 10 to 32 hectares (range = 6-35 ha) (Table 5-5). At different sites, population densities vary from 17.4 to 350 animals per km^2, and biomass has been estimated between 48.7 and 200 kg per km^2.

Larger home ranges and lower densities are found in drier or more disturbed habitats (Budnitz and Dainis 1975, O'Conner 1987, Sussman 1991). For example, at Beza Mahafaly groups living in drier habitats had home ranges averaging 32 ha, whereas those in wetter habitats averaged 17 ha. Group sizes in the two habitats were not significantly different. Since there is a higher density of large trees (>25 cm dbh) on wet soils, the small home ranges of groups living in wetter habitats contain a similar number of large trees to the larger ranges of groups in drier habitats (Sussman

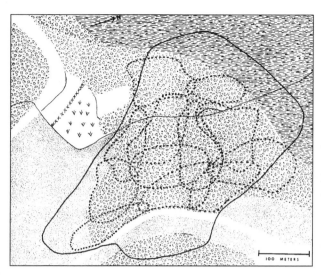

Figure 5-6 Antserananomby. Home ranges of *Lemur catta* and *Eulemur fulvus rufus*. This map includes the home range of one group of *L. catta* (19 animals) and 12 groups of *E. f. rufus* (112 animals).

———— *Lemur catta*   • • • • *Eulemur fulvus rufus*

and Rakotozafy 1994). However, groups with small home ranges expand their range during certain seasons to feed on fruit species absent from their normal range (Sussman 1991, Sauther 1998). Day ranges of groups at Antserananomby averaged 950 meters; at Berenty the average day range for 8 groups was 1377 meters (Jolly et al. 1993).

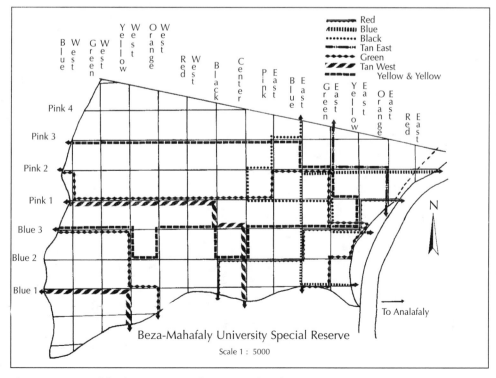

Figure 5-7 Map of the home ranges of groups within the reserve.

Table 5-5. Home Range Size, Population Density, and Group Structure of Some Diurnal Lemuriformes

Species	Home Range Size (hectares)	Day Range (meters)	Population Density (per sq. km.)	Social structure	Group Size	N	References
Lemur catta	6–35	1,000	17.4–350	multi-male	Avg. 13–15 (5–27)	66	Sussman 1991
Eulemur fulvus rufus (west)	1.0	150	1000	multi-male	Avg. 9 (4-17)	18	Sussman 1991
Eulemur fulvus rufus (east)	90–1,000	960	25	multi-male	Avg. 7 (6–12)	8	Overdorff 1991
Eulemur fulvus fulvus (west)	7 +	—	170	multi-male	Avg. 12	2	Harrington 1975 Ganzhorn 1988
Eulemur fulvus fulvus (east)	—	—	40–60	multi-male	(3–10)	—	Pollock 1979a
Eulemur fulvus mayottensis	—	800	1000	fission-fusion (multi-male "associations")	Avg. 9 (2–29)	—	Tattersall 1977
Eulemur fulvus albifrons	13.0	978	—	multi-male	7–11	1	Vasay 1997
Eulemur fulvus sanfordi	5–9	700–750	85	multi-male	Avg. 6.2	14	Freed 1995
Eulemur mongoz	0.5–1.0	450–750	350	pair bond	Avg. 2.75 (2-4)	8	Tattersall & Sussman 1975 Sussman & Tattersall 1976
Eulemur mongoz	3.0	1000–1400	100	multi-male	Avg. 3.2 (2–4)	4	Curtis 1997
Eulemur macaco	3.5–7.0	750–1000	200	multi-male	Avg. 10 (5–14)	4	Colquhoun 1993
Eulemur coronatus	6–9	750–1000	105	multi-male	Avg. 6.4	15	Freed 1995
Eulemur rubriventer	15–20	70–1020	15–30	pair bond	Avg. 3 (2–4)	12	Overdorff 1991

Table 5-5 *(continued)*

Species	Home Range Size (hectares)	Day Range (meters)	Population Density (per sq. km.)	Social structure	Group Size	N	References
Varecia variegata (Nosy Mangabe)	30	♂1905 ♀1022	29–43	fission-fusion (communities)	8–16	2	Moreland 1991
Varecia variegata (Masoala peninsula)	25	436	—	fission-fusion	5–6	2	Rigamonti 1993
Varecia variegata (Masoala peninsula)	58.0	♂969 ♀1074	—	fission-fusion (communities)	18–31	1	Vasey 1997
Hapalemur griseus griseus	6–15	425	47–62	multi-male	2–6	—	Pollock 1979 a Wright et al. 1987
Hapalemur griseus alaotrensis	2	—	~2.5?	multi-male	2–9	78	Mutschler 1998
Propithecus verreauxi	3–8.5	550–110	110–550	multi-male	Avg. 6 (2–13)	—	Richard 1978 a Jolly et al. 1982 b Richard et al. 1993
Propithecus tattersalli	8–12	460–1077	76–147	multi-male	Avg. 5.6 (3–10)	17	Meyers 1993
Propithecus diadema	100–200	350–850	15	multi-male	(3–9)	3	Overdorff 1991 Wright 1987, 1995 Powzyk 1997
Propithecus diadema	33–42+	1629	13	multi-male	3–8	2	Powzyk 1997
Indri indri	17–18	300–700	8–16	pair bond	Avg. 3.1 (2–5)	18	Pollock 1979 a
Indri indri	34–40	774	7	pair bond	2–3	2	Powzyk 1997
Avahi laniger	1–4	300–620	70	pair bond	(2–5)	—	Albignac 1981 Ganzhorn et al. 1985 Ganzhorn 1988 Harcourt 1991

There is some confusion in the literature as to how to define the spatial relationships of ringtailed lemurs. Using earlier reports from Berenty, Mitani and Rodman (1979) listed ringtails as the only primate species with multi-male groups that is territorial. However, Jolly (1985:151) considered them a "dubious case" for territoriality because groups have highly overlapping ranges. Currently, Jolly et al. (1993) characterize this species as territorial because overlap zones are highly contested, though an average of over 40% of the ranges are shared. I believe that this confusion is related to the lack of a reliable definition of territoriality.

Using the definition given in Chapter 1, i.e., territorial species defend home range boundaries *and* maintain essentially exclusive use of their home range, ringtailed lemurs would not be territorial. There is considerable and in some cases almost total overlap of home range boundaries among ringtailed lemur groups, and as Jolly (1972) has described, a number of groups may time-share sites for feeding, sleeping, or resting. There are core areas which groups use more intensively than others but these can change seasonally, and even core areas are often shared (Sauther and Sussman 1993, Sauther 1998, Sauther et al. in press).

DIET AND FORAGING BEHAVIOR

The diets of *L. catta* and *E. fulvus* were radically different. Both species ate leaves, shoots, flowers and fruits, but the proportion of these items and the number of species eaten differed. The brown lemur fed on a narrow range of food items; the ringtailed lemur, with its diverse vertical and horizontal ranging pattern, fed on a wider range of items. The brown lemur ate only 8 and 11 species of plants at the two forests; a total of 13 different plant species. The ringtailed lemur fed on 24 plant species at both forests, with a total of 45 species eaten, which covered one season at each forest. Over an entire year, two groups of ringtailed lemur at Beza Mahafaly fed on 44 and 50 species of plant (Sauther 1992). At Antserananomby, neither brown or ringtailed lemurs were observed feeding on animal prey. However, they have been observed feeding on small quantities of invertebrates at other sites (O'Conner 1987, Sauther 1992).

For the brown lemur a few species of plants made up a large proportion of the diet and leaves of the tamarind or kily tree (*Tamarindus indica*, the most common species of tree in all three forests) were the main staple. Three species of plant accounted for over 80% of the observed feeding. The *leaves* of the kily alone made up 42% of the diet at Tongobato (studied during the wet season), and over 75% at Antserananomby (studied during the dry season). In fact, during the dry season almost 90% of its diet consisted of leaves (Table 5-6).

The diet of the ringtailed lemur was more varied with eight species accounting for 70%–80% of the diet. In both forests, the dominant kily tree provided 23–24% of the diet but only about half (11–12%) of this was on kily leaves. The pods of the kily accounted for 10–12% of the diet of *L. catta* in the two forests. Indeed, a high proportion of fruit (34%) was consumed by *L. catta, e*ven in the dry season when fruit was very scarce (Table 5-6).

The diets were related to very different patterns of foraging. Brown lemur groups moved little throughout the day and had small home ranges. They rarely came to the ground or crossed open areas. Ringtailed lemurs, on the other hand, moved mainly on the ground and spent most of their time in open areas on the edge of the canopy forest. They moved constantly, and over relatively long distances to obtain food. Ringtails used one part of their home range for 3–4 days,

Table 5-6 Feeding Patterns of Diurnal Species Inhabiting Southwestern Madagascar. Data from Sussman (1977) and Richard (1978).

	Total Number Species Eaten	% Most Eaten Species	Number of Spp. Over 80% of Diet	Wet Season			Dry Season		
				% Fruit and Flowers	% Leaves	% Other	% Fruit and Flowers	% Leaves	% Other
Eulemur fulvus	13	50–75	3	47	50	bark <1	10	90	bark <1
Lemur catta	45	23–24	8	66	24	herbs 6	42	44	herbs 15
Propithecus verreauxi	85	12	26	60–70	<20	bark <20	<20	ca. 80	dead wood 5

then changed to another, covering the whole range in 7–10 days. Even when a feeding site was particularly favorable, the group moved continually while feeding, and movement was not related to the depletion of resources. Ringtails often fed on more plant species in one feeding session than did brown lemurs during the entire study.

The parts of plants eaten and the seasonal variation in the diets of the two species were determined to a great extent by the foraging patterns. The restricted ranging pattern of the brown lemur corresponded to a dependence on the kily tree, and in the dry season the brown lemur was able to subsist on a diet made up almost entirely of mature leaves. The foraging pattern of the ringtailed lemur allowed groups to exploit a number of different resources over a wide area: trees that were in blossom or fruit could be located and utilized, including many resources which were not available to the brown lemur (e.g., those outside the canopy or on the ground). Thus, even during the dry season ringtails had access to and ate considerably more fruit and flowers than did brown lemurs.

Thus, there is an interrelationship between foraging pattern, diversity of diet, and seasonal variation in diet. The ringtailed lemur has a very diverse diet and its foraging pattern ensures a certain amount of variety and a consistent dietary pattern. The brown lemur in these forests, had a restricted diet, but its ability to subsist mainly on mature leaves for part of the year provides it with abundant and predictable resources all year round. Ganzhorn (1985, 1986a,b) found similar patterns of feeding in semi-free ranging, sympatric groups of the same species living in a natural habitat enclosure at Duke University. These feeding patterns appear to be species-specific, and determined by the ecology, distribution, and evolutionary history of the two species.

Even though the foraging pattern of the ringtailed lemur is quite eclectic, at Beza Mahafaly, Sauther (1998) found them to be finely tuned to the seasonality of specific food resources. She also found that several plant species may function as keystone species and may be critical for the survival of this lemur. At Beza Mahafaly, these plant species either provided resources throughout the year (e.g., the kily tree) or produced large amounts of food during short but critical periods when other foods were not available.

Female ringtails depended on different food species which were available during different reproductive states (Sauther 1998). At Beza Mahafaly, early lactation coincides with an initial peak in fruit availability; late lactation/weaning occurred during a second peak in fruit. Young leaves also were available during lactation and weaning. During the dry season, however, the ringtails focus on a few critical plant species, including small pockets of ground herbs, and the leaves, fruit or flowers of a few tree species, including the kily. During the birth season, which occurs at the end of the dry season, only the flowers of two plants and the fruit of one tree (*Salvadora augustifolia*) were available. This is the time when the ringtails in the reserve expanded their home ranges to visit the stand of Salvadora trees for fruit.

These plants and their seasonality and overall production are dependent upon the yearly amount and patterns of rainfall (Sauther 1998). In years when drought occurs in the dry forests of Southern Madagascar, fruit and flower production can be seriously reduced and this has a direct effect on the lemur population. In fact, after two years of drought in 1991-92, the adult population of ringtails at Beza

Mahafaly declined by 31%. The infant mortality reached 80% and the mortality of females reached 29%, with most females lactating at the time of their death (Gould et al. in 1999). Thus, in such seasonal habitats as Southern Madagascar, the reduction of production in keystone species can have an enormous impact on ringtailed lemur demography and survival.

ACTIVITY CYCLES

Although both *L. catta* and *E. fulvus* are diurnal and have morning and evening peaks of activity, they exhibit different overall activity patterns. *E. fulvus* awoke very early in the morning and by 10:00 a.m. was already resting (Figure 5-8). After resting most of the day, they were active again late in the afternoon and often fed until after dark. The brown lemur was not active in direct sunlight and remained in the shade most of the time. *L. catta* awoke later and did not become very active until about 9:00–10:00 a.m. After this, *L. catta* moved out of the center of the forest to various feeding sites and was often active in the open, under direct sunlight. The ringtailed lemur had a short siesta during the hottest portion of the day, which was followed by an intense feeding bout. Then the animals moved back to the center of the forest and settled in a large tree just before sunset. At night, brown lemurs slept in clumps, the group divided into indistinguishable "masses" of 2–5 individuals each. The large group of ringtails slept in one or two subgroups of individuals lined up "choo-choo" fashion on a large branch.

Both species exhibit "sunning" behavior, in which animals sit, spread-eagle, exposing the light undersurface of their stomachs to the direct sunlight (Figure 5-9). This occurs mostly in the morning and especially during cold mornings. Ringtails sun twice as long as brown lemurs and this may be an important thermoregulating mechanism in the former species. Brown lemurs were active much less than ringtails throughout the day and there is some evidence that they may be active, to some extent, during the night. In the southwest, however, nocturnal activity was neither regular nor frequent (Sussman 1972).

SOCIAL STRUCTURE AND ORGANIZATION

Both species live in multi-male groups with approximately 1:1 sex ratios. Brown lemur group size ranges from 4–17 (N=18) animals, and averages 9 animals (Sussman 1975); ringtail groups are larger, ranging from 5–27 individuals (N=83, this includes some groups recensused over more than one year at Berenty and Beza Mahafaly) and averaging 13–15 animals (Sussman 1991) (Table 5-5).

Among groups of ringtails there is a dominance hierarchy (determined by the outcome of agonistic encounters and priority of access to certain resources). There is no noticeable hierarchy in groups of brown lemurs (Sussman 1977, Pereira et al. 1990, Pereira and McGlynn 1997). Among ringtails adult females are dominant to males (Jolly 1966, Taylor 1986, Kappeler 1990, Sauther 1993). Female dominance is rare among primates and mammals in general, but is characteristic of a number of Malagasy lemurs (Kappeler 1993, see below). The hierarchical structure facilitates the division of the large group into subgroups. As the group moves, the females, juveniles, and a central core of dominant males usually move together and ahead of less dominant males. The latter subgroup generally is composed of older juvenile and subadult males, subordinate males, and some males in the process of

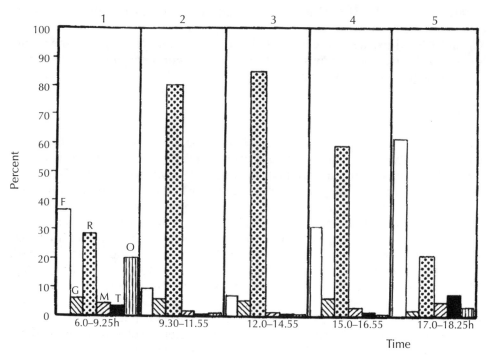

a. *Eulemur fulvus rufus* at Antserananomby.

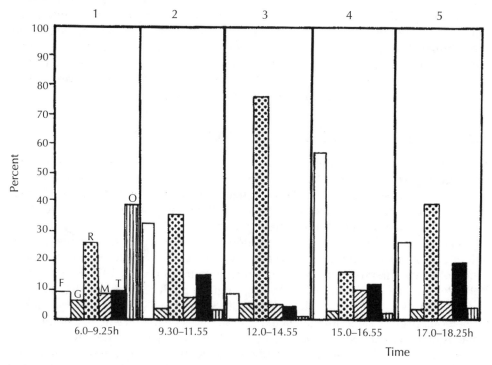

b. *Lemur catta* at Antserananomby.

Figure 5-8 Mean percentage of individual activity records for each of six activities during five periods of the day. F=Feeding, G=Grooming, R=Resting, M=Moving, T=Travel, O=Other.

migrating into or out of the group (see below). These males tend to feed and rest together (Jolly 1966, Sauther and Sussman 1993, Sauther et al. in press).

Like most monkeys of Africa and Asia, ringtailed lemurs display male-biased dispersal and female philopatry (Pusey and Packer 1987). That is, males migrate between groups and most females remain in their natal group and home range throughout their lifetime. Since females remain in the same group, they tend to interact more with close female relatives throughout their lives, and dominance and affiliative patterns are formed around matrilines (Taylor

Figure 5-9 *L. catta* sunning. [Photo by Robert W. Sussman]

and Sussman 1985, Taylor 1986). Subgroups form between more closely related females within the group. Kinship is an important component in group life of primates, and various aspects of this will be discussed throughout the book.

Males first migrate from their natal group between 3–5 years of age. Older males also transfer between groups on the average of every 3.5 years. Males form partnerships while transferring although these affiliations do not usually continue once group membership is attained (Sussman 1992, Gould 1994, 1997). Over his lifetime a male may live in a number of groups, and his dominance status and affiliative patterns change also. In all cases of male transfer, the males hold low-ranking and peripheral positions upon entering a new group and are challenged by other group members of both sexes for many months (Gould 1997). At Beza Mahafaly, after 6 years of study of 9 ringtail groups, only 7 adult males of 43 (16%) were still in their original group, and only 14 males (33%) were still in the census population, whereas 76% of the females were still in their original group (Sussman 1992). After 10 years, only one original male was still in the study population but not in his original group. By this time 5 of the 29 originally marked females were still in the population (many had died during the drought mentioned above), all in their original group (Sussman unpublished data). Migration is part of the mating strategy of the males.

In all group-living primate species studied to date, individuals of one or both sexes migrate between groups during some stage of their lives. Generally, in mammals males show a much greater tendency to leave their natal group and to transfer between groups later in life. Fewer gregarious species exhibit female biased dispersal or dispersal by both sexes, though this is not generally so for primates (see Strier 1994). The proximate causes for migration are still not well understood, nor are the factors that determine which sex migrates. However, ultimately dispersal ensures genetic interchange within the total population and thus is one mechanism which helps to maintain species coherence (see Pusey and Packer 1987). Theoretical issues related to kinship, dominance, and patterns of migration will be discussed in Volume II and Volume III.

Ringtail groups frequently divide into subgroups which separate while moving, foraging, or resting. This is facilitated by dominance hierarchies and by matrilineal kinship. Groups of brown lemurs, on the other hand, are very cohesive and regular subgroups are not seen. As yet, the dispersal patterns of western Madagascar brown lemurs is unknown and we know nothing about their kinship relationships.

As in all Lemuriformes studied to date except the aye-aye and *Hapalemur,* mating and births in *L. catta* and *E. fulvus* are seasonal and restricted to a very brief period of the year. The mating season lasts for about 1-3 weeks, however, during this time each female is receptive for no more than 24 hours. Among ringtails, females within each group enter into estrus asynchronously and thus each female within a group is receptive on different days during the mating season (Pereira 1991, Sauther 1991a). During the short breeding period, there is a great increase in male activity and in male-male fighting as males compete for estrus females. Jolly (1966) observed agonistic encounters to rise in frequency from 1-2 per hour to 21 per hour. In brown lemurs, during the mating season, there is an increase in the frequency of scent-marking, but agonistic encounters do not increase in frequency or intensity (Sussman and Richard 1974, Harrington 1975). Little is known about mating behavior in the brown lemurs of Western Madagascar.

A ringtailed lemur adult female mates with a number of males during her one day estrus period. Top ranking males usually are first to mate (Sauther 1991a) but this is not always the case (Koyama 1988, Gould 1994). Female choice of mating partners is paramount (Van Horn and Resko 1977, Taylor 1986, Sauther 1991a). Despite vigorous fighting between males, no male is successful in mating without the female's acquiescence. Females do not mate with natal males (Taylor 1986). However, besides mating with dominant males, they mate with subordinates, with migrating males, and also with males from other groups who "roam" during the mating season. In fact, during the most detailed study of mating behavior, of the 17 matings observed, 35% involved mating with males from other groups in spite of vigorous harassment by group members. Furthermore, initial approach by females occurred in approximately half of the matings (Sauther 1991a, Sussman 1992). Given these data, Sauther (1991a) has derived hypothetical mating strategies for both male and female ringtailed lemurs. These strategies are depicted in Figure 5-10.

The gestation period in captive *L. catta* and *E. fulvus* is from 120–135 days (Table 5-7). However, at Beza Mahafaly, the gestation period for ringtails was somewhat longer, 136–144 days (mean=141) (Sauther 1991a). Births occur at each site during a one month period between August and October (depending on locality in Madagascar). Single births are common but twinning occurs about 4% of the time in *E. fulvus* and 16% in *L. catta* (Tattersall 1982). From birth, infants are carried by the mother (the infant actually clings to the mother's fur) (Figure 5-11). Ringtailed lemur infants are much more precocious than infant brown lemurs, moving from the ventral surface of the mother to her back after one week and locomoting independently by 10–12 weeks of age. These events occur at 4 weeks and 14–16 weeks, respectively, in *E. fulvus* (Sussman 1977, Harrington 1978, Klopfer and Boskoff 1979). Among captive old world monkeys, it has been shown that infants of arboreal species become independent of their mothers at a later age than do infants of terrestrial species (Chalmers 1972). At Beza Mahafaly, 80–85% of females give birth annually, and 30–50% perish each year (Sussman 1991). Although more needs to be done on this subject, the causes of death appear to be

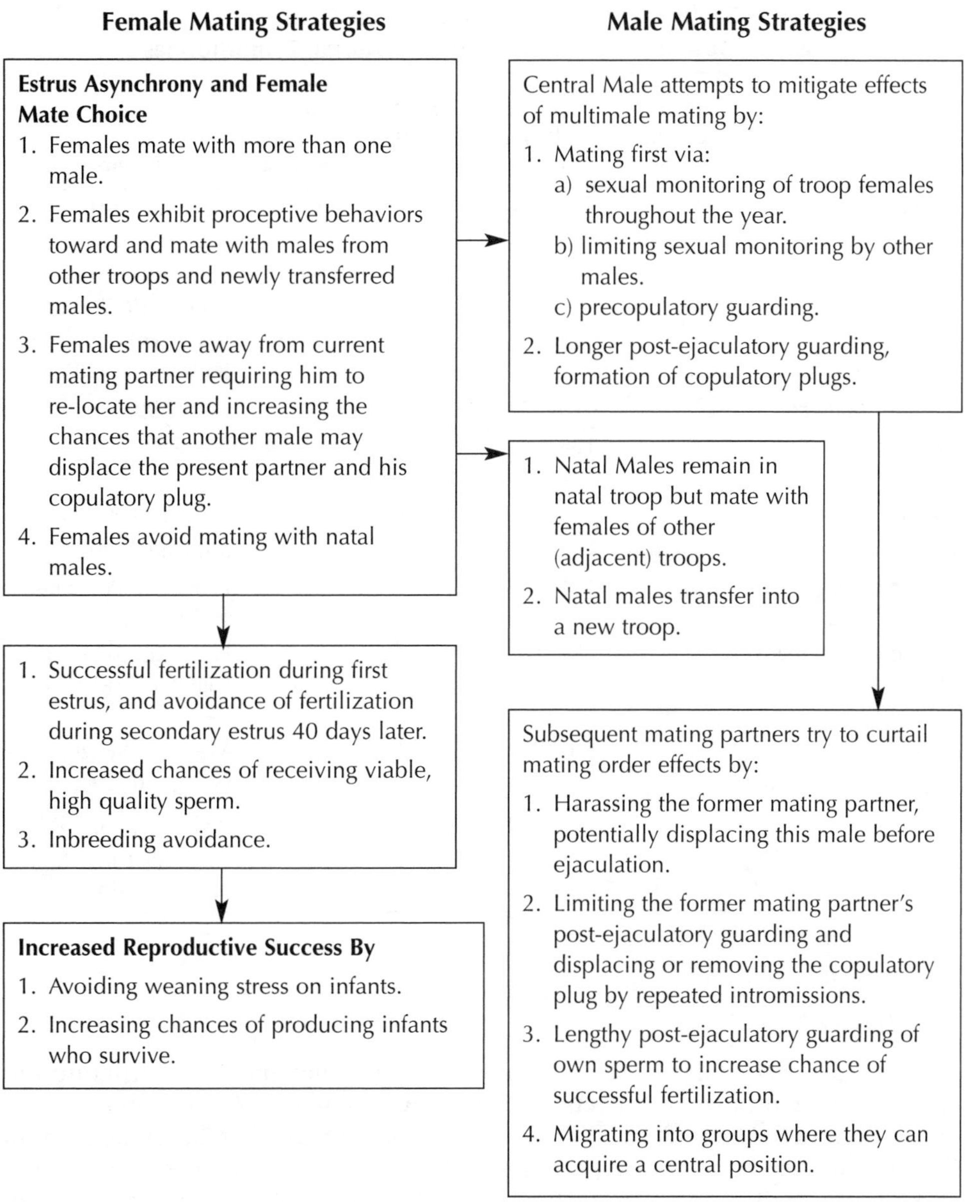

Female Mating Strategies   **Male Mating Strategies**

Estrus Asynchrony and Female Mate Choice
1. Females mate with more than one male.
2. Females exhibit proceptive behaviors toward and mate with males from other troops and newly transferred males.
3. Females move away from current mating partner requiring him to re-locate her and increasing the chances that another male may displace the present partner and his copulatory plug.
4. Females avoid mating with natal males.
	Central Male attempts to mitigate effects of multimale mating by:
	1. Mating first via:
	  a) sexual monitoring of troop females throughout the year.
	  b) limiting sexual monitoring by other males.
	  c) precopulatory guarding.
	2. Longer post-ejaculatory guarding, formation of copulatory plugs.

	1. Natal Males remain in natal troop but mate with females of other (adjacent) troops. 2. Natal males transfer into a new troop.
1. Successful fertilization during first estrus, and avoidance of fertilization during secondary estrus 40 days later. 2. Increased chances of receiving viable, high quality sperm. 3. Inbreeding avoidance.	Subsequent mating partners try to curtail mating order effects by: 1. Harassing the former mating partner, potentially displacing this male before ejaculation. 2. Limiting the former mating partner's post-ejaculatory guarding and displacing or removing the copulatory plug by repeated intromissions. 3. Lengthy post-ejaculatory guarding of own sperm to increase chance of successful fertilization. 4. Migrating into groups where they can acquire a central position.
Increased Reproductive Success By 1. Avoiding weaning stress on infants. 2. Increasing chances of producing infants who survive.	

Figure 5-10 A model of mating strategies for ringtailed lemurs [modified from Sauther and Sussman, 1993]

mainly predation, parasites, illness, falls, and during some years the effects of drought (Gould et al. 1999). This rate of infant mortality is not unusual among free-ranging primates. It has been suggested that infanticide may be a major cause of infant death among ringtails, however, there is little evidence to support this contention (see Sauther et al. in press and section on infanticide). In fact, unrelated males exhibit an interest in infants, engage in affiliative behavior with them, and provide alloparental care (Gould 1992, 1996, 1997).

Table 5-7 Gestation Periods, Litter Sizes, and Ages of Sexual Maturation in Various Diurnal Lemuriformes

Species	Cycle Length in Days	Gestation Period in Days	Modal Litter Size	Female Age of Sexual Maturation (months)	References
Lemur catta	39	141 (136–144)*	1	36*	Sussman 1991 Sauther 1991
Eulemur fulvus (captive)	30	120–135	1	20–24	Petter-Rousseaux 1962 Boskoff 1978
Eulemur macaco (captive)	33–34	125–129	1	20–24	Bogart et al.1977 Balke et al. 1988
Eulemur mongoz	31	125 (121–129)	1	18	Clark 1993
Eulemur coronatus (captive)	34	125	1	20–24	Kappeler 1987
Eulemur rubriventer			1		
Varecia variegata	30–44	102.5 (99–06)	2	18-20	Moreland 1991
Hapalemur griseus		140	1	—	Petter and Peyrieras 1970
Propithecus verreauxi		150–162	1	60*	Petter-Rousseaux 1962 Eaglen & Boskoff 1978 Richard et al. 1991
Propithecus tattersalli		@180*	1	—	Meyers & Wright 1993
Propithecus diadema		179*	1	48*	Wright 1995
Indri indri		120–150*	1	210?*	Pollock 1975, 1977
Avahi laniger		155?	1	—	Tattersall 1982 Harcourt 1991

*Data from wild populations

SUMMARY

Differences were found in the utilization of space, the diet, the social structure and organization and in some aspects of infant development of *L. catta* and western populations of *E. fulvus*. These differences are related to the habitat preferences of the two species, and are independent of the presence or absence of the other species. In a study of semi-free ranging populations of the same species, Ganzhorn (1985, 1986a) found a number of behaviors that paralleled those found in natural populations. These studies indicate that the differences are not caused by the interaction between the populations but by phylogenetic adaptations to environmental conditions which occur where the species are allopatric.

   L. catta is a highly terrestrial, frugivorous species. It is the only extant diurnal species of lemur inhabiting the dry brush and scrub vegetation of the south and southwest of Madagascar and utilizing the dry, rocky and mountainous areas of the south where only patches of deciduous forest remain. The ranging and foraging pattern of *L. catta* is related to its ability to cope with these semi-arid environments in which resources are sparse and unevenly distributed. The population of *E. fulvus* living in Western Madagascar on the other hand, is arboreal and highly folivorous and is able to subsist mainly on mature leaves during certain times of the year. It is adapted to exploit resources which are abundant and uniformly distributed. The western brown lemur shows ecological and behavioral adaptations which conform to those found in many arboreal, leaf-eating species of New and Old World monkeys, whereas adaptations of the ringtailed lemur compare to those of many terrestrial primates living in forest edges (ecotones).

Propithecus verreauxi: ONE SPECIES LIVING IN TWO DIFFERENT HABITATS

Propithecus verreauxi (Verreauxi's sifaka), the third species of diurnal lemur in the west and south, is larger than *Lemur catta* and *Eulemur fulvus* (weighing 3–5 kg). It is a vertical clinger and leaper (see Chapter 3) (Figure 5-12). This mode of locomotion is characteristic of the Indriidae, as are morphological and physiological adaptations for a folivorous diet. These include: large salivary glands, capacious stomach and long, convoluted caeca (Hill 1953).

Figure 5-11 *L. catta* infant. [Photo by A. Schilling.]

The first long-term study of Verreauxi's sifaka was conducted at Berenty by Jolly (1966), and research has continued at this site (e.g., Jolly et al. 1982a,b, O'Conner 1987, Jolly et al. 1993). In 1971–1972, Richard (1974, 1978a,b, 1985) did a comparative study of two populations of *P. verreauxi* living in vastly different habitat types. One study site (Ampijoroa) is a deciduous forest found in the northwest; the other site, a desert-like forest in the south (Hazafotsy) (Figure 5-13). In her study, Richard focused on the way in which this species adapts to vastly different habitats. More recently, Richard has led a team of researchers at the Beza Mahafaly Reserve, focusing on the demography and social system of the sifaka.

METHODOLOGY

In her earlier study, Richard employed a modification of the scan-sampling method. Rather than scanning the behavior of all individuals, she recorded the behavior of a single individual each minute. This technique is known as focal animal sampling (Altmann 1974, Lehner 1996). It is a strenuous and difficult method but, when possible, is one of the most productive methods of

Figure 5-12 Leaping *Propithecus*. [Photo by Robert W. Sussman]

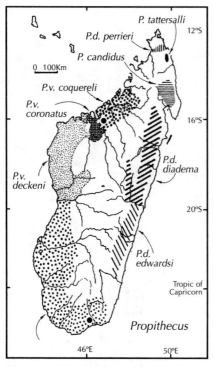

Figure 5-13 Map of *Propithecus*. [From Mittermeier, et al., 1994] ●=northern and southern study sites of Richard.

data collection. Because of the small group size, the ability to individually recognize animals and to follow an individual almost continually throughout the day, Richard was successful in employing this methodology.

Data were collected on two adjacent groups in each forest; with seventy-two hours collected for each group, in each month, evenly distributed in time and between different age/sex classes. Observations were made for three months during the wet season and three months during the dry season in both study sites. Thus, Richard could determine variation in behavior regionally (between groups in the different forests), seasonally, and locally (between groups in the same forest). At Beza Mahafaly, since 1984, Richard and her colleagues have captured, marked and censused animals from 43 social groups to determine population and group structure, patterns of dispersal and social behavior (e.g., Richard et al. 1991, 1993, Kubzdela 1995, 1997, Brockman 1994, Brockman et al. 1995, 1998). Comparative studies of social organization are currently in progress at this site.

VERTICAL AND HORIZONTAL RANGING

Because of its VCL mode of locomotion, the sifaka moves through the forest mainly on the vertical trunks below the canopy (level 3) or high in the emergent layer (level 5, Table 5-4, Figure 5-5). This mode of locomotion also is ideal for travel in the desert-like didierea forests of the south. *P. verreauxi* is the only diurnal lemur that is common in didierea forests. The anatomical adaptations associated with VCL locomotion, and the relatively large body size of the sifaka, were once thought to restrict this species to a relatively narrow range of arboreal supports and postures (Napier and Walker 1967). However, *Propithecus* is very agile and uses an amazing array of postures. It is completely at home in the terminal branches of the canopy; over 70% of its feeding takes place in small branches and twigs (Richard 1978a). Most feeding on small supports is done using suspensory postures, and sifakas climb a great deal during feeding (Vasey 1992). Richard found that substrate use and time spent in different postures in each study area was extremely similar despite the considerable differences in the physical structure of the study forests.

At Ampijoroa and Hazafotsy, home ranges of groups were 6.75–8.50 ha. Home ranges were determined by noting 50 meter quadrants in which the animals entered (Figure 5-14). This method overestimates home range size because groups may not use the whole quadrant. However, even given this overestimate, the home ranges of these groups were larger than those studied

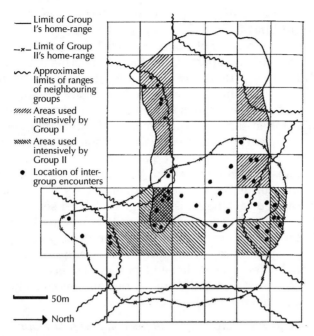

Figure 5.14a Areas of intensive use, areas of exclusive use, and the location of inter-group encounters within the home-range of the northern study groups. [From Richard, 1978a]

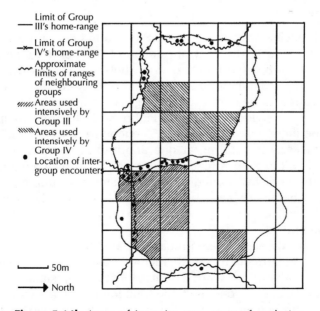

Figure 5.14b Areas of intensive use, areas of exclusive use, and the location of inter-group encounters within the home-range of the southern study groups. [From Richard, 1978a]

by Jolly (1966, Jolly et al. 1982b) at Berenty (3 ha) and by Richard et al. (1993) at Beza Mahafaly (4 ha).

Much like ringtails, groups of sifakas followed the same general route for 3–4 days and moved to a different part of their range, covering the total range in 10–20 days (Jolly 1972, Richard 1977). By monitoring the amount of time each group spent in each 50 m. quadrant, Richard could investigate the way groups used various portions of their range. Groups used some parts of the range more than others; at least 60% of their time was spent in only 10 quadrants, representing about 20% of the home range. However, there was no significant difference between the types of activity performed in different quadrants.

Northern groups moved further each day than the southern groups, and day ranges were longer during the wet season than the dry season. In the north, the mean distance traveled each day was 1,100 m. in the wet season and 750 m. in the dry season. In the south the day ranges were 1000 m. and 550 m. in the wet and dry seasons. Home ranges were not expanded appreciably during the wet season, the animals just moved more rapidly through the range. These differences were related to seasonal differences in diet and food availability.

At Berenty, groups defended the boundaries of their small home ranges. Jolly describes ritualized defense in sifakas as follows:

> *Territorial "battles" are formal affairs composed of leaping, staring, and scent-marking with very low growls or in silence. . . . A territorial battle may be very beautiful, since everything depends on a fast, formal pattern of movement, each animal occupying sections of tree rather than opposing individuals of the other troop. Each tense leap, therefore, carries an attacker toward a particular undefended area of tree, not into contact with an enemy. The troops, therefore, move in reciprocal formations (1966:49–50).*

Ritualized defense was observed by Richard in her southern site but not in the north. In the south, the central portion of the home range of each group formed an exclusively used block; overlap between ranges was minimal. Outside of the mating season, 20 of 29 encounters between groups culminated in ritualized "battles." In the north, home ranges overlapped extensively and areas of exclusive use were widely scattered (Figure 5-14). Intergroup interactions did not occur at boundaries and only 18 of 59 encounters were ritualized "battles." In most instances, after staring at one another, scent-marking and growling, groups moved off in opposite directions. As in brown lemurs, encounters between groups here functioned as mechanisms maintaining group integrity and cohesion, whereas those between groups in the southern site can be considered "territorial" encounters, since they involved the actual defense of an exclusive home range.

These regional differences are probably related to resource distribution in the two forests. At Ampijoroa in the north, Richard postulates that home range size is related to the patchy and scattered distribution and small size of the food resources. However, at any one time, the total food available was in excess of the requirements of any one group, thus allowing extensive overlap between groups. In the desert-like vegetation of the south, she believes that food was limited toward the end of the dry season, and the year-round territoriality was an adaptive response to the minimum foraging area required at times of reduced food availability (Richard 1978a).

DIET AND FORAGING BEHAVIOR

P. verreauxi has a diverse diet. The two northern groups ate 85 and 98 different species and the two southern groups ate 65 and 77 species. Sifakas had similar dietary patterns in both study areas, regardless of the extreme differences in vegetation type and species composition. The animals fed primarily on a few species, despite the overall diversity of the diet: all four groups spent over 65% of total feeding time eating 12 species (Table 5-6), and few species were eaten more than 1% of the time (Table 5-8). Thus, the animals spent a large amount of time (35%) eating a very small amount (<1%) of many different species.

Table 5-8 Twelve Species on Which Each Group Spent Most Time Feeding and Percentage of Total Feeding Time Spent Eating Each

Food Species	Time Spent %	Food Species	Time Spent %
Group I (North)		Group II (North)	
Drypetes sp. no. 1*	12.4	*Drypetes* sp. no. 1	11.7
Cedrelopsis sp. no. 1	5.4	*Cedrelopsis* sp. no. 1	8.9
Combretum sp. no.1	5.3	*Bussea perrieri* R. Vig.	4.5
Dead wood	5.3	*Commiphora pervilleana* Perr	4.4
Capurodendron microlobum (Baker)		*Bathiorhamnus louveli* Perr	3.9
Aubreville	5.1	Dead wood	3.9
Rheedia arenicola Jerm and Peff	4.5	*Rhovalocarpus simitis* Hemsley	3.7
Commiphora pervilleana Perr	4.2	*Combretum* sp. no. 1	3.6
Apaloxylon madagascariense Drake	4.0	Liana sp. no. 1	3.3
Liana sp. no. 1	3.7	*Boscia* sp.	3.1
Protorhus deflexa Perr	3.4	*Protorhus deflexa* Perr	2.4
Hippocratea sp.	3.3	*Capurodendron microlobum* (Baker)	
Mundulea sp.	2.9	Aubreville	2.3
Group III (South)		Group IV (South)	
Terminalia sp. no. 1	21.5	*Terminalia* sp. no. 1	21.0
Mimosa sp. no. 1	16.8	Liana sp. no. 2	15.4
Liana sp. no. 2	12.3	*Mimosa* sp. no. 1	11.2
Grewia sp. no. 2	7.7	*Grewia* sp. no. 1	9.9
Terminalia sp. no. 2	5.0	*Hagunta modesta* (Baker) M.	
Diospyros humbertiana Perr	4.1	Pichon	4.8
Grewia sp. no. 1	5.2	*Terminalia* sp. no. 2	4.3
Euphorbia plagiantha Drake	2.3	*Grewia* sp. no. 2	2.6
Liana sp. no. 4	2.1	Liana sp. no. 3	2.5
Liana sp. no. 3	2.1	*Diospryos humbertiana* Perr	1.9
Hagunta modesta (Baker) M.		*Rothmannia decaryi*	1.6
Pichon	1.8	*Commiphora* sp. no. 1	1.6
Liana sp. no. 5	1.7	*Grewia* sp. no. 3	1.5

* Where species identification could not be made, samples were numbered serially.

[Adapted from Richard, 1977]

The tendency to feed on a few species was more pronounced in the south than in the north (Table 5-8). In both study sites, the number of food species eaten declined in the wet season. This trend was more pronounced in the south. Furthermore, within each study area, the composition of the diet changed almost completely between seasons (Table 5-6). The only plant species that were eaten by sifakas throughout the year were those that produced leaves continuously: animals ate both mature and immature leaves of these plants. Plants that produced flowers and fruit during short periods and that shed their leaves during the dry season dominated the diet for only short periods.

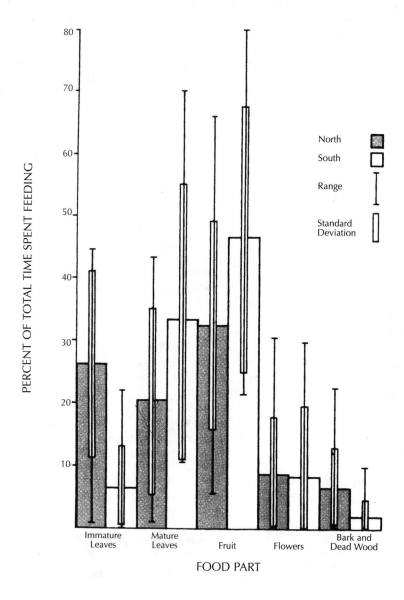

Figure 5-15 Amount of time spent eating different food parts, irrespective of season, by the groups in each study area. [From Richard, 1978a]

Although *Propithecus* has morphological adaptations for feeding on leaves, a large proportion of the year-round diet was composed of fruit; @ 30% in the north and 50% in the south (Figure 5-15). However, the food parts eaten changed seasonally and closely mirrored phenological changes in the plants. Generally, flowers and fruits made up the greatest portion of the diet in the wet season, young leaves did so during the beginning of the wet season, and mature leaves were eaten mainly during the dry season. In fact, mature leaves accounted for @ 70% of the diet in the south in the dry season. Some local differences (between neighboring groups) were observed, primarily in species composition of the diet. In both study sites, only eight of the twelve most commonly eaten species were the same in both groups. This may have been because of differences in plant availability and distribution, or differing traditions between groups. A small proportion of the diet was made up of bark and dead wood. It is likely that the animals obtain water or certain nutrients from these foods. Tooth combs and premolars were used to exploit these resources.

In sum, *P. verreauxi* is morphologically specialized for a diet of leaves (see Yamashita 1998). It has a highly variable diet and spends a large proportion of time feeding on small quantities of many species. In the dry season, Verreauxi's sifaka can subsist almost entirely on mature leaves. However, unlike brown lemurs whose diet becomes more monotonous in the dry season, the sifakas diet becomes more diverse at this time.

ACTIVITY CYCLES

Daily activity cycles of *P. verreauxi* differed seasonally. During the dry season, activity began 1–2 hours after sunset with sunning behavior, which sometimes lasted for over an hour. They then fed until early afternoon, before entering large sleeping trees for the night. In the hot, wet season, the animals usually fed before dawn and began a prolonged siesta by mid-morning. They began feeding again in the late afternoon and fed until well after sunset.

SOCIAL STRUCTURE AND ORGANIZATION

As in *Lemur catta* and *Eulemur fulvus*, *Propithecus verreauxi* lives in multi-male groups, though *P. verreauxi* groups are generally smaller. Group sizes range from 2–14 individuals, with most groups having 4–6 members (Richard 1978a, Jolly et al. 1982b, Richard et al. 1993, Kubzdela 1995, 1997). Sifaka groups appear to be fairly fluid foraging parties consisting mainly of philopatric females, young natal males, and frequently transferring adult males (Jolly 1966, Richard 1974, Jolly et al. 1982b, Richard et al. 1991). The sex ratio of groups vary widely, as does individual male membership. Male intergroup transfer occurs frequently, with males usually transferring between neighboring groups with adjacent boundaries. At Beza Mahafaly, where individuals are identified, males left their natal group between 3 and 6 years of age, and were likely to transfer two to three times during this period. Transfer rates dropped among 7–8 year olds and fluctuated thereafter. There appeared to be no particular seasonality to male transfer (Richard 1992, Richard et al. 1993).

Young adult males were not forced out of their group and transferred slowly between groups, spending some time in both the old and new group. Sometimes older males were evicted by females and other males. Younger males often transferred into groups with a higher proportion and number of females than their group of origin. Success in entering a new group depended upon a number of factors, and seemed to be related to previous knowledge of the approaching male by members of the new group. Some males were able to join females and establish new groups. It appeared to be more difficult for older males to successfully transfer.

In fact, at Beza Mahafaly, Kubzdela (pers. comm.) observed that all of her focal groups had at least one "prime" male with fully developed throat glands. Less developed males associated with one another and sometimes were able to transfer between groups without fighting. Prime males, however, fought one-another and only one successful transfer of a prime male occurred in 19 months of study. Richard (1992, Richard et al. 1993) believes that the mechanisms used by males, and outcome of their dispersal is determined by age, condition, kinship network, and previous history in the neighborhood. Some female migration also occurs and this appears to be related to the size of the group and the age of the females. Kubzdela (1997) found that, in larger groups with 4 or more females, the younger females were likely to attempt to migrate. These females would attempt to establish a new group, or would enter groups with fewer adult females.

Groups tend to contain more males than females (Richard et al. 1991, Kubzdela 1997), and this is unusual among primates. In most groups of monkeys and apes, there are more females than males. The sex ratio of males to females is referred to as the *socionomic sex ratio* and it is normally expressed as the fraction of females to males. In five years at Beza Mahafaly, the average ratio was 0.45 for sifaka groups. However, this ratio varied greatly from group to group, and group membership, especially of adult males, was not stable (Richard et al. 1991, 1993).

At any one time, *P. verreauxi* group members move cohesively and, although a dominance hierarchy exists in relation to agonistic encounters over access to food, this hierarchy does not pervade other aspects of behavior. In fact, there are few social interactions outside of the mating season; so few that it is difficult to define a dominance hierarchy (Richard 1992). For example, Richard found no correlation between the feeding hierarchy and ranks established according to frequency of aggression, direction or frequency of grooming, or preferential access to females during the mating season (Richard 1978a). As in ringtailed lemurs, however, adult females are always dominant to males, and it appears that older females may be dominant to younger ones (Kubzdela 1997).

Mating season at any one site lasts from approximately 10 to 21 days, with individual females being receptive for 36 to 42 hours (Jolly 1966, Richard 1978a). There are increases in scent-marking by adult males, intragroup fighting, intergroup encounters, and "roaming" by adult males. At Beza Mahafaly, Richard (1992) observed two distinct patterns among groups during the mating season. In some groups, males from other groups visited, and males engaged in fights and chases. In these groups, estrus females sometimes mated with the winner, and other times with a non-resident and non-combative male. Fights between males can be quite severe at this time (Jolly 1966, Richard 1992). Females can influence the occurrence and intensity of fights by not mating with group males and precipitating fights in a number of ways (Richard 1992). In some groups, however, no visiting males were seen, and there was no increase in aggression.

Jolly (1966) and Richard (1978a, Richard et al. 1993) have referred to sifaka study populations as socioreproductive *neighborhoods* because of the frequency of intergroup transfer by males, the fluidity of groups, the frequent intergroup encounters, the visits by males between groups, and the regularity of mating by non-group males. This is a very useful concept for many primate species, since individuals in different groups within a population usually know one another to a greater or lesser degree. This knowledge affects many aspects of the behavior of both the individual and the group.

The gestation period of *P. verreauxi* is around 150–162 days (Petter-Rousseaux 1962; Eaglen and Boskoff 1978) and females usually give birth to one infant. The mother and young infant provide a focal point for the group and the frequen-

Figure 5-16 *Propithecus* with infant. [Photo by Robert W. Sussman]

cy of grooming increases during the birth season (Jolly 1966; Richard 1976). The infant is first carried on the belly and then on the back of the mother for six or seven months (Figure 5-16). At Beza Mahafaly, females did not reproduce until they were 5 or 6 years old. This delayed reproduction is quite unusual for prosimians and for relatively small primates. Male sifakas have been observed to mate at 3 years, though most first mate at 6 years of age (Richard et al. 1991).

Reproduction is an energetically costly period for females, and it may be especially so for *P. verreauxi*. Females may give birth to infants yearly, but most successful births occur at 24 month intervals. At Beza Mahafaly, on average 60% of the infants survived to one year but over a four year period, this rate varied from 33–80%. There is a complex relationship, within sifaka groups, among female age, dominance rank, and reproductive success (Kudzdela 1997). In 11 years at Beza Mahafaly, no female lower than 3rd highest in dominance rank has given birth. Furthermore, very few females under six years of age give birth and those between 7 and 9 years old have fewer offspring than do females 10-24 years old. However, since rank in this study was directly related to age, 4th ranking females also would be in these younger age categories. Furthermore, all females that attempted to emigrate in Kubzdela's study were 3-8 years old and all were among the youngest females in groups with many females.

In sum, *Propithecus verreauxi* occupies a unique niche in Southwestern and Southern Madagascar and differs, in a number of ways, from *Lemur catta* and *Eulemur fulvus*. All three species live in multi-male groups with promiscuous mating, however, *P. verreauxi* has smaller, foraging groups, which frequently change in male membership and in sex ratio. The sifaka is morphologically adapted to feed on leaves, yet its diet is much more diverse than the more monotonous diet of western *E. fulvus*. *L. catta* is generally frugivorous throughout the year. These three species also have different locomotor adaptations and abilities, and move in different forest strata. It is these adaptations which are probably most related to their distribution in different forest types in regions where their populations overlap.

All three species coexist in the rich deciduous, canopy forests of Southwestern and Southern Madagascar. However, *E. fulvus* is strictly arboreal and is found only in canopy forests. *L. catta* is highly terrestrial and can also exploit forest edges, brush and scrub forests common to this region, as well as mountainous rocky regions of the central plateau. These areas necessitate terrestrial locomotion. The vertical clinging and leaping of *P. verreauxi* is a sufficient mode of locomotion for the canopy forests, but it is also a necessary adaptation for exploiting the desert-like, didierea forests of the south; *Propithecus* and the nocturnal *Lepilemur* are the only lemurs commonly found in these types of forest.

OTHER DIURNAL SPECIES OF THE OCCIDENTAL VEGETATION ZONE

Moving from south to north in the occidental zone of Madagascar, the amount of rainfall increases, there are large areas of rich deciduous forest, and a larger number of vegetation types. The rich forest vegetation was even more widespread and diverse in the recent past, before the arrival of man (Burney and MacPhee, 1988, Burney et al. 1991, 1997), and deforestation is occurring at a very rapid pace in this region. On the northwest coast there is one small area, called the Sambirano Domain, containing rain forest similar to that found in the east but with a pronounced dry season lasting for three to four months. The northern tip of Madagascar is extremely arid except for Mt. d'Ambre which lies in the center of this region and is covered with evergreen rain forest.

The diversity of lemurs in the northwest is greater than that of the southwest. *Lemur catta* is not found here but the ranges of *Propithecus verreauxi* and *Eulemur fulvus* extend to the north of the island. In fact, there are seven subspecies of *E. fulvus,* each with a different pelage coloration (Figure 5-17). Three of these are found in the northwest and north, and one on the Comoro Island of Mayotte (Tattersall 1992a, Tattersall and Sussman 1998). There are three other species of *Eulemur, E. mongoz* (the mongoose lemur), *E. macaco* (the black lemur), and *E. coronatus* (the crowned lemur), which replace one another as one moves from the northwest to the extreme north of Madagascar (Figure 5-18), though each of these is sympatric with *E. fulvus. E. macaco* is the same size as *E. fulvus* (@ 2,000–2,500 gm.), and is similar in general morphology (Figure 5-19); *E. coronatus* and *E. mongoz* are slightly smaller (@ 1,000–1,500 gm.) (Tattersall 1982, Smith and Jungers 1997,

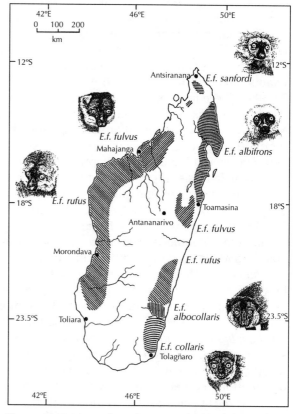

Figure 5-17 Map of Madagascar showing approximate areas of distribution of the subspecies of *Eulemur fulvus.* [From Tattersall and Sussman, 1998, with modified nomenclature]

182

Rasmussen pers. comm.). *E. mongoz* also is found on the islands of Anjuoan and Dzaoudzi in the Comoro Islands.

The only population of the newly discovered species *P. tattersalli* also occurs in Northern Madagascar. Furthermore, there are localized populations of *P. diadema,* of the bamboo specialist, *Hapalemur griseus,* and of the nocturnal indriid *Avahi laniger.* Since the ranges of these three species are more extensive in the east, I will discuss their ecology in the next section.

Until recently, studies of *E. fulvus, E. mongoz, E. macaco,* and *E. coronatus* in Northern Madagascar were of quite brief duration. However, now, relatively long-term research has been conducted on *E. macaco* (Andrews in prep., Colquhoun 1993, 1995, 1998, Birkinshaw 1995, Andrews and Birkinshaw 1998), *L. mongoz* (Curtis 1997, Rasmussen

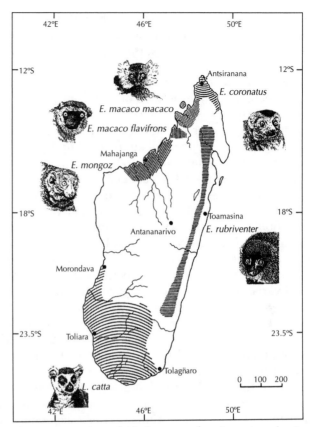

Figure 5-18 Map of Madagascar showing approximate areas of distribution of the subspecies of *Eulemur,* excluding *E. fulvus.* [From Tattersall and Sussman, 1998, with modified nomenclature.]

1998a, b, in prep.) and on sympatric populations of *E. fulvus* and *E. coronatus* (Freed 1996). Tattersall (1976, 1977) studied the Comoro populations of *E. mongoz* and *E. fulvus. P. tattersalli* has been the subject of a 13 month study by Meyers (1993).

Bird watchers usually look for colorful and unusual birds and often are unimpressed with the many little brown, nondescript species found in a diversity of habitats, filling many different niches. They often refer to these birds as little brown jobs (LBJ's), while going to great lengths to specifically identify other birds. The four species of *Eulemur* in Northern Madagascar could be called the LBJ's of the lemur world. They are all brownish or black, quite similar in size and morphology, relatively small, mainly frugivorous, arboreal quadrupeds, and are quite adaptable. As Tattersall (1992b: 33) notes, if we look at *Eulemur*

Figure 5-19 *E. macaco.* [Photo by Robert W. Sussman]

in its various guises simultaneously with the other species in its genus, we find representatives in virtually all forested regions of Madagascar except the extreme south. At least superficially, they appear to be relatively nondescript, and similar in behavior and ecology.

How do these closely related species coexist? Is this the result of millenia of stable ecological interactions or of recent population movements, with competition occurring between the various pairs? Are these species so adaptable that no predictable patterns of habitat preferences can be found between pairs? Preliminary research indicates that subtle but important distinctions exist between these species. Even though each species is quite adaptable, differences are found between coexisting pairs, especially in strata of the forest used, in specific food items chosen, and in activity cycle. There also are differences in social structure and organization (Tattersall and Sussman 1998).

Populations of brown lemur (both *Eulemur fulvus rufus* and *E. f. fulvus*) living in Northern Madagascar have not been the subject of detailed study. However, as the study conducted by Tattersall (1977) on brown lemurs in Mayotte indicates, various populations and subspecies of this taxon may show a great deal of behavioral variation (see also Harrington 1975, Vasey 1997a). Although Mayotte lemurs are canopy dwellers, and have substrate preferences and locomotor habits similar to those of brown lemurs in the west, other aspects of their behavior are quite different. Rather than being mainly folivorous, the Mayotte lemur is highly frugivorous and has a relatively diverse diet. It also has a mean day range of 800 meters, similar to that of ringtailed lemurs (see above).

Even though Mayotte lemurs are highly gregarious, individuals do not live in stable groups; there seems to be a constant shuffling of individuals into fluid and temporary "associations." These associations constantly change in size and composition, even during the course of a day. Interestingly, the average size of these associations was 9.1 individuals and the density of the study population was 1000 per sq. km., in both cases figures similar to those found for brown lemurs in the southwest. At Ampijoroa, Northern Madagascar, Harrington (1975) observed them to live in more permanent groups similar to those of brown lemurs in the southwest.

Mayotte brown lemurs have a cathemeral activity period, that is they are active both during the hours of daylight and darkness throughout the year (Tattersall 1979, 1987). The same appears to be true for *E. macaco* (Andrews and Birkinshaw 1998, Colquhoun 1998) and other observations suggest that activity during the night is not infrequent in other members of the *Eulemur* group (Sussman 1972, Harrington 1975, 1978, Conley 1975, Meyers 1988, Overdorff 1991).

Figure 5-20 *E. mongoz.* [Photo by I. Tattersall]

In some forests, there is good reason to believe that coexistence between the *Eulemur fulvus* and *E. mongoz* is related to major differences in activity cycles during the dry season. For example, at Ampijoroa, the mongoose lemurs (Figure 5-20) were mainly diurnal during the wet season when fruit is abundant, and often rest, feed and move with brown lemurs for up to 4 hours at a time (Harrington 1975, Andriatsarafara 1985). However, during the dry season when fruit is scarce, in some forests, mongoose lemurs change activity cycle completely and become nocturnal (Tattersall and Sussman 1975, Sussman and Tattersall 1976, Rasmussen in prep. see also Andriatsarafara 1985, Curtis 1997), sleeping throughout the day in densely foliated sites.

Nocturnal activity allows mongoose lemurs to take advantage of flowers that open only during the night and may provide protection from certain predators when leaf cover is absent (Sussman and Tattersall 1976, Rasmussen in prep). As stated above, brown lemurs are cathemeral, with most activity taking place during the day, and nocturnal activity appearing to be sporadic. There is no indication that brown lemurs change their activity cycles seasonally or feed on nocturnal flowers. Recently, *E. fulvus* and *E. mongoz* have been studied in sympatry (Rasmussen 1998a,b, in prep.). In her study area, Rasmussen (1998a, in prep.) found that the mongoose lemur was almost exclusively active during the night in the dry season but was mainly active during the day in the rainy season. In contrast, the brown lemur was active during the day throughout the year and only active during the night during the dry season. When both species were active at night their patterns of activity differed.

In other circumstances it appears that physiological limitations rather than competition may dictate certain flexibility in the activity cycles of the mongoose lemurs. On the Comoro Islands, where mongoose lemurs do not co-occur with brown lemurs, they have been observed to be diurnal in cool highland areas and nocturnal in warm, dry, seasonal environments (Tattersall 1976, 1978). These differences may be related to thermoregulation. In another forest in northern Madagascar, mongoose lemurs were not found to change there diet appreciably in relation to season even though they were sympatric with brown lemurs (Curtis 1997). In this forest, Curtis believed the advantages of changes in activity cycle among the mongoose lemurs most likely were related to thermoregulation and possibly predator avoidance.

E. mongoz groups contain only one adult male, one adult female, and one or two young individuals (Sussman and Tattersall 1976, Curtis 1997, Rasmussen in prep.). During nocturnal activity, the members of the group remain together throughout the night, coordinating movement and activity precisely. While travelling through the forest at night, group members constantly exchange vocalizations.

E. mongoz is quadrupedal and moves mainly high in the continuous canopy, only coming lower in the trees to feed. It is rarely seen on the ground. Day ranges are small, between 460-1800 m. Home ranges are also small (1–5.5 hectares) and overlap extensively. In fact, different groups use many of the same resources at different times during the night (Sussman and Tattersall 1976, Curtis 1997, Rasmussen in prep.). Intergroup encounters were infrequent and very similar to those of *E. fulvus* in that they were not associated with defense of boundaries but, rather, seemed to function in maintaining group integrity and spacing within shared ranges (Sussman and Tattersall 1976).

The mongoose lemur mainly eats fruit, making up approximately 60–65% of its diet year-round, but other foods are important at various times of the year (Andriatsafara

**Table 5-9. Percentage of Time *Eulemur mongoz*
Was Seen Feeding on Various Plant Parts**

Plant Species	Nectar		Fruits	Leaves	Total
	Flowers	Nectarines			
C. pentandra	64.0		17.5		81.5
H. crepitans		14.0			14.0
K. madagascariensis	2.0				2.0
C. phaneropetalum	1.0			1.0	2.0
T. indica				0.5	0.5
Total		81.0	17.5	1.5	

[From Sussman and Tattersall, 1976]

1985, Curtis 1997, Rasmussen in prep.). For example, during the dry season at Ampijoroa it becomes a specialized feeder, eating few species of plants (Table 5-9). Sussman and Tattersall (1976) found that 81% of its diet at this time consisted of the nectar-containing components of four species. Nectar from the flowers of the kapok tree *(Ceiba pentandra)* made up the major portion of the diet, 64%, and extra floral nectaries of Hura crepitans, 14% (Figure 5-21). The kapok tree is utilized by many flower-visiting bats in both the Old and New Worlds and is adapted to attract bats as pollinating agents (see Chapter 1). However, the plant-pollinator relationship exhibited by mongoose lemurs and the angiosperms they visit are suggestive of possible coevolutionary relationships of the past (see Sussman and Raven 1980, Kress 1993, Nilsson et al. 1993, Kress et al. 1994, Birkenshaw and Colquhoun 1998).

However, although Rasmussen (in prep.) observed that *H. crepitans* nectaries constituted between 82–100% of the diet during a portion of the dry season, the kapok flowers were not a major source of food. Curtis (1997) also did not find the same type of seasonal dependence on floral nectar in her study site. During her 10 month study, leaves, flowers, and animal matter were consumed more frequently during the dry season and nectar was consumed predominantly during the wet season. Thus, the behavior of the mongoose lemur is quite variable in different regions throughout its small geographic range.

Further to the north, *Eulemur macaco* overlaps with *E. fulvus fulvus* in the Sambirano Domain (Figure 5-18). The black lemur is widely distributed in this region of overlap and it is relatively easy to find, whereas the brown lemur is more restricted, both geographically and ecologically (Tattersall 1976, Andrews 1990). *E. macaco* is found in primary forest, secondary forest, and in timber and crop plantations; in fact, in some areas it is considered an agricultural pest (Andrews in prep., Colquhoun, 1997, 1998). In contrast, *E. fulvus* is found mainly in canopy and rain forest. The brown

Figure 5-21 "Flowers eaten by *E. mongoz*", but in this photo it is a nocturnal lemur visiting the flowers. [Photo by D. Baum]

lemurs also appear to live in smaller groups than the black lemurs, and where they occur their density is considerably lower (Tattersall 1976, Andrews 1990).

E. macaco appears very similar in behavior and ecology to *E. fulvus*. Black lemurs live in groups of around 7 to 10 individuals, with equal numbers of adult males and females (Andrews 1990, in prep., Colquhoun 1997). Smaller groups are cohesive but larger groups (up to 15 animals) tend to split up and reassemble, especially during the dry season and in disturbed habitats. It is not yet known if both males and females transfer between groups.

Like the brown lemurs, black lemurs are active both during the day and night, notably on moonlit nights (Colquhoun 1998). Nocturnal feeding tends to be more predominantly on fruit than during the daytime; more time is spent in the canopy at night, because of reduced risk of avian predation, or because of lowered canopy temperatures (Andrews and Birkinshaw 1998). In areas where black lemurs raid crops, their activity is more *crepuscular* (around sunrise and sunset) (Petter 1962, Birkel 1987, Andrews 1990). When fruit is scarce these lemurs reduce activity levels and spend more time resting (Colquhoun 1993, 1997).

Black lemurs spend most of their time in the middle and upper level of the canopy, whether in primary or secondary forest. Their diet varies seasonally. They are mainly frugivorous; at Lokobé (Nosy Bé), for example, fruit predominates in all months but May, when leaves are much the largest dietary component (Andrews and Birkinshaw 1998). Flowers and nectar also are exploited in some months. Black lemurs are documented to be important seed dispersers (Birkinshaw 1995) and pollinators, especially from the endemic travelers palm (*Ravenala*, Figure 5-22) (Birkinshaw and Colquhoun 1998).

Home ranges of black lemurs average @ 5.0 ha, and overlap considerably. Population density reaches around 200 animals per km^2. In the wet season, day ranges cover several hundred meters, whereas in the dry season groups move only 50 meters on some days. Group encounters occur at shared resources and seem to function to maintain group and not spatial integrity. As in brown lemurs, intra-group agonism is rare, but unlike the brown lemurs, black lemur females are dominant to males (Colquhoun 1993, Fornasieri and Roeder 1993). During the mating season, there is an increase of agonism between males and they "roam" between groups. Females first give birth at 2 years of age, and gestation is 128–129 days.

Although the brown lemur has not been studied in this region, and the behavior and ecology of these two species appear to be broadly similar, there are indications that the two species differ in a number of ways (Koenders 1989).

In the extreme north, *E. fulvus sanfordi* (Sanford's lemur) coexists throughout most of its range with *E. coronatus* (the crowned

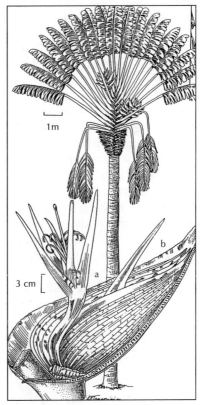

Figure 5-22 Traveler's palm. [After A.R. Tangezini, 1993]

lemur) (Figure 5-23). From early surveys of these two species, it was thought that crowned lemurs preferred lowland habitats and spent much of the time on the ground, like ringtailed lemurs. Sanford's lemur was thought to be more prevalent in humid highland forests (Rand 1935, Tattersall 1977, Arbelot-Tracqui 1983). However, Petter et al. (1977) found both species in the highland forests of Mt. d'Ambre. In the strange limestone massif of Ankarana, Wilson et al. (1989) found the crowned lemur mainly in undisturbed canopy forest

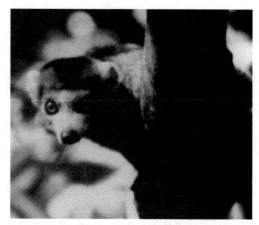

Figure 5-23 *E. coronatus.* [Photo by Ben Freed]

and Sanford's lemur most often in edge and degraded forest.

Noting that the surveys just quoted left somewhat unclear the question as to differences in habitat preferences between these two species, Freed (1996) conducted extensive surveys of the two species, and studied them in sympatry for 13 months. He found crowned lemurs to be more widely distributed than Stanford's lemurs, both geographically, and in types of forest inhabited. The two species coexist in some dry and wet, low and high elevation, and disturbed and undisturbed forests. However, crowned lemurs were found in some areas with discontinuous forest, whereas Sanford's lemurs only inhabited forests with closed, continuous canopy. Although crowned lemurs are found alone in some areas, this was not true of the Sanford's lemurs. The question remained as to how did these two species partition shared forests? As with the other LBJ's of the north, Freed (1996) found some subtle but important distinctions.

Both Arbelot-Traqui (1983) and Wilson et al. (1989) reported that both species were mainly frugivorous and had similar dietary preferences. Furthermore, Arbelot-Traqui did not observe any differences in horizontal or vertical habitat preferences. However, neither study was conducted over an entire annual cycle, and seasonal differences remained undetermined. Freed's (1996) 13-month study corrected for this deficiency with instructive results. Although both species spent most of their time in the understory and middle story of the canopy, the crowned lemur mainly used the understory, whereas Sanford's lemur was found preferentially in the middle canopy. These differences were significant in both the cooler, wetter and the hotter, dryer seasons, and during foraging and feeding activities (Figure 5-24).

Both lemur species ate mostly ripe fruit from the same plant species, in similar proportions, and had comparatively diverse diets. However, the most frequently consumed plant parts were found at different forest levels. Crowned lemurs fed and foraged mainly in the understory and in smaller trees. These smaller plants had less abundant crops of fruit and flowers, which made it necessary for crowned lemurs to visit more trees, and to spend significantly more time foraging than the Sanford's lemur. Finally, during the dry season the crowned lemurs ate more unripe fruit and flowers, whereas Sanford's lemur continued to feed on whatever ripe fruit was available. The crowned lemurs' ability to subsist on food from understory plants leads to differences in ranging behavior, and may be the major factor that allows these two closely related species to coexist.

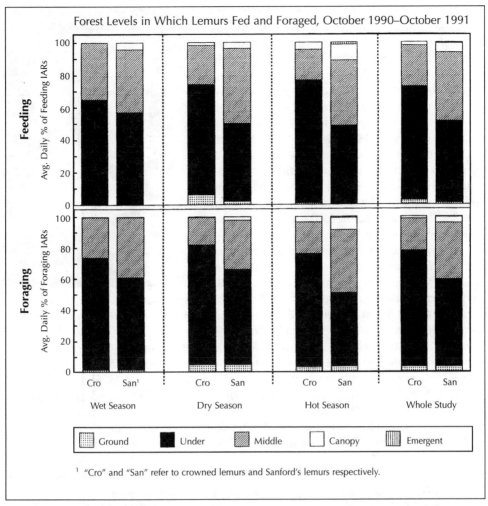

Figure 5-24 *E. coronatus* and *E. fulvus* use of forest strata. [Adapted from *Co-occurrence Among Crowned Lemurs (Lemur coronatus) and Sanford's Lemur (Lemur fulvus sanfordi) of Madagascar*, by B. Freed, Ph.D Thesis. Washington University, St. Louis, MO, 1997.]

The social organization of the two forms was similar. Both species lived in groups of from 6–8 individuals, with approximately equal sex ratios. At times, there was a tendency for larger groups to break into subgroups. Freed (1996) could discern no tendency for females to dominate males in either species. Crowned and Sanford's lemurs spent a small amount of time in mixed species groups, but mainly remained apart. However, conspecific groups were more agonistic toward one another than were groups of the different species. Like most brown lemurs, agonistic encounters appeared to maintain group integrity. There is little difference between reproductive parameters in these species (Table 5-7).

In sum, even though these four species of LBJ's of the north are quite similar in a number of ways, and are quite adaptable in ecology and behavior, it appears that subtle but important distinctions can be found in their habitat preferences and, in some cases, their social behavior. Since very few long-term, or detailed studies have been done, and studies have not been done in a variety of habitats for any species, the extent of behavioral flexibility is unknown. We still do not know, for example, if the distinctions so far described are consistent in different localities throughout a species range, and related to real species-specific morphophysiological and behavioral adaptations. Or, as suggested by Tattersall (1992b:36) in referring to the LBJ's of northern Madagascar "that habitat preference—ecology—cannot be viewed as a property of any of these taxonomic groups . . . we cannot use aspects of ecology to help us define or recognize lemur species or subspecies." Much remains to be done on these relatively ignored animals.

In Northeastern Madagascar, *Eulemur fulvus* and *Eulemur coronatus* coexist with *Propithecus tattersalli*. Recently, Meyers (1993) conducted a year-long study of this indriid in three separate forests (Figure 5-25). He found that it was in many ways very similar to *P. ver-*

Figure 5-25 *P. tattersalli.* [Photo by Ben Freed]

reauxi studied by Richard in the west. As in the west, forests in the extreme northeast are seasonal with distinct wet and dry seasons. In the dry season, young leaves, flowers, and fruit are scarce.

The three study forests were slightly different in structure, species composition, and phenology. However, even though there were differences in some of the plant species eaten at each forest, Meyers found that the general pattern of the diet was similar at all three sites. Tattersall's sifaka ate a wide variety of plants;

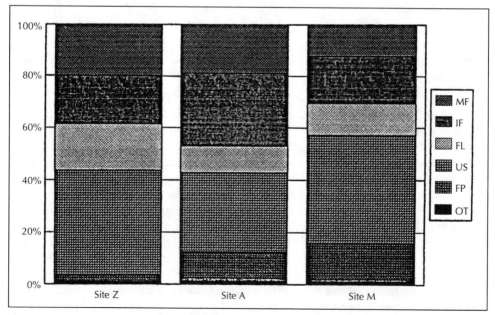

Figure 5-26 Annual diet composition of the three primary study groups of P. tattersalli [Adapted from Meyers, 1993]

over 80 species at each site. Immature leaves, unripe fruit and seeds, especially from leguminous pods, made up @ 60% of the diet at all sites (Figure 5-26). Less than 50% of the diet consisted of leaves.

Common food items changed seasonally in relation to their availability. Unripe fruit and seeds were eaten mainly in the mid-wet through early dry season, mature fruit in the mid-dry season, flowers in the late dry season, and immature leaves in the late dry to early wet season. When all food items were abundant, animals preferred to eat immature leaves and flowers. When young leaves became scarce, unripe fruit and seeds were eaten more. Only when new leaves and fruit were not available did the sifakas eat mature leaves. Many folivorous primates use mature leaves as dry season alternatives to more preferred food items. Also, as in some other folivores, *P. tattersalli* was a seed predator rather than a seed disperser. Seeds were not a major dietary item of *P. verreauxi.* This was one of the major dietary differences between the two species of sifaka. The use of seeds by Tattersall's sifaka may be facilitated by detoxification capabilities of stomach microbes, as found in some colobines (Davies et al. 1988, Ganzhorn 1989, Davies 1991, Meyers 1993).

The sifakas were able to maintain a similar dietary pattern in the three different forests by using different foraging patterns in each. Meyers found that the daily number of feeding bouts, bout duration, and food tree crown diameters varied significantly across sites. Patterns of range use also differed significantly, with monthly ranges at one forest being consistently smaller than at the others. Generally, when and where food patches were smaller, more patches were used, animals moved more, and each feeding bout was shorter. However, dietary diversity, types of food items chosen, and total daily feeding time did not vary among populations. Meyers (1993) suggests that the necessity to maintain a highly diverse diet in sifakas may be necessary to provide a minimum mix of nutrients, and to protect this primary consumer from high levels of specific plant toxins (see also Freeland and Janzen 1974, Glander 1982). He further suggests that the stability of daily feeding time is a result of the constraints of food passage in herbivores (Westoby 1974, Milton 1984).

The social structure and organization of *P. tattersalli* and *P. verreauxi* are very similar. Tattersall's sifaka groups average between 5 and 6 individuals, and range from 3 to 9. Each group contains 1 or 2 adult of each sex, and the adult sex ratio is equal.

Adult females were dominant over adult males and aggressive interactions within groups were mostly associated with access to food. Females had priority of access to feeding sites. Most feeding agonism was related to the use of small food patches with high quality immature foliage, which was available mainly during the wet season.

Home ranges at Meyer's sites ranged from 8 to 12 ha and were highly overlapping. Meyers found that the amount of overlap in home range boundaries was greater in the site with higher population density. This finding is similar to that on the western sifaka by Richard, where the dry forest population had a lower density and lower range overlap than the population living in the semi-deciduous site of Ampijoroa. In Meyer's study, intergroup interactions were significantly correlated with daily path length. He considered the frequency of intergroup interactions to be a byproduct of spatial and temporal foraging strategies, population density, and home range overlap. In seasons when daily path lengths were long because of the necessity of high foraging effort, groups were more likely to encounter one another, and the number of encounters would be related to population density and home range size. Thus, he believes that intergroup encounters were directly related to the distribution of food resources but not caused by between-group competition.

THE ECOLOGY OF THE DIURNAL LEMURIFORMES OF THE ORIENTAL ZONE

Along the east coast of Madagascar there is a thin coastal plain of littoral forest and, further inland, a higher and colder montane forest. Seasonality is not as marked in this region as in the west and north, and different species of plants, as well as different individuals of the same species, may flower, fruit, and produce new leaves at widely different times of the year. When humans arrived on Madagascar, there were approximately 11 million ha of rain forest along the eastern coast. In 1985, only 4 million ha remained and deforestation was proceeding at a pace of 111,000 ha per year (Green and Sussman 1990). Clearly, these rates must be slowed if the unique flora and fauna of Madagascar is to be saved (see Ganzhorn et al. 1996/1997).

There are eight species of large, mainly diurnal Lemuriformes and one secondarily nocturnal species that are widely distributed in the eastern rain forests. These include: three species of Indriidae, *Indri indri, Propithecus diadema,* and *Avahi laniger;* and six species of Lemuridae, *Eulemur fulvus, E. rubriventer, Varecia variegata, Hapalemur griseus, H. simus,* and the recently discovered *H. aureus.* The indriids are mainly folivorous, *Eulemur* and *Varecia* are highly frugivorous, and *Hapilemur* species are bamboo and grass specialists.

Although there are a number of studies recently completed or still in progress on eastern rain forest species, there is relatively little actually published on projects of at least one year duration. Furthermore, at present few species have been studied at more than one locality. All except *Indri* exist at the newly created Ranomafana National Park (Wright "in prep."), and studies of various length and intensity have been conducted on the Ranomafana primates. Reports on long-term research at this site have been published on *E. fulvus* and *E. rubriventer,* and on *P. diadema.* Outside Ranomafana, published reports on year-long studies are available only on *H. griseus, I. indri, P. diadema,* and *V. variegata.* Preliminary studies have been published on the other species.

THE FRUGIVOROUS LEMURS OF THE EASTERN RAIN FOREST

Eulemur fulvus (the brown lemur) and *E. rubriventer* (the red-bellied lemur) (Figure 5-27) are the little brown jobs of the eastern rain forest. These two species show little difference in weight, measurements, dentition, or cranial and skeletal morphology. There is no sexual dimorphism in either species. The red-bellied lemur does

Figure 5-27 *E. rubriventer.* [Photo by Robert W. Sussman]

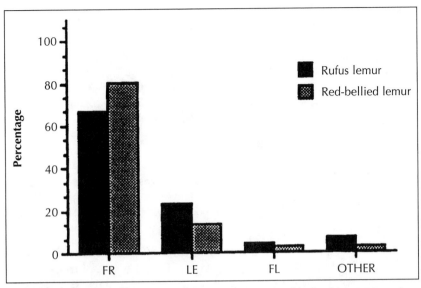

Figure 5-28 Overall composition of diet for rufus and red-bellied lemurs based on (A) total number of feeding bouts and (B) percentage of observed feeding time in earch food category. FR, fruit; LE, leaves; FL, flowers; OTHER, miscellaneous food material. [Adapted from Overdorff, 1993a.]

have two morphological features that are peculiar, however; pointed, keeled nails, for which the function is as yet unknown (Gebo 1987, Tilden 1990, Overdorff 1991), and a brush-like feathered tongue which may be adapted for extracting nectar and pollen (Overdorff 1992). The geographical ranges of the two species generally overlap over much of the east (Figures. 5-17,18), and both are arboreal quadrupeds restricted to relatively undisturbed forest. Population densities of red-bellied lemurs are very low throughout their range and this species may be the most rare of the genus (Harcourt 1990).

Overdorff (1991) did a one year ecological study of coexisting red-bellied and brown lemurs (*E. fulvus rufus*) at Ranomafana. She also conducted a 10 year study of the demography of this species (Overdorff et al. in press). As with the LBJs of the north, both species were highly frugivorous; 70-80% of their diet consisted of fruit, and fruit made up over 50% of the diet in all months but one. They supplemented this with young and old leaves, flowers, and a small number of insects. Both ate approximately 100 plant species, and the proportion of food items in their diets were very similar (Figure 5-28). The two lemurs fed on the same species of plants 80% of the time, with the top 5 species making up 40-50%, and the top ten 60-70% of the diet. However, there also were important differences. There was only @ 40% overlap in plant parts eaten. Ripe fruit was the most favored item, but brown lemurs ate more unripe fruit and mature leaves than red-bellied lemurs; red-bellied lemurs ate more young leaves.

Flowers were fed upon only 5% of the time but were important resources between peaks of fruit and new leaf production. The two lemur species consumed flowers in a very different manner. Brown lemurs consumed the whole flower 70% of the time. Red-bellied lemurs licked flowers 86% of the time, using their specialized tongues (Figure 5-29) to gather the nectar, and did not destroy the flowers (Overdorff 1992). The red-bellied lemur is a pollinator of some flower species, and its tongue may be an adaptation for nectar feeding. One species of leguminous liana, *Strongylodon cravieniae* appears to have successfully evolved to exclusively exploit lemurs as pollinators (Nilsson et al. 1993). Given its ability to eat unripe fruit, mature leaves, and whole flowers, it appears that *E. f. rufus* is better able to tolerate potentially toxic, secondary plant compounds than is *E. rubriventer*. This may be an important factor in niche separation between these two lemur species (Ganzhorn 1988, 1989, Overdorff 1991, 1993a, 1996a,b).

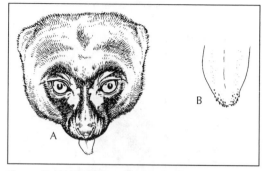

Figure 5-29 These two drawings depict the brush-like feathered tongue of the red-bellied lemur in proportion to the animal's head (A) and in detail (B). [Adapted by Overdorff, 1992]

Differences also were noted in group size and in the size of trees fed upon. At Ranomafana, brown lemur groups averaged around 10 animals. Red-bellied lemurs lived in smaller, "family" groups, averaging 3 animals (Overdorff 1993b, 1996b, 1998, Overdorff et al. in press), though larger groups have been reported (see Harcourt 1990). During peaks of fruit abundance, the brown lemur fed in small trees, visiting many per hectare. While doing this, the group divided into small parties with some animals eating in different but nearby trees. Red-bellied lemurs fed in larger trees and visited fewer trees per hectare with all group members eating in the same tree. During this period, both lemurs ate common plant species. When fruit became scarce, brown lemurs fed on large trees of rare plant species. The whole group fed together and stayed in the tree for twice as long as in periods of fruit abundance, thus traveling less. Red-bellied lemurs fed in smaller, common trees when fruit was scarce, but visited more trees per hectare. The group was small enough to allow all members to feed together.

Even though the brown lemur fed on a greater variety of food items than the red-bellied lemur, the larger brown lemur groups searched for scarce fruit patches over a wider area, especially when fruit was scarce. Their day ranges were twice as large as those of red-bellied lemurs (mean = 961 vs 444 m), and the home ranges were 5 times as large (100 vs 20 ha) (Table 5-5). In fact, brown lemurs made long forays to areas outside of their home range to feed at groves of introduced guava trees during a period of extreme fruit scarcity. However, on average this species traveled less when fruit was scarce than when females were lactating (Sept.–Dec.), at which time they ate a greater variety of food items, moved further, and had more feeding bouts. Brown lemur groups at Ranomafana were similar in size to those at other sites, and there was a 75% overlap in home range boundaries.

At Ranomafana, red-bellied lemurs lived in pair bonds and shared a number of traits that are commonly found in monogamous primates (see Fuentes 1999). They were territorial and monitored home range borders on a regular basis. Home range boundaries overlapped only slightly (8% of total home range). Also, males participated in infant care; sometimes holding the infant when the mother fed and carrying it when the group traveled. Males and females spent equal amounts of time caring for infants for the first two months, and after this the infant moved independently or was carried by the male. Red-bellied lemur group members rarely were more than

10 meters apart. No dominance hierarchy was discernable in either species, and females were not dominant over males.

Varecia variegata (the ruffed lemur), of which there are two subspecies (Figure 5-30), is the third mainly frugivorous, diurnal species in the eastern rain forests. Like the other two species, it is widely distributed in this region. The ruffed lemur lives in low densities in high altitudes (Pollock 1975, Harcourt 1990, White et al. 1995) but at higher densities in lowland forests (Vasey 1997a). The ruffed lemur is an evolutionary enigma in that it is the largest extant species within the Lemuridae (2.5–5 kg), and yet retains a number of features normally found in small, nocturnal prosimians. For example, ruffed lemurs have shorter gestation periods (mean = 102 days) and larger litter sizes (typically 2–3) (Rasmussen 1985) than other diurnal lemurs (Table 5-7). They also have an absentee parental system. Females build nests for their newborn infants. Later

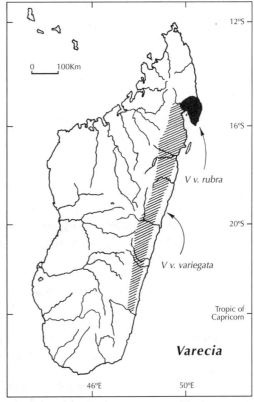

Figure 5-30 Map of *Varecia* distribution. [From Mittermeier, et al., 1994]

they carry the young by mouth and leave them unattended while they engage in other activities (Vasey 1997b). The infants are altricial (i.e., they mature quickly). In captivity, they have been observed to move off the mother within 4 days (Klopfer and Klopfer 1970). In the wild, they travel independently by 70 days (Morland 1991, Vasey in prep.). Infants achieve 70% of adult weight within 4 months and are adult size by 6 months (Pereira et al. 1987, Vasey 1997b).

Figure 5-31 *V. variegata.* [Photo by Robert W. Sussman]

Until recently reports were available only on captive and semi-free ranging groups of ruffed lemurs (e.g., Klopfer and Dugard 1976, Kress et al. 1978, Pereira et al. 1987, 1988). However, now data are available from a number of field studies. On the island of Nosy Mangabe, Morland (1991, 1993) spent 12 months over a 30 month period observing black and white ruffed lemurs on the Masoala Peninsula, Rigamonti (1993) studied the red ruffed lemur for 6 months over an 11 month period, and Vasey (1997a,b) recently completed a 13 month study at Andranobe also on

Table 5-10. Total Diet Composition of *Varecia* at Nosy Mangabe

	Fruit	Nectar	Leaves	Other	Unident
Percentage of feeding time spent on					
Mean	74.01	21.51	5.11	1.79	0.88
Median	70.53	25.26	3.79	3.80	0.61
Sample Points	2762	821	162	80	30

Notes: Figures show the percentage of total feeding 5-min, sample points, recorded during instantaneous sampling, in which focal animal fed on each item. [From Morland, 1991]

the Masoala Peninsula. Vasey also studied sympatric groups of brown lemurs *(E. fulvus albifrons)* at this site. Finally, White (1991, White et al. 1995) studied a pair of black and white ruffed lemurs for three months at Ranomafana.

Varecia (Figure 5-31) is an arboreal quadruped. It spends most of the time above 10 meters in the upper canopy (Morland 1991, Vasey 1997b). Between 70–86% of the ruffed lemur diet consists of ripe fruit, supplemented by flowers and leaves (Table 5-10). Few plant species make up a major portion of the diet. On Nosy Mangabe and on the Masoala Peninsula, ruffed lemurs spend over 70% of the time feeding on just 8 plant species (Morland 1991, Rigamonti 1993). On Nosy Mangabe, 2–3 species account for over 50% of the diet in 11 of the 12 months studied, and most feeding is on common tree species. On the Masoala Peninsula, two fig species *(Ficus* spp.) accounted for 58% of the fruit eaten, and four species made up 78% of the fruit diet (Rigamonti 1993). Animals feed mainly in large trees of around 45 cm diameter at breast height (DBH) (Vasey 1997b). In all activities, they spend most of their time above 15 meters (Vasey 1997b) and over 85% in trees with 41–80 cm DBH (Rigamonti 1993). Ruffed lemurs also depend on large fruit trees at Ranomafana (White et al. 1995).

The social system of ruffed lemurs is still not fully understood and is the subject of some debate. On Nosy Mangabe, they were observed to live in fission-fusion communities (see Chapter 2) with home ranges of around 30 ha. At Ranomafana, an adult male and female pair were territorial; they shared and defended a home range of 197 ha. Several factors may explain the differing results. Ruffed lemurs were introduced to Nosy Mangabe in 1930. Since that time the population has grown, and densities on the island generally are high (Table 5-5). In the study site at Ranomafana, selective timbering of large trees has occurred and, perhaps due to this factor, ruffed lemur densities are extremely low. *Varecia* seems to be very sensitive to logging, and is thought to be one of the most vulnerable rain forest lemurs (White et al. 1995). Therefore, using these two studies, it is difficult to say which home range size and social system might be more typical, or indeed, if this is normal variation within the species.

On the Masoala Peninsula, where Rigamonti (1993) and Vasey (1997a,b) conducted their studies, the sites have not been selectively logged, and ruffed lemurs occur naturally. Vasey's is the only relatively long-term study covering all seasons. Vasey captured and radio-collared 3 individuals and followed them and their community members for the duration of the study. Her results were similar to those of Morland and Rigamonti in that she found that ruffed lemurs did not live in spacially cohesive groups. Instead, they were found in dispersed social networks of animals that regularly interacted with one another.

Following Morland (1991, 1993), Vasey referred to these stable social networks as communities (Figure 5-32). She divided these networks into three levels of organization: core groups, subgroups, and communities. Core groups consisted of individuals who shared the same core area and used it throughout the year. Subgroups were individuals that associated with one another and were formed either from members of the same core group or different core groups depending on the season. They varied daily in size, composition, and duration (Figure 5-33). Communities were animals that affiliated with one another regularly but rarely with conspecifics outside of the community. Females affiliated with animals within and outside of their core area but did not normally interact with animals from

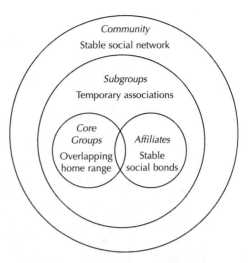

Figure 5-32 Schematic diagram of *Varecia* social system on Nosy Mangabe showing a nested series of four levels of grouping associations. [Adapted from Moreland, 1991]

outside of the community range, except during the mating season. Outside of the short mating season, males affiliated almost exclusively with members of their own core group.

Core areas within communities did not overlap but there was no defense of boundaries. However, there was ritualized defense at boundaries between adjacent communities. Non-community members only were allowed to cross these boundaries during the mating season. The community studied by Vasey (1997b) consisted of at least 5 core groups. Throughout the study, the community size ranged from 18–31 individuals. There were many reproductively active males and females in the community (at least 3 males and 8 females), and as many as two adult females within a single core group. Vasey defined the social structure as multi-male/multi-female and the social organization as fission-fusion, with community members defending an exclusive communal home range.

At Vasey's site, males and females displayed different ranging patterns during different seasons. These patterns appear to be tightly linked to the different reproductive investments of the males and females and to the ruffed lemurs' reliance on a spatially and temporally patchy diet. Generally, males resided in their core area of between 7–13 hectares throughout the year. Female ranges, on the other hand, varied seasonally in direct relationship to their reproductive cycle (from 2–27ha).

The females reduced home range size and distances traveled during the cold rainy season and during gestation relative to lactation and the hot rainy season. This was a strategy for conserving energy when less food was available and during extremely short and costly pregnancies. Directly after litters were born, females expanded the distances but stayed mainly within their own core areas. The reproductive burden decreases in December, during infant stashing period, when infants are left alone or guarded by other group members. During this time, males and other community members provided extensive care for infants, guarding them while the mothers were out feeding and socializing. Once the infants

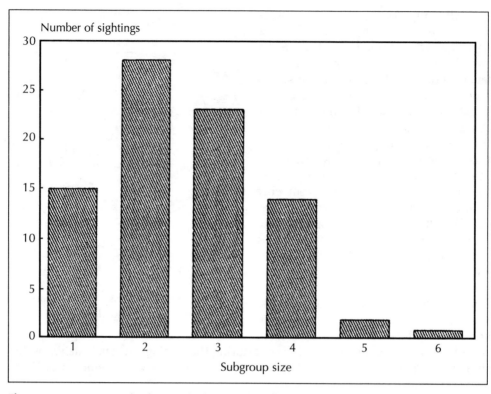

Figure 5-33 Frequency of sightings of subgroups of different sizes. Bars show the number of times that subgroups of each size were counted when ruffed lemurs were sighted for the first time per day before beginning the daily observation session. Data represent 84 sightings of community I, July–December 1987. [Adapted from Moreland, 1991]

began to travel independently, in the hot rainy season (January-March), females ranged long distances and over much larger areas. At this time they ranged throughout a large portion of the community range of 57.7 ha, and affiliated with members of other core groups as well as their own. These subgroups varied daily in size, composition, and duration.

Affiliative interactions that occurred in subgroups included female greeting displays, feeding, calling, resting, grooming, and traveling together. The greeting behavior of the ruffed lemur females is quite spectacular, involving anogenital marking of one another's backs, jumping over each other, writhing together, and emitting squealing vocalizations. These activities were not observed during the cold rainy season. During the hot rainy season, individual male home range sizes were less than half the size of the female ranges. This season had not been well studied at other research sites.

At Nosy Mangabe, Morland (1991, 1993) also described a fission-fusion social structure, but she recognized four levels of organization: communities, core groups, affiliates, and subgroups.

At her site, just as at Andranobe (Vasey's site), males were subordinate to females and did not take part in intercommunity encounters. Most spacing was accomplished by olfactory marking and vocal communication. Ruffed lemurs have loud calls that are often given spontaneously and then answered by neighboring communities or by subgroups of the same community. Morland noticed that some individuals interacted more frequently and had persistent social bonds. She termed these individuals "affiliates." Most but not all affiliates were in the same core group. Adult females had many affiliates, including all or most of the other females. Adult males were more solitary, and only interacted with one or two conspecifics.

On Nosy Mangabe, females were quite sedentary during the first day of sexual receptivity. They mated mainly with males that were affiliative partners outside of mating season and both males and females mated with more than one partner. At this site and at Vasey's, some matings occurred with roaming males from other communities. When the female was receptive, she continued to be agonistic during copulation, grappling, cuffing, and biting the male. The females either approach the male while in estrus (Morland 1993, Vasey in prep.), or allowed him to approach, inviting copulation.

Pregnant females began to build nests as much as 3 weeks prior to birth. After the short gestation period, females spent most of the time feeding or resting with their infants and became more solitary. During the first two months of lactation, females greatly reduce time spent traveling (Morland 1993) and spend most of their time near the nest (Vasey 1997b). These differences were the result of climatic conditions, seasonal food supply, and female reproductive status. During the cool months, besides low temperatures, there was continuous daily rainfall, and fruit was relatively scarce. The animals, thus, conserved energy and minimized intraspecific competition.

The flexible social organization of the ruffed lemur may be interpreted as an adaptive response to reproductive constraints and feeding ecology. Their reproductive pattern and reliance on spatio-temporally patchy fruit resources necessitates the fission-fusion social structure and general flexibility in social behavior. Given this spatial pattern and the allo-parenting traits, Vasey (1997b, in prep.) describes the red ruffed lemur as a cooperative breeder. She includes the following traits as part of this breeding system: (1) mothers stash infants with other mothers; (2) mothers leave infants to be guarded by other group members including both breeding and non-breeding males and females; (3) adult males guard nests of several different females and help care for infants who are probably fathered by other males; (4) during infant stashing season, adult males effectively lighten the reproductive burden on mothers by spending time with yearlings, including playing, traveling, and feeding with them; (5) breeders and non-breeders participate in coordinated vigilant activities during the infant nesting and stashing seasons, and respond to infant distress calls; (6) non-maternal infant transport; (7) cooperative nursing; and (8) infants are adopted.

The unique features of reproduction and development found in the ruffed lemur, along with their highly frugivorous diet and dependence on large trees, form an integrated system which may help us to understand their unusual social system. One reason that most diurnal primates remain in cohesive groups is the protection this affords them from predation (see Chapter 1 and Chapter 2). *Varecia* is the largest of the diurnal lemurids and it has very loud alarm vocalizations which are used to warn others of the presence of predators. Furthermore, unlike other diurnal primates, ruffed lemurs do not constantly carry infants. Infants grow very quickly and, within a short period, they are almost adult size.

Morland hypothesizes that these characteristics of ruffed lemurs probably reduce predation pressure and allow them to live in fluid social groups, with adults spending much of their time alone. Furthermore, because there is more than one infant, and infants are large and grow rapidly, female competition for food may become quite severe. This may necessitate solitary foraging. Fission-fusion groups are rare among primates (though less rare than was once thought). Among forest dwelling primates, this social system is found in spider monkeys and in common chimpanzees, both of which are relatively large frugivores. As in the ruffed lemur, subgroups of these species are flexible and subgroup size is related to resource availability. When more fruit is available, subgroups are larger and when fruit is scarce individuals are more solitary. Thus, it seems that reduced predator pressure and dependence on widely distributed, patchy, and sometimes scarce fruit resources are also important factors related to the fluid social system of the ruffed lemur.

In comparing the sympatric populations of *Varecia* with that of *E. fulvus albifrons,* Vasey (1997b) found that these two species divide the environment by using different forest strata, forest heights, and by having different diets. *Varecia* was found mainly in the tree crowns, above 15 m high and ate mainly ripe fruit throughout the year. The white-fronted brown lemur used a variety of sites in the forest and moved mainly in the understory and lower canopy, below 15 m. The brown lemur supplemented a diet of fruit with substantial amounts of young and old leaves, flowers, and other foods, depending upon the seasonal availability of these items.

FOLIVOROUS LEMURS OF THE EASTERN RAIN FOREST

There are three species of Indriidae living in the eastern rain forest: *Indri indri, Propithecus diadema,* and *Avahi laniger.* Published reports on year-long or longer research are available on indris (e.g., Pollock 1975, 1977) and diademed sifakas (Wright 1987, 1995, Hemingway 1995, 1996, 1998), and a recently completed study of sympatric populations of these two species (Powzyk 1997). The results of short-term studies have been published on woolly lemurs. None of these folivores survives well in captivity, although the Duke Primate

Figure 5-34 *Indri.* [Photo by I. Tattersall]

Facility in North Carolina has been very successful with *Propithecus.*

Indri (Figure 5-34) is found in a 500 km stretch of eastern rain forest (Figure 5-35). Its geographic range does not include the Masoala Peninsula, nor is it found south of Mangoro River. Its range has been reduced even within the past few decades (Petter et al. 1977, Tattersall 1982, Harcourt 1990, Jungers et al. 1995, Mittermeier et al. 1994, Godfrey et al. in press). *Indri* occurs in forests from sea level to 1500 m. Like all indriids, *Indri* is morphologically adapted for a VCL mode of locomotion and a folivorous diet. It is the largest living lemur weighing between 6,000–7500 gm (Table 5-3). Like *Propithecus, Indri* is highly arboreal, usually remaining in or near the canopy, around 14 meters high. However, all levels

(1–40 m.) of the forest are utilized, especially during feeding (Pollock 1977, Powzyk 1997).

Indri lives in small groups of 2–5 individuals, containing an adult pair and their young (Table 5-5). Home ranges of *Indri* were about 18 hectares at Pollock's site (Figure 5–36) and 35–40 hectares at Powzyk's. Population densities at various sites ranged from 5 and 16 individuals per sq. km. Home ranges are defended by ritualized "singing" battles. Besides giving these vocalizations during border encounters, *Indri* are very vocal and groups constantly exchange beautiful, but eerie, loud calls which resound throughout the forest. Some of these singing exchanges may last as long as 30 minutes, and can be heard as far as 1200 m away in dense forest, thus providing vocal contact and communication between at least 20 groups.

The call has many functions (Pollock 1975, 1986). It is probably mainly used as an inter-group distance maintaining signal. However, calls also give information on age, sex, and possibly breeding condition of neighboring individuals, as well as

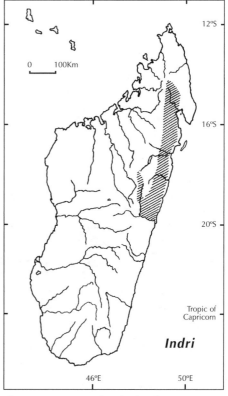

Figure 5-35 Map of *Indri* distribution. [From Mittermeier, et al., 1994]

density of groups in the area. The duration of songs may provide clues as to the size of neighboring groups. Olfactory communication is an important means of communication for most prosimians. *Indri* anogenital and cheek mark throughout the year and males increase marking behavior immediately preceding and during the mating season (Powzyk 1997).

I. indri awakes late and retires early and is active for only 5–6 hours daily, spending about 42% of this time feeding and half of it resting. Day ranges at Pollock's site were between 300-700 m and two distinct patterns of daily ranging behavior were observed: (1) When certain plant species were in flower, fruit, or contained new leaves, groups visited them to feed for 1-3 hours per day. These long, concentrated feeding bouts were then followed by a series of short feeding bouts on a diverse array of plants. (2) When there were no concentrated sources of food available, the groups ranged in a less predictable manner with short feeding progressions scattered throughout the day (Pollock 1979a). Using this pattern of daily ranging, in any one day, a noticeably constant number of species, from 5-12, was consumed by each group. Because of this, Pollock had the impression that "a precise control of dietetic variety existed" (1977:54). At Powzyk's site *Indri* daily path length was 330-1540 meters (avg.= 774 m).

Indri was observed to feed on 62 species during Pollock's 14 month study and 79 species during the 15 months studied by Powzyk. Pollock found that groups spent between 14 and 19% of their feeding time on the top ranking species and 50% on the top ranking 5 species. Fruit was preferred when available but immature

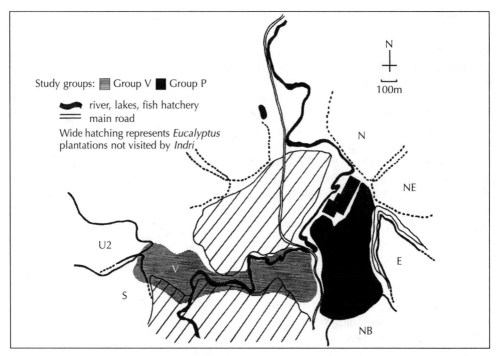

Figure 5-36 Local distribution of *Indri* groups in the forest of Analamazoatra near Perinet. Known (continuous lines) and approximate (discontinuous) boundaries are shown for ten *Indri* groups. [Adapted from Pollock, 1975]

leaves provided the bulk of the diet at both studies. Mature leaves were fed on regularly but in small amounts (Table 5-11). For example, Powzyk found that her two groups spent 71% of the time eating immature leaves, 16% fruit (mainly seeds), 7% flowers, and 5% on other parts, including mature leaves. Pollock found that most foods eaten by *Indri* came from the commonest trees of the forest.

Indri groups consistently consume small quantities of earth, often from sites with upturned trees. An analysis of one soil sample by Pollock failed to show high concentrations of specific minerals. Feeding on earth is referred to as geophagy and it is common in many non-human primate species and some human populations. Its function is not yet fully understood, though it has been related to folivory and to low nutrient soils (McKey 1978, Oates 1978, Davies et al. 1988, Browman and Gundersen 1993, Kay and Davies 1994).

The mean size of 18 groups censused by Pollock was 3.1 animals and 7 groups censused by Powzyk contained an average of 2.7 individuals. *Indri* appear to live in

Table 5-11. The Percentage of Feeding Observations on Each Food Part Category by Each Study Group Over the Whole Year.

Group	Leaf Shoots and Young Leaves	Fruit	Flowers	Mature Leaves	Unidentified
P	36.1	26.4	2.3	0.9	34.3
V	32.2	23.8	0.0	0.2	43.8

[From Pollock, 1977]

pair bonded groups, containing an adult male and female with their offspring. Although the group is very cohesive in its activity and daily ranging behavior, intra-group social activities are limited in variety and are infrequent. Only about 1.5% of the time being spent in any type of social behavior (Pollock 1975, Powzyk 1997). Females are dominant to males. Pollock observed mating to occur in January-February and only the adult pairs were involved in sexual behavior. The gestation period was between 17–22 weeks, and infant development is relatively slow. At 4–5 months of age, the infant moves from the stomach to the back of the mother and at 6 months of age the infant often moves alone but may fall 30 feet to the ground several times a day!

The second diurnal rain forest species of foli-vore is *Propithecus diadema* (Figure 5-37). This is the largest of the sifaka species (5–8 kg), and is variable in coat color from all white to all black, or with various shades of gold, gray or brown patches. There are four subspecies, with one of these being located in the extreme north (Figure 5-13) (Petter et al. 1977, Tattersall 1982). Population densities of this species are low (Table 5-5), and are declining mainly due to forest destruction. In 1990, Harcourt noted that there was little known about the ecology or behavior of this species. In 1986, Wright (1987, 1995, Meyers and Wright 1993) began a long-term study of *P. diadema edwardsi* at Ranomafana, and Hemingway (1995) completed a 13 month study (6/91–7/92) in the same forest. Powzyk (1997) studied *P. diadema diadema* for 18 months in the Mantadia National Park, north of Ranomafana.

Figure 5-37 *P. diadema.* [Photo by Robert W. Sussman]

The diademed sifaka is similar in many ways to its dry forest counterparts, Verreauxi's and Tattersall's sifakas. It has a very diverse diet. At Ranomafana the sifakas fed on 70–80 species of plants/year, with 16–22 species being eaten over 1% of the time and accounting for over 80% of the diet. At Mantadia, during Powzyk's 18 month study, 172 species were eaten by two groups, with an average of 26 species eaten per day. Overall, at both sites, leaves were eaten less than 50% of the time. Like Tattersall's sifaka, the major portion of the diademed diet was seeds and immature leaves (Meyers and Wright 1993, Hemingway 1995, Powzyk 1997). *P. diadema* fed at all forest levels, including the ground. As observed in the indri, soil was consumed regularly in small amounts.

There were interesting distinctions noted by Hemingway between the diet of diademed and Tattersall's sifakas. Both of these species are seed predators. However, the most important item in the diet of *P. diadema* was seeds, accounting for a third of the diet. The diademed sifakas fed on seeds whenever sources were available and, unlike Tattersall's, generally ate leaves when seeds were unavailable. *P. diadema* also differs from *P. tattersalli* in that leaves from vines were a dietary staple throughout the year at Ranomafana. At Mantadia, vines were important also, but 79% of feeding was from trees.

Seasonal variation in the diet of *P. diadema* was considerable. A few key plant species were preferred and were exploited during their peak availability throughout the year. For example, the top 5 species eaten each month made up between 60–95% of the diet, and many plant species were consumed for 2 to 3 months.

Hemingway (1995) notes that the percent of seeds and leaves in the diet of *P. diadema* at Ranomafana most closely resembles the folivore-gramnivore pattern seen in several species of colobine monkey.

As with the diet and ecology, the social system of the diademed sifaka is similar to that of other species of sifaka. Wright (1995) studied three groups over a nine year period. The groups varied in size from 3–9 individuals, each containing 1–2 adult males and females. The average adult male/female sex ratio was 1.5:1.8. Groups had home ranges of around 38 ha, much larger than their dry forest counterparts. Day ranges, however, were only around 670 m (350–850 m). There was little overlap of range boundaries and the group visited the entire home range every 5–6 days (Wright 1987). Females are dominant to males and appear to be philopatric, whereas males emigrate from their natal group at around 4–6 years of age (Wright 1995). Powzyk (1997) found similar group size and composition in her population. However, home ranges were smaller (33 and 42 ha in two study groups) and home ranges overlapped to some extent. Day ranges at Mantadia averaged 1650 m (600–2575 m).

The diademed sifaka has a very slow reproductive turnover. Females do not give birth until they are 4 years old and males first mate at 5 years of age. Average interbirth interval is 2 years and there is high infant mortality. In Wright's groups, 67% of the infants born over the 9 years of study did not reach adulthood (Wright 1995). Reproductive parameters are given in Table 5-7.

In comparing sympatric groups of *Indri* and *Propithecus,* Powzyk (1997) found that both species mainly ate immature leaves, fruit, and flowers. However, the indris spent significantly more time feeding on leaves, whereas the sifakas spent more time eating fruits and flowers. The top ten items in the diet of the sifakas contained high quantities of simple sugars and fats while those of the indris were low in these nutrients. There was little overlap in the main species eaten by the two prosimian species and *Propithecus* had a more diverse diet.

These dietary differences could be related to other behavioral differences that correlated with a more folivorous, less energenically expensive activity pattern in the *Indri* than in *Propithecus*. For example, *Indri* was active for shorter periods and spent more time resting than *Propithecus*. The latter species, in turn, spent more time traveling, feeding, socializing and playing, and had day ranges that were twice as long as *Indri*.

The third species of indriid found in the eastern rain forests is *Avahi laniger* (the woolly lemur) (Figure 5-38). There are two subspecies of *Avahi: A. l. laniger,* the eastern rain forest form, and *A. l. occidentalis,* with a restricted distribution in deciduous forests of the northwest (Figure 5-39). Both subspecies were more widely distributed in the recent past (Tattersall 1982, Mittermeier et al. 1994, Godfrey et al. in press), and a new population of the western form was recently discovered in the

Figure 5-38 *Avahi.* [Photo by I. Tattersall]

west (Mutschler and Thalmann 1990). Like others in this family the woolly lemur is folivorous and a VCL. It is the smallest of the indriids (@ 900–1200 gm). In fact, folivory is unusual for such a small primate, a characteristic *Avahi* shares with *Lepilemur*. The woolly lemur also is the only indriid that is nocturnal. There has been no long-term field research on this species but a number of short studies have been conducted.

Woolly lemurs are active mainly during the beginning and end of the night and spend most of the night resting (Ganzhorn et al. 1985, Harcourt 1991). At Ranomafana, for example, woolly lemurs spent 22% of the time feeding and 65% resting and grooming (Harcourt 1991). As with most folivores, low levels of activity are related to the low quality of the diet, and the difficulty in digesting fibrous and toxic foods (see Chapter 1). *Avahi* diet consists almost entirely of leaves, with only small amounts of flowers and fruits (Albignac 1981, Ganzhorn et al. 1985, Harcourt 1991).

Ganzhorn (1988) determined that the leaves eaten by *Avahi* were relatively high in protein and sugar, but did not contain alkaloids.

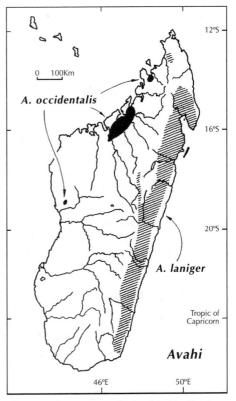

Figure 5-39 Map of *Avahi* distribution. [From Mittermeier, et al., 1994]

Alkaloids are secondary plant compounds that are toxic to many animals and are deterrents to leaf eating. Woolly lemurs were not deterred by high fiber content and probably are able to digest these structural carbohydrates and transform them to resorbable substances in the hindgut. Tannins, another type of oftent toxic secondary compound, did not deter the *Avahi,* and substantial amounts were consumed with their food. Although many primates do not eat items containing tannins, *Indri* also ingests substantial amounts of tannins (Ganzhorn 1985, 1988), as do howler monkeys and gorillas.

Lepilemur living sympatrically with *Avahi* did not avoid alkaloids but seemed to avoid tannins. The differences in choice of chemical contents of the leaves eaten may enable these two species to coexist. *Avahi* and *Indri* appear to choose leaves with similar chemical characteristics. However, *Indri* feeds higher in the trees and in larger trees than *Avahi,* whereas *Avahi* uses more uniform stands of younger trees (Ganzhorn 1988, 1989). The woolly lemur usually feeds at about 2–9 m high. It travels below the canopy approximately 5 m high, using vertical supports almost 90% of the time (Ganzhorn et al. 1985). *Avahi* lives in groups of 2–5 animals (Rand 1935, Petter 1962, Pollock 1975, Albignac 1981, Ganzhorn et al. 1985, Harcourt 1991). Group members often feed and travel alone but meet periodically during the night to rest and groom. At Ranomafana, group members were together 40% of the night (Harcourt 1991). The group also slept huddled together during the day. Sleeping sites were located in dense foliage 2-10 m high, or close to the ground hidden by tall grasses (Ganzhorn et al. 1985, Harcourt 1991). Over a three month period, one group at Ranomafana slept in the same site 80% of the days.

In the rain forest, *Avahi* has been observed to defend exclusive territories of 1–2 ha, with frequent spacing calls and agonistic defense of boundaries (Albignac 1981, Ganzhorn et al. 1985). Day ranges average 460 m (300–620 m) (Harcourt 1991). The western subspecies had larger, overlapping ranges (3–4 ha), called less, and had fewer intergroup agonistic encounters than the eastern form. It also was less active and moved less (180 m/day) (Razanahoera-Rakotomalala 1981, Albignac 1981). The population density of both subspecies has been estimated at around 70/km^2 (Ganzhorn 1988). Birth season occurs in August–September, with only one infant per female. Little is known about the reproductive behavior and biology of this species.

A. laniger, E. mongoz, and the New World monkey *Aotus* are the only nocturnal forms within their taxonomic group, and are assumed to be secondarily nocturnal (derived from a diurnal ancestor). There is still some question as to what factors may have led to nocturnal activity in these species (Wright 1985, 1989, Wright and Martin 1995, see Chapter 8). All live in monogamous family groups, and whether this is the result of ecological or phylogenetic factors, or some combination thereof, is currently unknown (Ganzhorn et al. 1985, Kinzey 1987).

BAMBOO AND GRASS EATERS OF THE EASTERN FORESTS

Hapalemur is specialized, both behaviorally (Petter and Peyrieras 1970, 1975, Glander et al. 1989) and morphologically (Milton 1978), for bamboo or grass eating. There are three species of bamboo lemur: *Hapalemur griseus, H. aureus,* and *H. simus. H. griseus* (the gentle lemur) is the smallest (700–1250 gm) and the most widespread of the three (Figure 5-40) (Tattersall 1982, Harcourt 1991, Smith and Jungers 1997). *H. aureus* (the golden bamboo lemur) is intermediate in size (1500–1800 gm), whereas *H. simus* (the greater bamboo lemur) is the largest (2300–2400 gm). Although their distribution in the recent past was much wider, the latter two species now are found only in a small region of the southeast (Godfrey and Vuillaumé-Randriamanantena 1986, Meier and Rumpler 1987). These two species are among the most endangered of all primates.

Three subspecies of the gentle lemur are recognized: *H. g. griseus* (Figure 5-41) has a wide distribution in the eastern forests; *H. g. alaotrensis* is located only in the dense reed beds on the shores of Lake Alaotra in the east; and *H. g. occidentalis* is found in two isolated populations in the west. There may be a fourth subspecies of gentle lemur in the southeast (Warter et al. 1987, see also Groves 1988). Within the forest, the distribution of *H. griseus griseus* is closely tied to grass

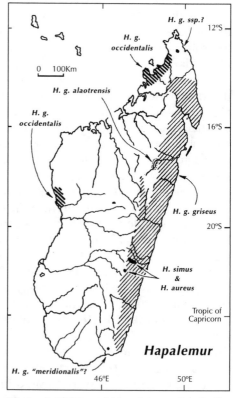

Figure 5-40 Map of *Hapalemur* spp. distribution. [From Mittermeier, et al., 1994]

or bamboo patches. Some bamboo patches are quite large and the individual plants can be up to 15 m high and 20 cm in diameter. *H. g. griseus* either runs horizontally through the dense vegetation or jumps to and from vertical stems of bamboo. It also frequently descends to the ground (Petter and Peyrieras 1970, 1975).

The diet of *H. g. griseus* has been studied most extensively at Ranomafana, where all three species of *Hapalemur* coexist. Ninety percent of the diet of this species consisted of new shoots, leaf bases, or stem pith of one species of bamboo, *Cephalostachyum perrieri* (Wright and Randrimanantena 1989, Glander et al. 1989). Small quantities of tree leaves, grasses, berries, and soil also were eaten (Wright 1986). These gentle lemurs fed in all habitats with bamboo, and from the ground to the canopy. They browsed continuously, eating rapidly, and almost half of the day was spent feeding. Gentle lemurs were quite active

Figure 5-41 *H. griseus.* [Photo by I. Tattersall]

throughout the day and group members slept together at night (Wright 1986).

This form of *Hapalemur* lives in small groups of from 3 to 6, and it appears that groups contain 1 adult male and 1–2 adult females. Home ranges of 6 to 15 ha have been reported and day ranges of 425 m (375–495) (Petter and Peyrieras 1970, Pollock 1986, Wright 1986). In one rain forest, Pollock (1979a) estimated a population density of about 47–62 individuals per sq. km. *H. g. griseus* usually gives birth to one infant and the gestation period is @ 140 days. The infant clings to the mother's back from birth and, in captivity, both the male and female carry the infant (Petter and Peryrieras 1970, 1975).

The subspecies *H. g. alaotrensis* differs from the other bamboo lemurs. In fact, in a 15 month study, Mutschler (1998) found that the Alaotran gentle lemur fed on a total of 11 different plant species and on soil. Dietary diversity was extremely low with focal animals eating a maximum of 6 plant species per month. Moreover, over 95% of feeding time was invested in four species of grass from the families of *Cyperaceae* and *Poaceae*. These gentle lemurs usually move in the dense vegetation or from one reed stem to another, often over the water (Petter and Peyriaras 1970, 1975). They are reported to be excellent swimmers.

Like other subspecies of gentle lemur, *H. g. alaotrensis* lives in small multi-male/multi-female groups of 2–9 individuals (Mutschler 1998). Median group size was 3 animals. No group contained more than 2 adult females and 3 adult males and approximately half of the groups contained 2 adult females. Although Petter and Peyriaras (1975) reported that, at certain times of the year, 30–40 animals often gather together, Mutschler (1998) did not observe these large gatherings. In fact, Nievergelt et al. (1998) found that these lemurs defended a large area of their home ranges and thought that they should be considered territorial. Mutschler (1998) observed both males and females emigrating, with females leaving their natal group as subadults and males transferring after reaching adulthood.

Unlike all other lemurs except *Daubentonia*, the Alaotran gentle lemur does not have a highly restricted birth season and births occur over a 6 month period. Forty percent of births were twins. Reproductive output was found to be significantly higher in groups with 2 adult females than in those with only one. This gentle lemur was found to have a crepuscular activity cycle which did not differ significantly between seasons (Mutschler 1998).

The marsh vegetation that supports the Alaotran gentle lemur has decreased dramatically over the past 50 years (Mutschler and Feistner 1995, Pidgeon 1996). This is mainly due to encroachment of agriculture on the natural marsh vegetation but this gentle lemur also is suffering from hunting pressure (Feistner and Rakotoarinosy 1993). In 1994, the total population size of this lemur was estimated to be approximately 7500 individuals (Mutschler and Feistner 1995).

H. aureus (the golden bamboo lemur) (Figure 5-42) was first discovered in 1985 (Meier et al. 1987). It has a very localized distribution in Southeastern Madagascar, around and including Ranomafana. This species feeds almost exclusively on giant bamboo, *C. cf viguieri;* 91% of the diet consists of shoots, leaf bases, pith and the viny portion of this bamboo. These particular food items are not eaten by the other bamboo lemurs. They are high in protein but also contain high quantities of cyanide, levels that

Figure 5-42 *H. aureus.* [Photo by B. Freed]

would kill most mammals (Wright and Randrimanantena 1989, Glander et al. 1989). In fact, the golden bamboo lemurs ingest 12 times the normal lethal dose of cyanide each day. The physiological mechanisms used to avoid cyanide poisoning are unknown.

The social system of this species appears to be similar to that of *H. griseus,* with groups containing 2–4 individuals. One study group had an exclusive home range of approximately 80 ha (Wright and Randrimanantena 1989). *H. aureus* is crepuscular and also may be active for part of the night. It has been reported that no more than 400 individuals of this species remain in the wild, though the means of determining this number was not specified (see Harcourt 1990).

Hapalemur simus (Figure 5-43) is also extremely rare and was thought to be extinct until rediscovered in 1964 and again in 1972 (Petter and Peyrieras 1970, Petter et al. 1977). Recently, preliminary surveys of this species have been conducted, and it was found in a few isolated localities in the southeast, including Ranomafana (Wright et al. 1987, Meier and Rumpler 1987). *H. simus* lives in quite low densities. Groups of up to seven individu-

Figure 5-43 *H. simus.* [Photo by Robert W. Sussman]

als have been observed to range over a home range of over 100 ha (Meier et al. 1987, Wright et al. 1987). Like the golden bamboo lemur, there may be only between 200-400 individuals still remaining, and *H. simus* is threatened by hunting and habitat destruction (Meier and Rumpler 1987).

At Ranomafana, the *H. simus* ate large quantities of giant bamboo, feeding mainly on the mature culms (stalks) (Wright et al. 1987, Glander et al. 1989). Unlike the portions of this plant fed on by *H. aureus,* the culm pith is low in protein, high in fiber, and contains no cyanide. The greater bamboo lemur was less specialized in food choice than the golden bamboo lemur, including small quantities of fruit and flowers in its diet (Meier et al. 1987).

The bamboo lemurs have exploited an open niche in Madagascar. Few mammals use this potentially abundant food resource. The giant panda feeds almost entirely on bamboo. There also are two species of Asian rodent, one South African rat, an African monkey *(Cercopithecus mitis),* and a South American monkey *(Callicebus moloch)* that eat bamboo to a more or less extent (Emmons 1981, Schaller et al. 1985, Wright 1985). It is quite remarkable that these three species of bamboo-eating lemurs coexist in some forests, and have divided this niche. This partitioning is accomplished by specializing on different species of bamboo, or on different parts of the same species, and food selection is mediated by the chemical and nutrient content of particular food items (Glander et al. 1989, see also Ganzhorn 1988).

PREDATION

Although larger than nocturnal lemurs, the same types of predators listed in the last chapter have been known to prey on diurnal lemurs. These include viverrids, especially the fossa *(Cryptoprocta ferox)* (Figure 5-44), feral cats, domestic dogs, snakes, and birds of prey. All diurnal lemurs react to the presence of potential predators, and they also react to one another's predator calls. Some species have particular reactions to different classes of predators. For example, *Lemur catta* has different reactions and alarm calls towards snakes, carnivores, and aerial predators (Sussman 1977, Macedonia and Polak 1989, Sauther 1989). Goodman et al. (1993) summarize the data available on predation on both nocturnal and diurnal lemurs (see also Rasoloarison et al. 1995, Powzyk 1997, Wright et al. 1997).

As summarized by Goodman, observed cases of predation by predatory birds have been reported on *L. catta, E. fulvus, Avahi,* and *P. verreauxi.* Many of the same species, and possibly *Hapalemur* have been identified in the scat of these birds. *L. catta, E. coronatus, E. fulvus, P. diadema, P. verreauxi* (and possibly *I. indri,* see Powzyk 1997) have been observed being preyed upon or in the feces of viverrids. Both the birds and viverrids eat lemurs of all ages. An adult *E. fulvus* has been observed being swallowed by a snake, *Acrantophis madagascariensis.*

Figure 5-44 Fossa. [Photo by R. Mittermeier]

It is interesting to note that the large subfossil eagle discovered by Goodman (1994, see Chapter 4) was a diurnal predator and modern interactions between predators and lemurs may have been more complex in the recent past. It is obvious, however, that predation has been an important selective pressure on both diurnal and nocturnal lemurs.

Humans are the most dangerous predators of most diurnal primates. Although hunting and eating of lemurs is taboo for many Malagasy, these animals are hunted in many regions (Richard and Sussman 1987). In fact, the extinct subfossil lemurs were all larger than the living forms, and it is likely that a combination of habitat disturbance and hunting led to their demise (Dewar 1984, 1997). It has been shown that low levels of indigenous human hunting pressure, carried out over a long period of time, can have devastating effects on primate population densities (Robinson and Redford 1994, Sussman and Phillips-Conroy 1995).

SUMMARY OF THE ECOLOGY AND SOCIAL BEHAVIOR OF DIURNAL LEMURIFORMES

As discussed in this and the last chapter, the suborder Lemuriformes contains 5 families endemic to Madagascar. There are 14 living genera and between 24 to 31 species, and possibly more. If one includes the 17 known species of subfossils, all of which coexisted with extant forms in the recent past, these animals exhibit an enormous range of body sizes, from the size of mice to that of gorillas. They also exhibit an extremely wide range of ecological and behavioral adaptations. Among the lemurs are the only diurnal forms of prosimians or Strepsirhini and the only primates that are cathemeral (mixed day and night activity). In this chapter, I discussed the families containing mainly diurnal and cathemeral species.

Madagascar is divided into two major climatic zones which differ radically in vegetation: the relatively dry occidental zone of the west and the rain forests of the eastern oriental zone. These two zones differ in fauna as well as vegetation and, for the most part, the species of lemur of Eastern and Western Madagascar also differ. These differences are more accentuated today than in the past because the reduction of the natural vegetation in the central plateau due to long-term climatic changes before the arrival of humans, and to those induced by humans over the past 2000 years (Richard and Dewar 1991). Major differences between faunal assemblages in the east and west are likely due to early, long-term isolation (Martin 1972). However, more recent changes, brought about by fragmented refugia isolating neighboring populations, have probably caused differentiation between some closely related species, or has led to subspeciation in a number of forms (Tattersall 1982).

Besides the major differences in ecological communities in which eastern and western forms are found the diurnal species of lemur are divided into two families, Indriidae and Lemuridae. The indriids are specialized in locomotor anatomy and gut morphology. They are vertical clingers and leapers and have digestive adaptations for a folivorous diet. *Propithecus verreauxi* is widely distributed in the west; *P. diadema* and *Indri indri* are eastern forms. The nocturnally active indriid, *Avahi laniger*, is found on both sides of the island.

Most Lemuridae are quadrupedal and do not have specialized digestive tract adaptations. They are mainly frugivorous. *Hapalemur* is an exception. It is often vertical in posture and locomotion, and is a bamboo and grass specialist. *Lemur catta* is the only lemurid found in the south. *Eulemur fulvus* is represented by subspecies in all coastal forested regions around the island except the south. In the north *E. mongoz, E. macaco,* and *E. coronatus* replace one another as one moves from the northwest to the northeast, and each is sympatric with *E. fulvus. E. fulvus, V. variegata,* and *E. rubriventer* are all relatively widely distributed in the east, as is *H. griseus. H. simus* and *H. aureus* are narrowly distributed in the east. *E. fulvus* and *H. griseus* are the only lemurids that occur in both the oriental and occidental zones.

As one moves from the dry forests of the south to wetter forests of the north, and finally, the rain forests of the east, the vegetation becomes more diverse. There is also greater faunal species diversity, and the lemur communities are more complex. The three diurnal species that coexist in the west are very different in their habitat preferences and diet. In the north of Madagascar, however, pairs of mainly frugivorous *Eulemur* species show much more subtle distinctions in habitat and food choices, as do coexisting, secondarily nocturnal and diurnal indriids.

In the eastern rain forests, some regions contain three sympatric species each of frugivorous lemurids, folivorous indriids, and bamboo lemurs. Studies available to date indicate that there are major differences in habitat selection between families and some genera, but that distinctions are much less noticeable within these taxonomic groups. In some cases, very subtle substrate preferences or seasonal dietary differences allow closely related taxa to coexist. In other cases, dietary preferences can be related to differences in the chemical properties of foods. The morphological and physiological correlates to these behavioral differences are not well understood and offer rich topics for future research.

Generally, diurnal lemuriformes fill similar ecological niches, and share a number of characteristics with diurnal monkeys and apes. In fact, the variability in ecology and behavior among lemurs is enormous. No other group of primates includes species that are cathemeral in activity cycle, or that switch from diurnal to nocturnal activity under different conditions. The locomotor and postural adaptations among the lemurs are extremely varied, especially when the subfossils are included. Among lemurs there are vertical clingers and leapers, slow and fast arboreal quadrupeds, partially terrestrial quadrupeds, and some subfossil forms were highly terrestrial, while others were sloth-like in their locomotion.

The diurnal lemurs eat many of the same food items fed upon by anthropoids such as fruit, leaves, flowers, seeds, gums, and insects. One fossil form, *Hadropithecus,* ate grass, much like living and fossil gelada baboons (Jolly 1970). Flower feeding might be more important for many lemurs than it is for any anthropoid, and some Malagasy flowering plants may have long-term coevolutionary relationships with some lemurs. Also, bamboo is an unusual food for mammals, and yet there are three coexisting species of bamboo lemurs in some eastern rain forests. In many lemurs, there are major seasonal differences in diet within the same population, and the diet of different populations of the same species may be quite different in different localities.

Diurnal lemurs live in similar types of groups as monkeys and apes. They are found in multi-male, pair bonded, and fission-fusion groups. In some species males migrate and females are philopatric, whereas in others both sexes migrate. Mating is monogamous or promiscuous, and often includes animals from outside of the group. As far as we know there is no propensity in any species for one male group structure, nor for the specialized polyandrous type of social system found among the callitrichids (see Volume II).

Even though the diurnal lemurs share the above characteristics with anthropoids, there are other lemurid traits which generally differ from those of monkeys and apes. Even with the great variability of social structure, lemurs are sexually monomorphic in both body size and in canine tooth morphology. Sexual dimorphism occurs in many anthropoids and is often related to different types of social systems. Lemurs generally live in small social groups of less than 15 individuals. They use their tooth combs for grooming instead of their hands, and olfactory communication is more important in lemurs than in anthropoids.

In many of the diurnal lemurs, females are dominant to males, a trait rare in mammals, generally. This may be related to high energetic costs of reproduction in female lemurs; lemurs have low basal metabolic rates, short gestation length in relation to body size, and high postnatal growth rates. These factors also may be related to the extremely short mating and births seasons of all lemurs (except the aye-aye and at least some bamboo lemurs), and to low survivership of adult females in relation to males in some species. This leads to sex ratios skewed in favor of adult males in some populations, another characteristic rare in other primates.

The juxtaposition of anthropoid-like characteristics with those unique to the lemuriformes, makes these animals ideal for testing many of the hypotheses concerning relationships between ecology and social structure and organization on the one hand, and individual behavior and morphophysiology on the other. However, only preliminary research has been conducted on many species, and most have not been studied in more than one locality. Long-term research on identified individuals has only begun, and so far only on a few populations. Furthermore, if deforestation continues at the same rate, we have predicted that very little forest will be left in 35 years (Sussman et al. 1994). What remains will be located on the steep mountain slopes of the east. If these predictions are correct, extinction rates will be very high, and there may be little time remaining to learn from these unique and wonderful animals.

ARCHAIC COEVOLUTIONARY POLLINATOR RELATIONSHIPS

Throughout the northwest of Madagascar there are a number of introduced populations of kapok trees and it is likely that *E. mongoz* acts as a major pollinator of these plants and other, similarly adapted plants. In Madagascar, there are only three species of bats and three of birds known to feed on nectar (Jenkins 1987, Langrand 1990, Kress et al. 1994). As we have seen in Chapter 4, a number of other nocturnal lemurs visit flowers and feed on nectar, at least during some months of the year. Outside of Madagascar, a number of species of non-flying mammals regularly visit flowers for food (see Nilsson et al. 1993 for bibliography). Besides lemurs, the most prominent groups are some of the phalangerid marsupials of Australia, rodents in Australia and South Africa, bushbabies in West Africa (see Chapter 3), and neotropical primates (see Volume II).

Just as *Daubentonia* may fill a woodpecker niche in the absence of these avian species in Madagascar, Sussman and Raven (1978) suggested that some lemurs, and possibly other non-flying mammal pollinators, may fill the role of bat-pollinators in communities in which flower visiting bats and birds are rare or absent. We also theorized that pollination by nonflying mammals may be an archaic diffuse coevolutionary system, and that some mammals may have been significant pollinators of angiosperms since the late Cretaceous. In this scenario bird and bat pollination would be derived. This would be consistent with the theory of primate-angiosperm coevolution developed in Chapter 2. Recent research by botanists suggests that several unrelated Malagasy angiosperms have evolved floral traits for lemur pollination (Kress 1993, Nilsson et al. 1993, Kress et al. 1994), and that some diurnal species of lemur also are major pollinators (Morland 1991, Overdorff 1992, Colquhoun 1997). These plant-pollinator relationships are consistent with the hypothesis that pollination by nonflying mammals is an archaic system which has persisted in some isolated geographic regions depauperate in significant flying vertebrate floral visitors (Sussman and Raven 1978, Kress 1993, Kress et al. 1994).

INFANTICIDE IN *Lemur catta*

It has recently been suggested that infanticide may be an adaptive aspect of the mating system of ringtailed lemurs (Pereira and Weiss 1991). The rationale behind such a strategy is that it may influence both male and female reproductive success. If a female loses her infant one year, it is suggested that her next infant has a better chance of survival. "By killing current infants, then, male lemurs may often advance by a full year the time of females' next successful reproduction . . . when infants die, for whatever reason, their fathers become unlikely to be chosen again as mates ("incompetent fathers")" (Pereira and Weiss 1991:149–150).

This hypothesis is problematic on a number of levels. First, if 80–85% of females give birth annually and approximately 50% of these infants perish each year, presumably a high proportion of these deaths should be due to infanticidal activities. However, infanticide has not been observed in ringtail lemurs either at Berenty or Beza Mahafaly. In fact, during two years of intensive observations of two groups (1988) and three groups (1992) at Beza Mahafaly, we failed to observe a single episode of male aggression directed toward any mother or infant (Sauther 1991, Sauther and Sussman 1993, Gould 1994). In fact, Gould observed male-infant affiliative behavior in all three study groups, as well as some male alloparental care.

Furthermore, many females successfully produce surviving infants two years in a row. As in many mammals, infant survival appears to be related to the age of the mother, with young prime and prime-aged females having above average infant survival rates and older females below average rates (Sussman 1991).

Second, a phenomenon of increased agonism directed by females toward males during the lactation period is hypothesized to be a strategy used by females to protect their infants. This has rarely been observed in studies of free-ranging or captive populations, despite hundreds of hours of observation on identified, habituated animals (Taylor 1986, Gould 1990, 1992, 1994, Lessnau pers. comm., Sauther 1992). Gould (1994) found no significant difference in female to male agonism during lactation and other times of the year.

Third, since females mate with a number of males both from within and outside of the group, it is difficult to conceive of either the female or the male recognizing the father of any particular infant. This pattern of multi-male mating has been observed both at Berenty and Beza Mahafaly and is clearly a part of the female mating strategy. It is therefore difficult to understand the advantages of infanticide in such a species.

In this species an interplay between stochastic events (predation, parasites, illness, falls, environmental perturbations) and the precocial development pattern are likely responsible for the high levels of infant death observed in ringtailed lemurs. Infanticide does not appear to be a viable reproductive tactic. The dramatic seasonal fluctuation of resource availability and the short mating season makes it essential that females mate with more than one male to ensure fertilization during the short estrous period. Female choice of non-group males, and intense competition results in the number of male mating strategies shown in Figure 5-12.

FEMALE DOMINANCE, REPRODUCTIVE STRESS, SEASON-ALITY, AND ADULT SEX RATIOS

Although in a number of species of primates females form coalitions against males (Smuts 1987), consistent dominance of adult females over males is rare in primates, and in mammals generally (Ralls 1976). In most mammals and primates, males are able to displace females, and are often larger and more aggressive. This is presumed to be a consequence of morphophysiological traits related to male competition for females, i.e., sexual selection (see Chapter 1). However, among many Malagasy lemurs, adult females consistently evoke submission from adult males. Female dominance has been reported in *Lemur catta, Eulemur macaco, Varecia variegata, Propithecus verreauxi, Propithecus diadema, Indri indri, Phaner furcifer, Microcebus murinus,* and *Daubentonia madagascariensis.*

Three hypotheses have been proposed to explain this characteristic: (1) In pair bonded species, the male may defer to the female as a parental investment, in order to ensure that their own young survive (Trivers 1972, Pollock 1979b); (2) Males deference may be a reproductive strategy in which males are conserving energy throughout the year for annual fights during the mating season (Hrdy 1981); (3) Finally, high reproductive costs of the females necessitate female priority of access to resources (Jolly 1984, Richard 1987, Young et al. 1990, Sauther 1992, 1993). This phenomena and its possible causes have been most thoroughly examined in *Lemur catta* (Sauther 1993) and *Propithecus verreauxi* (Richard 1987, Richard and Nicoll 1987, Richard et al. 1993).

The first of these hypotheses is not relevant here since few if any of these species has a monogamous mating system. Among ringtailed lemurs, Sauther (1992, 1993) found that the number of agonistic interactions in feeding situations did not change seasonally, and that there was no indication of male "chivalry." Males did not reduce feeding agonism during gestation and birth season, and feeding competition between the sexes significantly increased during lactation and weaning, the most stressful reproductive period. Females, however, always had feeding priority and could displace males from feeding sites. Thus, male deference does not appear to be a good explanation of female dominance, at least in ringtailed lemurs.

High reproductive costs, however, might explain the evolution of female dominance in these species. Among primates, Lemuriformes have relatively low basal metabolic rates and relatively high prenatal maternal investment (Richard and Nicoll 1987, Pollock 1989, Young et al. 1990, Ross 1992). Furthermore, they produce altricial infants which are fully supported by the mother during a rapid postnatal growth period. Finally, all species except the aye-aye and some Hapalemur, have very strict breeding seasonality tied to a seasonal availability of resources.

Sauther (1993:145) believes that, "given the close reproductive synchrony . . . . females will experience identical reproductive events and undergo similar reproductive stresses leading to high levels of interindividual resource competition." Thus, extra feeding competition from males would be a distinct disadvantage to females and their young, unless females had feeding priority. The larger *Propithecus* female has longer periods of gestation, lactation, and infant dependency, and this may exacerbate reproductive stress. Female sifakas may be particularly vulnerable during years of low resource availability. This in turn may explain the irregular pulses in mortality of females of reproductive age (Richard et al. 1991), and the male biased sex ratio among adults.

BIBLIOGRAPHY

Albignac, R. 1981. Variabilité de L'Organisation Territoriale et Ecologie de *Avahi laniger* (Lémurien Nocturne de Madagascar). *C.R. Acad. Sci. Paris 292* (Serie III):331–334.

Altmann, J. 1974. Observational Study of Behaviour: Sampling Methods. *Behaviour* 49:227–265.

Andrews, J.R. 1990. *A Preliminary Survey of Black Lemurs, Lemur macaco, in North West Madagascar.* Unpublished Report.

Andrews, J.R. In Prep. *Ecology of Black Lemurs in Primary Rainforest Versus Converted Forest.* Ph.D. Thesis, Washington University, St. Louis.

Andrews, J.R., Birkinshaw, C.R. 1998. A Comparison Between the Daytime and Night-Time Diet, Activity and Feeding Height of the Black Lemur, *Eulemur macaco* (Primates: Lemuridae), in Lokobe Forest Madagascar. *Folia Primatol.* 69, Suppl. 1:175–182.

Andriatsarafara, F.R. 1985. Notes sur les Rythmes D'activité et sur le Régime Alimentaire De *Lemur mongoz* Linnaeus, 1766 À Ampijoroa. Pp. 103–106 in *L'equilibre des Ecosystemes Forestiers: A Madagascar: Actes d'un Seminaire International.* L. Rakotovao; V. Barre ; J. Sayer, Eds., Gland, I.U.C.N.

Arbelot-Tracqui, V. 1983. *Etude Ethoecologique de Deux Primates Prosimiens:* Lemur coronatus *Gray et* Lemur fulvus sanfordi *Archbold.* These. L'Université De Rennes.

Balke, J., Porton, I., Boever, W., Junge, R., Birkel, R., Lasley, B., Colquhoun, I. 1988. *Physiologic and Behavioral Studies of Reproduction in the Black Lemur* (Lemur macaco macaco). Paper Delivered at Fifth Intl. Conference of Breeding Endangered Species in Captivity.

Birkel, R. 1987. *International Studbook for the Black Lemur, Linnaeus, 1966.* St. Louis, St. Louis Zoological Park.

Birkinshaw, C.R. 1995. *The Importance of the Black Lemur, Eulemur macaco, (Lemuridae, Primates), for Seed Dispersal in Lokobe Forest, Madadascar.* Ph.D. Thesis. University College London, London.

Birkinshaw, C.R., Colquhoun, I.C. 1998. Pollination of *Ravenala madagascariensis* and *Parkia madagascariensis* by *Euleumur macaco* in Madagascar. *Folia Primatol.* 69:252–259.

Bogart, M.H., Kumanamoto, A.T., Lasley, B.L. 1977. A Comparison of the Reproductive Cycle of Three Species of Lemur. *Folia Primatol.* 28:134–143.

Boskoff, K.J. 1978. The Oestrous Cycle of the Brown Lemur, *Lemur fulvus. J. Reproduct. and Fertility* 54:313–318

Brockman, D.K. 1994. *Reproduction and Mating System of Verreauxi's Sitaka*, Propithecus verreauxi, *at Beza Mahafaly, Madagascar.* Ph.D. Thesis. Yale University, New Haven.

Brockman, D.K., Whitten, P.L., Russell, E., Richard, A.R., Izard, M.K. 1995. Reproduction in Free-Ranging Male *Propithecus verreauxi:* Estrus and the Relationship Between Multiple Partner Matings and Fertilization. *Am. J. Primatol.* 36:313–325.

Brockman, D.K., Whitten, P.L., Richard, A.R., Schneider, A. 1998. Reproduction In Free-Ranging Male *Propithecus verreauxi:* The Hormonal Correlates of Mating and Aggression. *Am. J. Phys.Anthopol.* 105:137–151.

Browman, D.L., Gundersen, J.N. 1993. Altiplano Comestible Earths: Prehistoric and Historic Geophagy of Highland Peru and Bolivia. *Geoarchaeology* 8:413–425.

Budnitz, N., Dainis, K. 1975. *Lemur catta:* Ecology And Behavior. Pp. 219–236 in *Lemur Biology.* I. Tattersall; R.W. Sussman, Eds., New York, Plenum.

Burney, D.A. 1997. Theories and Facts Regarding Holocene Environmental Change Before and After Human Colonization. Pp. 75–89 in *Natural and Human-Induced Change in Madagascar.* B.D. Patterson; S. Goodman, Eds. Washington D.C., Smithsonian.

Burney, D.A., James, F.V., Grady, F.V., Rafamantananatsoa, J.-G., Ramilisonina, Wright, H.T., Cowart, J.B. 1997. Environmental Change, Extinction, and Human Activity: Evidence from Caves in NW Madagascar. *J. Biogeography* 24:755–767.

Burney, D.A., Macphee, R.D.M. 1988. Mysterious Island. *Nat. Hist.* 97:46–55.

Chalmers, N.R. 1972. Differences in Behaviour Between Some Arboreal and Terrestrial Species Of African Monkeys. Pp. 69–100 In *Comparative Ecology and Behaviour of Primates.* R.P. Michael; J.H. Crook, Eds., London, Academic.

Charles-Dominique, P. 1975. *Ecology and Behaviour of Nocturnal Primates.* New York, Columbia University Press.

Colquhoun, I.C. 1993. The Socioecology of *Eulemur macaco:* A Preliminary Report. Pp. 11–23 In *Lemur Social Systems and Their Ecological Basis.* P.M. Kappeler; J.U. Ganzhorn, Eds., New York, Plenum.

Colquhoun, I.C. 1997. *The Socioecology of the Black Lemur:* Eulemur macaco. Ph.D. Thesis, Washington University, St. Louis.

Colquhoun, I.C. 1998. Cathemeral Behavior of *Eulemur macaco macaco* at Ambato Massif, Madagascar. *Folia Primatol.* 69, Suppl. 1:22–34.

Conley, J.M. 1975. Notes on the Activity Pattern of *Lemur fulvus. J. Mammal.* 56:712–715.

Conroy, G.C. 1990. *Primate Evolution.* New York, Norton.

Crook, J.H., Aldrich-Blake, P. 1968. Ecological and Behavorial Contrasts Between Sympatric Ground-Dwelling Primates in Ethiopia. *Folia Primatol.* 8:192–227.

Curtis, D.J. 1997. *The Mongoose Lemur* (Eulemur mongoz): *A Study in Behaviour and Ecology.* Ph.D. Thesis. University of Zurich, Zurich.

Davies, A.G. 1991. Seed-Eating by Red Leaf Monkeys *(Presbytis rubicunda)* in Dipterocarp Forest of Northern Borneo. *Intl. J. Primatol.* 12:119–144.

Davies, A.G., Bennett, E.L., Waterman, P.G. 1988. Food Selection by Two South-East Asian Colobine Monkeys *(Presbytis rubicunda* and *Presbytis melalophos)* in Relation to Plant Chemistry. *Biol. J. Linn. Soc.* 34:33–56.

Dewar, R.E. 1984. Recent Extinctions In Madagascar: The Loss of the Subfossil Fauna. Pp. 574–593 in *Quarternary Extinctions: A Prehistoric Revolution.* P.S. Martin; R.G. Klein, Eds., Tucson, University of Arizona Press.

Dewar, R.E. 1997. Were People Responsible for the Holocene Extinctions in Madagascar, and How Will We Ever Know? Pp. 364–377 in *Natural Change and Human Impact in Madagascar.* S.M. Goodman; B.D. Patterson, Eds., Washington, D.C., Smithsonian Institution Press.

Eaglen, R.H., Boskoff, K.J. 1978. The Birth and Early Development of a Captive Sifaka, *Propithecus verrauxi coquereli. Folia Primatol.* 30:206–219.

Emmons, L.H. 1981. Morphological, Ecological, and Behavioral Adaptations for Arboreal Browsing in *Dactylomys dactylinus* (Rodentia, Echimyidae). *J. Mammol.* 62:183–189.

Feistner, A.T.C., Rakotoarinosy, M. 1993. Conservation of Gentle Lemur *Hapalemur griseus alaotrensis* at Lac Alaotra, Madagascar: Local Knowledge. *Dodo J. Widl Preserv. Trusts* 29:54–65.

Fornasieri, I., Roeder, J-J. 1993. Dominance Relationships in a *Lemur fulvus* Group. *Rev. Ecol.(Terre Vie)* 48:155–161.

Freed, B.Z., 1996. *A Comparative Ecological Study of Two Sympatric Species of Primates,* Lemur coronatus *and* Lemur fulvus sanfordi, *in Northern Madagascar.* Ph.D. Thesis. Washington University, St. Louis.

Freeland, W.J., Janzen, D.H. 1974. Strategies in Herbivory by Mammals: The Role of Plant Secondary Compounds. *Am. Natur.* 108:269–289.

Fuentes, A. 1999. Reevaluating Primate Monogamy. *Am. Anthropol.* 100:890–907.

Ganzhorn, J.U. 1985. Habitat Separation of Semifree-Ranging *Lemur catta* and *Lemur fulvus. Folia Primatol.* 45:76–88.

Ganzhorn, J.U. 1986a. Feeding Behavior of *Lemur catta* and *Lemur fulvus. Intl. J. Primatol.* 7:17–30.

Ganzhorn, J.U. 1986b. The Influence of Plant Chemistry on Food Selection by *Lemur catta* and *Lemur fulvus.* Pp. 21–29 in *Primate Ecology and Conservation.* J.G. Else; P.C. Lee, Eds., Cambridge, Cambridge University Press.

Ganzhorn, J.U. 1988. Food Partitioning Among Malagasy Primates. *Oecologia* 75:436–450.

Ganzhorn, J.U. 1989. Niche Separation of Seven Lemur Species in the Eastern Rainforest of Madagascar. *Oecologia* 79:279–286.

Ganzhorn, J.U., Abraham, J.P., Razanahoera-Rakotomalala, M. 1985. Some Aspects of the Natural History and Food Selection of *Avahi laniger. Primates* 26:452–463.

Ganzhorn, J.U., Langrand, O., Wright, P.C., O'Conner, S., Rakotosamimanana, B, Feistner, A.T.C., Rumpler, Y. 1996/1997. The State of Lemur Conservation in Madagascar. *Primate Conserv.* 17:70–86.

Garber, P. 1980. Locomotor Behavior and Feeding Ecology of the Panamanian Tamarin *(Sanguinus oedipus geoffroyi,* Callitrichidae, Primates). *Intl. J. Primatol.* 1:185–201.

Gebo, D.L. 1987. Locomotor Diversity in Prosimian Primates. *Am. J. Primatol.* 13:271–281.

Glander, K.E. 1982. The Impact of Plant Secondary Compounds on Primate Feeding Behavior. *Yrbk. Phys. Anthropol.* 25:1–18.

Glander, K.E., Wright, P.C., Siegler, D.S., Randrianasolo, V., Randrianasolo, B. 1989. Consumption of a Cyanogenic Bamboo by a Newly Discovered Species of Bamboo Lemur. *Am. J. Primatol.* 19:119–124.

Glander, K.E., Wright, P.C., Daniels, P.S. 1992. Morphometrics and Testicle Size of Rain Forest Lemur Species From Southeastern Madagascar. *J. Hum. Evol.* 22:1–17.

Godfrey, L.R., Jungers, W.L., Reed, K.E., Simons, E.L., Chatrath, P.S. 1997. Inferences About Past And Present Primate Communities In Madagascar. Pp. 218–256 in *Natural Change and Human Impact in Madagascar.* S.M. Goodman; B.D. Patterson, Eds., Washington, D.C., Smithsonian Institution Press.

Godfrey, L.R., Jungers, W.L., Simons, E.L., Chatrath, P.S., Rokotosamimanana, B. In Press. Past and Present Distributions of Lemurs in Madagascar. In *New Directions in Lemur Studies.* J. Ganzhorn; S. Goodman, Eds. New York: Plenum.

Godfrey, L.R., Vuillaume-Randriamanantena, M. 1986. *Hapalemur simus:* Endangered Lemur Once Widespread. *Primate Conserv.* 7:92–96.

Goodman, S.M. 1994. The Enigma of Antipredator Behavior in Lemurs: Evidence of a Large Extinct Eagle on Madagascar. *Intl. J. Primatol.* 15:129–134.

Goodman, S.M.; O'Conner, S.; Langrand, O. 1993. A Review of Predation on Lemurs: Implications for the Evolution of Social Behavior in Small, Nocturnal Primates. Pp. 51–66 in *Lemur Social Systems and Their Ecological Basis.* P.M. Kappeler; J.U. Ganzhorn, Eds., New York, Plenum.

Gould, L. 1990. The Social Development of Free-Ranging Infant *Lemur catta* at Berenty Reserve, Madagascar. *Intl. J. Primatol.* 11:297–318.

Gould. L. 1992. Alloparental Care in Free-Ranging *Lemur catta* at Berenty Reserve, Madagascar. *Folia Primatol.* 58:72–83.

Gould, L. 1994. *Patterns of Affiliative Behavior in Adult Male Ringtailed Lemurs* (Lemur catta) *at the Beza Mahafaly Reserve, Madagascar.* Ph.D. Thesis. Washington University, St. Louis.

Gould, L. 1996. Male-Female Affiliative Relationships in Naturally Occurring Ringtailed Lemurs *(Lemur catta)* at the Beza-Mahafaly Reserve, Madagascar. *Am. J. Primatol.* 39:63–78.

Gould, L. 1997. Intermale Affiliative Behavior in Ringtailed Lemurs *(Lemur catta)* at the Beza-Mahafaly Reserve, Madagascar. *Primates* 38:15–30.

Gould, L., Sussman, R.W., Sauther, M.L. 1999 Natural Disasters And Primate Populations: The Effects of a Two-Year Drought on a Naturally Occurring Population of Ringtailed Lemurs in Southwestern Madagascar. *Intl. J. Primatol.* 20:69–84.

Green, G.M., Sussman, R.W. 1990. Deforestation History of the Eastern Rainforests of Madagascar From Satellite Images. *Science.* 248:212–215.

Groves, C.P. 1988. Gentle Lemurs: New Species, and How They Are Formed. *Australian Primatol.* 2/3:9–12.

Hall, K.R.L. 1963. Some Problems in the Analysis and Comparison of Monkey and Ape Behaviour. Pp. 273–300 in *Classification and Human Evolution,* S.L. Washburn, Ed., New York, Aldine.

Hall, K.R.L. 1965. Experiment and Quantification in the Study of Baboon Behaviour in Its Natural Habitat. Pp. 29–42 in *The Baboon in Medical Research,* H. Vagtborg, Ed., San Antonio, University of Texas Press.

Harcourt, C. 1990. *Lemurs of Madagascar and the Comoros. The IUCN Red Data Book.* Gland, IUCN.

Harcourt, C. 1991. Diet And Behaviour of a Nocturnal Lemur, *Avahi laniger,* in the Wild. *J. Zool. Lond.* 223:667–674.

Harrington, J.E. 1975. Field Observations of Social Behavior of *Lemur fulvus fulvus* E. Geoffroy 1812. Pp. 259–279 in *Lemur Biology.* I. Tattersall; R.W. Sussman, Eds., New York, Plenum.

Harrington, J.E. 1978. Diurnal Behavior of *Lemur mongoz* at Ampijoroa, Madagascar. *Folia Primatol.* 29:291–302.

Hemingway, C. 1995. Feeding and Reproductive Strategies of the Milne-Edwards' Sifaka, *Propithecus diadema edwardsi.* Ph.D. Thesis. Duke University, Durham.

Hemingway, C. 1996. Morphology and Phenology of Seeds and Whole Fruit Eaten By Milne-Edwards' Sifaka, *Propithecus diadema edwardsi,* in Ranomafana National Park, Madagascar. *Intl. J. Primatol.* 17:637–660.

Hemingway, C. 1998. Selectivity and Variability in the Diet of Milne-Edwards' Sifaka, *Propithecus diadema edwardsi:* Implications For Folivory And Seed-Eating. *Intl. J. Primatol.* 19:355–378.

Hill, W.C.O. 1953. *Primates: Comparative Anatomy and Taxonomy. I. Strepsirhini.* Edinburgh, Edinburgh University Press.

Hrdy, S. 1981. *The Woman That Never Evolved.* Cambridge, Harvard University Press.

Jenkins, M.D., Ed., 1987. *Madagascar: An Environmental Profile.* Gland, IUCN.

Jolly, A. 1966. *Lemur Behaviour: A Madagascar Field Study.* Chicago, University Of Chicago Press.

Jolly, A. 1972. Troop Continuity and Troop Spacing in *Propithecus verrauxi* and *Lemur catta* at Berenty (Madagascar). *Folia primatol.* 17: 335–362.

Jolly, A. 1984. The Puzzle of Female Feeding Priority. Pp. 197–215 In *Female Primates: Studies by Women Primatologists.* M. Small, Ed., New York, Liss.

Jolly, A. 1985. *The Evolution of Primate Behavior.* New York, Macmillan.

Jolly, A., Oliver, W.L.R., O'Connor, S.M. 1982a. Population and Troop Ranges of *Lemur catta* and *Lemur fulvus* at Berenty, Madagascar: 1980 Census. *Folia Primatol.* 39:115–123.

Jolly, A., Gustafson, H., Oliver, W.L.R., and O'Conner, S.M. 1982b. *Propithecus verreauxi* Population and Ranging at Berenty, Madagascar, 1975 and 1980. *Folia Primatol.* 39:124–144.

Jolly, A., Rasamimanana, H.R., Kinnaird, M.F., O'Brien, T., Crowley, H.M., Harcourt, C.S., Gardner, S., Davidson, J.M. 1993. Territoriality in *Lemur catta* Groups During the Birth Season at Berenty, Madagascar. Pp. 85–109 in *Lemur Social Systems and Their Ecological Basis.* P.M. Kappeler; J.U. Ganzhorn, Eds., New York, Press.

Jolly, C.J. 1970. *Hadropithecus,* A Lemuroid Small-Object Feeder. *Man* 5–26.

Jungers, W.L., Godfrey, L.R., Simons, E.L., Prithijit, S.C. 1995. Subfossil *Indri indri* from Ankarana Massif of Northern Madagascar. *Am. J. Phys. Anthropol.* 97:357–366.

Kappeler, P.M. 1987. Reproduction in the Crowned Lemur *(Lemur coronatus)* in Captivity. *Am. J. Primatol.* 12:497–503.

Kappeler, P.M. 1990. The Evolution of Sexual Size Dimorphism in Prosimian Primates. *Am. J. Primatol.* 21:201–204.

Kappeler, P.M. 1993. Sexual Selection and Lemur Social Systems. Pp. 223–240 in *Lemur Social Systems and Their Ecological Basis.* P.M. Kappeler; J.U. Ganzhorn, Eds., New York,Plenum.

Kay, R.F. 1994. "Giant" Tamarin from the Miocene of Colombia. *Am. J. Phys. Anthropol.* 95:333–353.

Kay, R.N.B., Davies, A.G. 1994. Digestive Physiology. Pp. 229–249 in *Colobine Monkeys: Their Ecology, Behaviour and Evolution.* A.G. Davies; J.F. Oates, Eds., Cambridge, Cambridge University Press.

Kinzey, W.G. 1987. Monogamous Primates: A Primate Model for Human Mating Systems. Pp. 105–114 in *The Evolution of Human Behavior: Primate Models.* W.G. Kinzey, Ed., Albany, SUNY Press.

Klopfer, P.H., Boskoff, K.J. 1979. Maternal Behavior in Prosimians. Pp. 123–156 in *The Study of Prosimian Behavior.* G.A. Doyle; R.D. Martin, Eds., New York, Academic.

Klopfer, P.H., Dugard, J. 1976. Patterns of Maternal Care in Lemurs: III: *Lemur variegatus. Z. Tierpsychol.* 40:210–220.

Klopfer, P.H., Klopfer, M.S. 1970. Patterns of Maternal Care in Lemurs: I. Normative Description. *Z. Tierpsychol.* 27:984–996.

Koechlin, J. 1972. Flora And Vegetation of Madagascar. Pp. 145–226 in *Biogeography and Ecology Of Madagascar.* R. Battistini; G. Richard-Vindard, Eds., The Hague, W. Junk.

Koenders, L. 1989. An Eco-Ethological Comparison of *Lemur fulvus* and *Lemur macaco. Hum. Evol.* 4:187–193.

Koyama, N. 1988. Mating Behavior of Ring-Tailed Lemurs *(Lemur catta)* at Berenty, Madagascar. *Primates* 29:163–175.

Kress, J.H., Conley, J.M., Eaglen, R.H., Ibanez, A.E. 1978. The Behavior of the *Lemur variegatus,* Kerr 1972. *Z. Tierpsychol.* 48:87–99.

Kress, W.J., Schatz, G.E., Andrianifahanana, M., Moreland, H.S. 1994. Pollination of *Ravenala madagascariensis* (Strelitziaceae) by Lemurs in Madagascar: Evidence for an Archaic Coevolutionary System? *Am. J. Botany* 81:542–551.

Kress, W.J. 1993. Coevolution of Plants and Animals: Pollination of Flowers by Primates in Madagascar. *Current Science:*253–257.

Kubzdela, K.S., Enafa, Esoahere. 1995. Group Size, Social Structure, and Population Dynamics In *Propithecus v. verreauxi.* Pp. 32–33 in *Environmental Change in Madagascar.* B.D. Patterson; S.M. Goodman; J.L. Sedlock, Eds., Chicago, The Field Museum.

Kubzdela, K.S. 1997. *Sociodemography in Diurnal Primates: The Effects of Group Size and Female Dominance Rank on Intra-Group Spatial Distribution, Feeding Competition, Female Reproductive Success, and Female Dispersal Patterns in White Sifaka,* Propithecus verreauxi verreauxi. Ph.D. Thesis. University of Chicago, Chicago.

Langrand, O. 1990. *Guide to the Birds of Madagascar.* New Haven, Yale University.

Lehner, P.N. 1996. *Handbook of Ethological Methods, Second Edition.* New York, Garland.

Macedonia, J.M., Polak, J.F. 1989. Visual Assessment of Avian Threat in Semi-Captive Ringtailed Lemurs *(Lemur catta). Behavior* 111:291–304.

Martin, R.D. 1972. Adaptive Radiation and Behavior of the Malagasy Lemurs. *Phil. Trans. Roy. Soc. Lond. (Series B)* 264:295–352.

Martin, R.D. 1990. *Primate Origins and Evolution: A Phylogenetic Reconstruction.* Princeton, Princeton University Press.

Martin, R.D. 1992. Goeldi and the Dwarfs: The Evolutionary Biology of the Small New World Monkeys. *J. Hum. Evol.* 22:367–393.

Mitani, J.C., Rodman, P.S. 1979. Territoriality: The Relation of Ranging Pattern and Home Range Size to Defendability, with an Analysis of Territoriality Among Primates. *Behav. Ecol. Sociobiol.* 5:241–251.

Mittermeier, R.A., Tattersall, I., Konstant, W.R., Meyers, D.M., Mast, R.B. 1994. *Lemurs of Madagascar.* Washington, D.C., Conservation International.

Meier, B., Rumpler, Y. 1987. Preliminary Survey of *Hapalemur simus* and of a New Species of *Hapalemur* in Eastern Betsileo, Madagascar. *Primate Conserv.* 8:40–43.

Meier, B., Albignac, R., Peyrieras, A., Rumpler, Y., Wright, P. 1987. A New Species of *Hapalemur* (Primates) from South-East Madagascar. *Folia Primatol.* 48:211–215.

Meyers, D. 1988. Behavorial Ecology of *Lemur fulvus rufus* in Rain Forest in Madagascar. *Am. J. Phys. Anthropol.* 75:250.

Meyers, D.M. 1993. *The Effects of Resource Seasonality on Behavior and Reproduction in The Golden-Crowned Sifaka* (Propithecus tattersalli, *Simons, 1988*) *in Three Malagasy Forests.* Ph.D. Thesis. Duke University, Durham.

Meyers, D.M., Wright, P.C. 1993. Resource Tracking: Food Availability and *Propithecus* Seasonal Reproduction. Pp. 179–192 in *Lemur Social Systems and Their Ecological Basis.* P.M. Kappeler; J.U. Ganzhorn, Eds., New York: Plenum.

McKey, D.B. 1978. Soils, Vegetation, and Seed-Eating by Black Colobus Monkeys. Pp. 423–438 in *The Ecology of Arboreal Folivores.* G.G. Montgomery, Ed., Washington, DC, Smithsonian Institution.

Milton, K. 1978. Role of the Upper Canine and p^2 in Increasing the Harvesting Efficiency of *Hapalemur griseus* Link, 1795. *J. Mammol.* 59:188–190.

Milton, K. 1984. The Role of Food-Processing Factors in Primate Food-Choice. Pp. 241–251 In *Adaptations for Foraging in Nonhuman Primates: Contributions to an Organismal Biology of Prosimians, Monkeys, and Apes.* P.S. Rodman, J.G.H. Cants, Eds., New York, Columbia University Press.

Mitani, J.C., Rodman, P.S. 1979. Territoriality: The Relation of Ranging Pattern and Home Range Size to Defendability, with an Analysis of Territoriality Among Primate Species. *Behav. Ecol. Sociobiol.* 5:241–251.

Moreland, H.S. 1991. *Social Organization and Ecology of Black and White Ruffed Lemurs* (Varecia variegata variegata) *in Lowland Rain Forest, Nosy Mangabe, Madagascar.* Ph.D. Thesis. Yale University, New Haven.

Moreland, H.S. 1993. Seasonal Behavorial Variation and its Relationship to Thermoregulation In Ruffed Lemurs *(Varecia variegata).* Pp. 193–203 in *Lemur Social Systems and their Ecological Basis.* P.M. Kappeler; J.U. Ganzhorn, eds., New York, Plenum.

Mutschler, T. 1998. *The Alaotran Gentle Lemur* (Hapalemur griseus alaotrensis): *A Study in Behavioural Ecology.* Ph.D. Thesis. Zurich University, Zurich.

Mutschler, T., Feistner, A.T.C. 1995. Conservation Status and Distribution of the Alaotran Gentle Lemur *Hapalemur griseus alaotrensis. ORYX* 29:267–274.

Mutschler, T., Thalmann, U. 1990. Sighting of *Avahi* (Woolly Lemur) in Western Madagascar. *Primate Conserv.* 11:15–17.

Napier, J.R. Walker, A.C. 1967. Vertical Clinging and Leaping—A Newly Recognized Category of Locomoter Behaviour in Primates. *Folia Primatol.* 6:204–19.

Nievergelt, C, Mutschler, T., Feistner, A.T.C. 1998. Group Encounters and Territoriality in Wild Alaotran Gentle Lemur *(Hapalemur griseus alaotrensis). Am. J. Pirmatol.* 46:251–258.

Nillson, L.A., Rabakonandrianina, E., Pettersson, B., Grunmeier, R. 1993. Lemur Pollination in the Malagasy Rainforest Liana *Strongylodon craveniae* (Leguminosae). *Evol. Trends In Plants* 7:49–56.

Oates, J.F. 1978. Water-Plant and Soil Consumption by Guereza Monkeys *(Colobus guereza):* A Relationship With Minerals and Toxins in the Diet? *Biotropica* 10:241–253.

O'Connor, S.M. 1987. *The Effect of Human Impact on Vegetation and the Consequences to Primates in Two Riverine Forests, Southern Madagascar.* Ph.D. Thesis, Cambridge University, Cambridge.

Overdorff, D.J. 1991. *Ecological Correlates to Social Structure in Two Prosimian Primates:* Eulemur fulvus rufus and Eulemur rubriventer *in Madagascar.* Ph.D. Thesis. Duke University, Durham.

Overdorff, D.J. 1992. Differential Patterns in Flower Feeding by *Eulemur fulvus rufus* and *Eulemur rubriventer* in Madagascar. *Am. J. Primatol.* 28:191–203.

Overdorff, D.J. 1993a. Similarities, Differences, and Seasonal Patterns in the Diets of *Eulemur rubriventer* and *E. fulvus rufus* in the Ranomafana National Park, Madagascar. *Intl. J. Primatol.* 14:721–753.

Overdorff, D.J. 1993b. Ecological and Reproductive Correlates to Range Use in Red-Bellied Lemurs *(Eulemur rubriventer)* and Rufous Lemurs *(Lemur fulvus rufus).* Pp. 167–178 in *Lemur Social Systems and Their Ecological Basis.* P.M. Kappeler; J.U. Ganzhorn, Eds., New York, Plenum.

Overdorff, D.J. 1996a. Ecological Correlates to Social Structure in Two Prosimian Primates in Madagascar. *Am. J. Phys. Anthropol.* 100:487–506.

Overdorff, D.J. 1996b. Ecological Correlates to Activity and Habitat Use in Two Prosimian Primates: *Eulemur rubriventer* and *Lemur fulvus rufus* in Madagascar. *Am. J. Primatol.* 40:327–342.

Overdorff, D.J., Merenlender, A.M., Talata, P., Telo, A., Forward, Z.A. in Press. Life History of L*emur fulvus rufus* from 1988–1997 in Southeastern Madagascar. *Am. J. Phys. Anthropol.*

Pereira, M.E. 1991. Asynchrony Within Estrous Synchrony Among Ringtailed Lemurs (Primates: Lemuridae). *Physiol. Behav.* 49:47–52.

Pereira, M.E., Klepper, A., Simons, E.L. 1987. Tactics of Care for Young Infants by Forest-Living Ruffed Lemurs *(Varecia variegata variegata):* Ground Nests, Parking, and Biparental Guarding. *Am. J. Primatol.* 13:129–144.

Pereira, M.E., Seeligson, M.L., Macedonia, J.M. 1988. The Behavioral Repertoire of the Black-and-White Ruffed Lemur, *Varecia variegata variegata* (Primates, Lemuridae). *Folia Primatol.* 51:1–32.

Pereira, M.E., Kaufman, R., Kappeler, P.M., and Overdorff, D.J., 1990. Female Dominance Does Not Characterize All of the Lemuridae. *Folia Primatol.* 55:96–103.

Pereira, M.E., McGlynn, C.A. 1997. Special Relationships Instead of Female Dominance for Redfronted Lemurs, *Eulemur fulvus fulvus. Am. J. Primatol.* 43:239–258.

Pereira, M.E. , Weiss, M.L., 1991. Female Mate Choice, Male Migration, and the Threat of Infanticide in Ringtailed Lemurs. *Behav. Ecol. Sociobiol.* 28:141–152.

Petter-Rosseaux, A. 1962. Recherches Sur La Biologie De La Réproduction Des Primates Inférieurs. *Mammalia* 26 Suppl. 1:1–88.

Petter, J.-J. 1962. Recherches sur L'écologie et L'éthologie des Lémuriens Malgaches. *Mem. Mus. Natl. Hist. Nat. Paris* 27:1–146.

Petter, J.-J. 1965. The Lemurs of Madagascar. Pp. 292–319 in *Primate Behaviour: Field Studies of Monkeys and Apes.* I. Devore, ed., New York, Holten, Rinehart, and Winston.

Petter, J.-J., Peyrieras, A. 1970. Observations Éco-Éthologiques sur les Lémuriens Malgaches du Genre *Hapalemur. Terre Vie* 24:356–382.

Petter, J.-J., Peyrieras, A. 1975. Preliminary Notes on the Behaviour and Ecology of *Hapalemur griseus.* Pp. 281–286 in *Lemur Biology.* I. Tattersall; R.W. Sussman, Eds., New York, Plenum.

Petter, J.-J., R. Albignac, Y. Rumpler. 1977. *Faune De Madagascar 44: Mammiferes Lemuriens (Primates, Prosimiens).* Paris, Orstom/Cnrs.

Pidgeon, M. 1996. An Ecological Survey of Lake Alaotra and Selected Wetlands of Central and Eastern Madagascar in Analysing the Demise of Madagascar Pochard *Aythya innotata* L. Wilmé. Antananarivo, *WWF/MBG.*

Pollock, J.I. 1975. Field Observations in the *Indri indri:* A Preliminary Report. Pp. 287–311 in *Lemur Biology.* Tattersall, I.; R.W. Sussman, Eds., New York, Plenum.

Pollock, J.I. 1977. The Ecology and Sociology of Feeding in *Indri indri.* Pp. 37–69 in *Primate Ecology: Studies of Feeding and Ranging Behavior in Lemurs, Monkeys, and Apes.* T.H. Clutton-Brock, Ed., New York, Academic.

Pollock, J.I. 1979a. Spatial Distribution and Ranging Behaviour in Lemurs. Pp. 359–409 in *The Study of Prosimian Behaviour.* G.A. Doyle; R.D. Martin, eds., New York, Academic.

Pollock, J.I. 1979b. Female Dominance in *Indri indri. Folia Primatol.* 31:143–164.

Pollock, J.I. 1986. The Song of the Indris *(Indri indri,* Primates: Lemuroidea) Natural History, Form and Function. *Intl. J. Primatol.* 225–264.

Pollock, J.I. 1989. Intersexual Relationships Amongst Prosimians. *Hum. Evol.* 4:133–143.

Powzyk, J. 1997. *The Socio-Ecology of Two Sympatric Indriids:* Propithecus diadema diadema *and* Indri indri, *a Comparison of Feeding Strategies and Their Possible Repercussions on Specis-Specific Behaviors.* Ph.D. Thesis. Duke University, Durham.

Pusey, A.E., Packer, C. 1987. Dispersal and Philopatry. Pp. 250–266 in *Primate Societies.* B.B. Smuts; D.L. Cheney; R.M. Seyfarth; R.W. Wrangham; T.T. Struhsaker, Eds., Chicago, University of Chicago Press.

Rabinowitz, P.D., Coffin, M.G., Falvey, D. 1983. The Separation of Madagascar and Africa. *Science* 220:67–69.

Rand, A.L. 1935. On The Habits of Some Madagascar Mammals. *J. Mammal.* 16:89–104.

Ralls, K. 1976. Mammals in Which Females Are Larger Than Males. *Quart. Rev. Biol.* 51:245–276.

Rasmussen, D.T. 1985. A Comparative Study of Breeding Seasonality and Litter Size in Eleven Taxa of Captive Lemurs *(Lemur* and *Varecia). Intl. J. Primatol.* 6:501–517.

Rasmussen, M.A. 1998a. Variability in the Cathemeral Activity of Two Lemurid Primates at Ampijoroa, Northwest Madagascar. *Am. J. Phys. Anthropol.* Suppl. 26:183.

Rasmussen, M.A. 1998b. Ecological Influences on Cathemeral Activity in the Mongoose Lemur *(Eulemur mongoz)* At Ampijoroa, Northwest Madagascar. *Am. J. Primatol.* 45:202

Rasmussen, M.A. In Prep. Ph.D. Thesis. Duke University, Durham.

Razanahoera-Rakotomalala, M. 1981. *Les Adaptations Alimentaires Comparees de Deux Lemuriens Folivores Sympatriques:* Avahi *Jourdan, 1834,* Lepilemur *I. Geofferey, 1851.* These Doctorat. University De Madagascar, Tananarive.

Richard, A.F. 1973. *Social Organization and Ecology of* Propithecus verreauxi *Grandidier 1867.* Ph.D. Thesis. University of London, London.

Richard, A.F. 1974. Patterns of Mating in *Propithecus verrauxi.* Pp. 49–74 in *Prosimian Biology.* R.D. Martin; G.A. Doyle; A.C. Walker, Eds., London, Duckworth.

Richard, A.F. 1976. Preliminary Observations on the Birth and Development of *Propithecus verrauxi* to the Age of Six Months. *Primates* 17:357–66.

Richard, A.F. 1977. The Feeding Behaviour of *Propithecus Verrauxi.* Pp. 72–96 in *Primate Ecology: Studies of Feeding and Ranging Behavior in Lemurs, Monkeys, and Apes.* T.H. Clutton-Brock, ed., London, Academic.

Richard, A.F. 1978a. *Behavioral Variation: Case Study of a Malagasy Lemur.* Lewisburg, Bucknell University Press.

Richard, A.F. 1978b. Variability in the Feeding Behavior of the Malagasy Prosimian, *Propithecus verrauxi:* Lemuriformes. Pp. 519–553 in *The Ecology of Arboreal Folivores.* G.G. Montogomery, ed., Washington D.C., Smithsonian Institution.

Richard, A.F. 1985. Social Boundaries in a Malagasy Prosimian, The Sikafa *(Propithecus verreauxi). Intl. J. Primatol.* 6:553–568.

Richard, A.F. 1987. Malagasy Prosimians: Female Dominance. Pp. 25–33 in *Primate Societies.* B.B. Smuts; D.L. Cheney; R.M. Seyfarth; R.W. Wrangham; T.T. Struhsaker, eds., Chicago, University of Chicago Press.

Richard, A.F. 1992. Aggressive Competition Between Males, Female-Controlled Polygyny and Sexual Monomorphism in A Malagasy Primate, *Propithecus verrauxi. J. Hum. Evol.* 22:395–406.

Richard, A.F., Dewar, R.E. 1991. Lemur Ecology. *Ann. Rev. Ecol. Syst.* 22:145–175.

Richard, A.F., Nicoll, M. 1987. Female Social Dominance and Basal Metabolism in a Malagasy Primate, *Propithecus verrauxi. Am. J. Primatol.* 12:309–314.

Richard, A.F., Rakotomanga, P., Schwartz, M. 1991. Demography of *Propithecus verreauxi* at Beza Mahafaly: Sex Ratio, Survival, and Fertility, 1984–1988. *Am. J. Phys. Anthropol.* 84:307–322.

Richard, A.F., Rakotomanga, P., Schwartz, M. 1993. Dispersal by *Propethicus verrauxi* at Beza Mahafaly, Madagascar. *Am. J. Primatol.* 30:1–20.

Richard, A.F., Rakotomanga, P., Sussman, R.W. 1987. Beza Mahafaly: Formation et Mesures de Conservation. Pp. 41–43 in *Priorites en Matiere de Conservation des Especes a Madagascar.* R.A. Mittermeier; L.H. Rakotovao; V. Randrianasolo; E.J. Sterling; D. Devitre, Ed., Gland, IUCN

Richard, A.F., Sussman, R.W. 1987. Framework for Primate Conservation in Madagascar. Pp. 329–341 in *Primate Conservation in the Tropical Rain Forest.* C.W. Marsh; R.A. Mittermeier, eds., New York, Alan R. Liss.

Rigamonti, M.M. 1993. Home Range and Diet in Red-Ruffed Lemurs *(Varecia variegata rubra)* on the Masoala Peninsula, Madagascar. Pp. 25–39 in *Lemur Social Systems and their Ecological Basis.* P.M. Kappeler, J.U. Ganzhorn, eds., New York, Plenum.

Robinson, J.G., Redford, K.H. 1994. Measuring the Sustainability of Hunting in Tropical Forests. *Oryx* 28:249–256.

Ross, C. 1992. Basal Metabolic Rate, Body Weight and Diet in Primates: An Evaluation of the Evidence. *Folia Primatol.* 58:7–23.

Sauther, M.L. 1989. Antipredator Behavior in Troops of Free-Ranging *Lemur catta* at Beza Mahafaly Special Reserve, Madagascar. *Int. J. Primatol.* 10:595–606.

Sauther, M.L. 1991. Reproductive Behaviour of Free-Ranging *Lemur catta* at Beza Mahafaly Special Reserve, Madagascar. *Am. J. Phys. Anthropol.* 84:463.

Sauther, M.L. 1992. *The Effect of Reproductive State, Social Rank and Group Size on Resource Use Among Free-Ranging Ringtailed Lemurs* (Lemur catta) *of Madagascar.* Ph.D. Thesis. Washington University, St. Louis.

Sauther, M.L. 1993. Resource Competition in Wild Populations of Ringtailed Lemurs *(Lemur catta):* Implications for Female Dominance. Pp. 135–152 in *Lemur Social Systems and Their Ecological Basis.* P.M. Kappeler; J.U. Ganzhorn, eds., New York, Plenum.

Sauther, M.L. 1998. Interplay of Phenology and Reproduction in Ring Tailed Lemurs: Implications for Ring Tailed Lemur Conservation. *Folia Primatol.* 69, Suppl. 1:309–320.

Sauther, M.L., Sussman R.W. 1993. A New Interpretation of the Social Organization and Mating System of the Ringtailed Lemur *(Lemur catta).* Pp. 111–121 in *Lemur Social Systems And Their Ecological Basis.* P.M. Kappeler; J.U. Ganzhorn, eds., New York, Plenum.

Sauther, M.L., Sussman, R.W., L. Gould. In Press. The Socioecology of the Ringtailed Lemur: 35 Years of Research. *Evol. Anthropol.*

Schaller, G., Jinchu, H., Wenshi, P., Jing, Z. 1985. *Giant Pandas of Wolong.* Chicago, University of Chicago Press.

Simons, E.L. 1972. *Primate Evolution: An Introduction to Man's Place in Nature.* New York, Macmillan.

Simons, E.L. 1997. Lemurs: Old and New. Pp. 142–166 in *Natural Change and Human Impact in Madagascar.* S.M. Goodman; B. Patterson, eds., Washington, D.C., Smithsonian Institution.

Smith, R.J., Jungers, W.L. 1997. Body Mass in Comparative Primatology. *J. Hum. Evol.* 32:523–559.

Smuts, B.B. 1987. Sexual Competition and Mate Choice. Pp. 385–399 in *Primate Societies.* B.B. Smuts; D.L. Cheney; R.M. Seyfarth; R.W. Wrangham; T.T. Struhsaker, eds., Chicago, University of Chicago Press.

Strier, K. 1994. Myth of the Typical Primate. *Yrbk. Phys. Anthropol.* 37:233–271.

Sussman, R.W. 1972. *An Ecological Study of Two Madagascan Primates: Lemur fulvus rufus Audebert and Lemur catta linnaeus.* Ph.D. Thesis, Duke University, Durham.

Sussman, R.W. 1974. Ecological Distinctions in Sympatric Species of Lemur. Pp. 75–108 in *Prosimian Biology.* R.D. Martin; G.A. Doyle; A.C. Walker, Eds., London, Duckworth.

Sussman, R.W. 1975. A Preliminary Study of the Behavior and Ecology of *Lemur fulvus rufus* Audebert, 1800. Pp. 237–258 in *Lemur Biology.* I. Tattersall; R.W. Sussman, eds., New York, Plenum.

Sussman, R.W. 1977. Socialization, Social Structure, and Ecology of Two Sympatric Species of Lemur. Pp. 515–529 in *Primate Social Development: Biological, Social, and Ecological Determinants.* S. Chevalier-Skolnikoff; F. Poirier, eds., New York, Garland.

Sussman, R.W. 1988. The Adaptive Array of the Lemurs of Madagascar. *Monogr. Syst. Bot. Missouri Bot. Gard.* 25:215–226.

Sussman, R.W. 1991. Demography and Social Organization of Free-Ranging *Lemur catta* in the Beza Mahafaly Reserve, Madagascar. *Am. J. Phys. Anthropol.* 84:43–58.

Sussman, R.W. 1992. Male Life History and Intergroup Mobility Among Ringtailed Lemurs *(Lemur catta). Intl. J. Primatol.* 13:395–413.

Sussman, R.W., Green, G.M., Sussman, L.K. 1994. Satellite Imagery, Human Ecology, Anthropology, and Deforestation in Madagascar. *Hum. Ecol.* 22:333–354. 64:419–449.

Sussman, R.W., Phillips-Conroy, J. 1995. A Survey of the Diversity and Density of the Primates of Guyana. *Intl. J. Primatol.* 16:761–791.

Sussman, R.W., Rakotozafy, A. 1994. Plant Diversity and Structural Analysis of a Tropical Dry Forest in Southwestern Madagascar. *Biotropica* 26:241–254.

Sussman, R.W., Raven, P.H. 1978. Pollination by Lemurs and Marsupials: An Archaic Coevolutionary System. *Science:* 200:731–736.

Sussman, R.W., Richard, A. 1974. The Role of Aggression Among Diurnal Prosimians. Pp. 49–76 in *Primate Aggression, Territoriality, and Xenophobia.* R. Holloway, ed., New York, Academic.

Sussman, R.W., Tattersall, I. 1976. Cycles of Activity, Group Composition, and Diet of *Lemur mongoz mongoz* Linnaeus, 1766 in Madagascar. *Folia Primatol.* 26:270–283.

Tattersall, I. 1976. Group Structure and Activity Rhythm in *Lemur mongoz* (Primates, Lemuriformes) on Anjouan and Moheli Islands, Comoro Archipelago. *Anthropol. Pap. Am. Mus. Nat. Hist.* 53:369–380.

Tattersall, I. 1977. Ecology and Behavior of *Lemur fulvus mayottensis* (Primates, Lemuriformes). *Anthrop. Pap. Amer. Mus. Nat. Hist.* 54:421–482.

Tattersall, I. 1978. Behavorial Variation in *Lemur mongoz* (= *L. m. mongoz).* Pp. 127–132 in *Recent Advances In Primatology.* D.J. Chivers; K.A. Joysey, Eds., London, Academic.

Tattersall, I. 1979. Patterns of Activity In The Mayotte Lemur, *Lemur fulvus mayottensis. J. Mammal.* 60:314–323.

Tattersall, I. 1982. *The Primates of Madagascar.* New York, Columbia University Press.

Tattersall, I. 1987. Cathemeral Activity In Primates: A Definition. *Folia Primatol.* 49:200–202.

Tattersall, I. 1992a. Systematic Versus Ecological Diversity: The Example of the Malagasy Primates. Pp. 25–39 in *Systematics, Ecology and the Biodiversity Crisis.* N. Eldridge, Ed., New York, Columbia University Press.

Tattersall, I. 1992b. The Mayotte Lemur: Cause for Alarm. *Primate Conserv.* 10:26–27.

Tattersall, I. 1997. Malagasy Primates. Pp. 636–639 in *Encyclopedia of the History of Physical Anthropology.* F. Spencer, Ed., New York, Garland.

Tattersall, I., Sussman, R.W. 1975. Observations on the Ecology and Behavior of the Mongoose Lemur *Lemur mongoz mongoz* Linnaeus (Primates, Lemuriformes), at Ampijoroa, Madagascar. *Anthropol. Pap. Am. Mus. Nat. Hist.* 52:195–216.

Tattersall, I., Sussman, R.W. 1998. "Little Brown Lemurs" of Northern Madagascar. Phylogeny and Ecological Role Iin Resources Partitioning. *Folia Primatol.* 69, Suppl. 1:379–388.

Taylor, L.L. 1986. *Kinship, Dominance, and Social Organization in a Semi-Free Ranging Group of Ringtailed Lemurs.* Ph.D. Thesis. Washington University, St. Louis.

Taylor, L.L., Sussman, R.W. 1985. A Preliminary Study of Kinship and Social Organization in a Semi-Free Ranging Group of *Lemur catta. Int. J. Primatol.* 6:601–614.

Tildon, C. 1990. A Study of Locomotor Behavior in a Captive Colony of Red-Bellied Lemurs *(Eulemur rubriventer). Am. J. Primatol.* 22:87–100.

Trivers, R.L. 1972. Parental Investment and Sexual Selection. Pp. 136–179 in *Sexual Selection And The Descent of Man.* B.G. Campbell, Ed., Chicago, Aldine.

Vasey, N. 1992. *Positional Behavior of the White Sifaka* (Propithecus verreauxi) *at Beza Mahafaly, Southwestern Madagascar.* M.A. Thesis. SUNY, Stonybrook, New York.

Vasey, N. 1997a. *Community Ecology and Behavior of* Varecia variegata rubra *and* Lemur fulvus albifrons *on the Masoala Peninsula, Madagascar.* Ph.D. Thesis. Washington University, St. Louis.

Vasey, N. 1997b. How Many Ruffed Lemurs Are Left? *Intl. J. Primatol.* 18:207–216.

Van Horn, R.N., Resko, J.A. 1977. The Reproductive Cycle of the Ring-Tailed Lemur *(Lemur catta):* Sex Steroid Levels and Sex Receptivity Under Controlled Photoperiods. *Endocrinology* 101:1579–1586.

Ward, S.C., Sussman, R.W. 1979. Correlates Between Locomotor Anatomy and Behavior in Two Sympatric Species of Lemur. *Am. J. Phys. Anthropol.* 50:575–90.

Warter, S., Randrianasolo, G., Dutrillaux, B., Rumpler, Y. 1987. Cytogenic Study of a New Subspecies of *Hapalemur griseus. Folia Primatol.* 48:50–55.

Westoby, M. 1974. An Analysis of Diet Selection by Large Generalist Herbivores. *Am. Natur.* 108:290–304.

White, F.J. 1991. Social Organization, Feeding Ecology, and Reproductive Strategy of Ruffed Lemurs, *Varecia variegata.* Pp. 81–84 in *Primatology Today: Proceedings of the XIII Congress of the International Primatological Society.* A. Ehara; T. Kimura; O. Takenaka; M. Iwamoto, Eds., Amsterdam, Elsevier.

White, F.J., Overdorff, D.J., Balko, E.A., Wright, P.C. 1995. Distribution of *Variecia variegata* in Ranomafana National Park, Madagascar. *Folia Primatol.* 64:124–131.

Wilson, J.M., Stewart, P.D., Ramangason, G.-S., Denning, A.M., Hutchings, M.S. 1989. Ecology and Conservation of the Crowned Lemur, *Lemur coronatus,* at Ankara, N. Madagascar. *Folia Primatol.* 52:1–26.

Wright, P.C. 1985. *Costs And Benefits of Nocturnality of* Aotus, *The "Night Monkey."* Ph.D Thesis. CUNY, New York.

Wright, P.C. 1986. Ecological Correlates To Monogamy in *Aotus* and *Callicebus* Pp. 159–167 in *Primate Ecology and Conservation.* J.G. Else, P.C. Lee, Eds., Cambridge: Cambridge University Press.

Wright, P.C. 1987. Diet and Ranging Patterns of *Propithecus diadema edwardsi. Am. J. Phys. Anthropol.* 72:271.

Wright, P.C. 1989. The Nocturnal Monkey Niche in the New World. *J. Hum. Evol.* 18:635–658.

Wright, P.C. 1995. Demography and Life History of Free-Ranging *Propithecus diadema edwardsi* In Ranomafana National Park, Madagascar. *Intl. J. Primatol.* 16:835–854.

Wright, P.C., Daniels, P.S., Meyers, D.M., Overdorff, D., Rabesoa, J. 1987. A Census and Study of *Hapalemur* and *Propithecus* in Southeastern Madagascar. *Primate Conserv.* 8:84–88.

Wright, P.C., Heckscher, S.K., Duhham, A.E. 1997. Predation on Milne-Edward's Sifa-ka *(Propithecus diadema edwardsi)* by the Fossa *(Cryptoprocta ferox)* in the Rain Forest of Southeastern Madagascar. *Folia Primatol.* 68:34-43.

Wright, P.C., Martin, L.B. 1995. Predation, Pollination and Torpor in Two Prosimi-ans: *Cheirogaleus major* and *Microcebus rufus* in the Rain Forest of Madagascar. Pp. 45–60 in *Creatures of the Dark.* L. Alterman; G.A. Doyle; M.K. Izard, Eds. New York, Plenum.

Wright, P.C., Randrimanantena, M. 1989. Comparative Ecology of Three Sympatric Bamboo Lemurs In Madagascar. *Am. J. Phys. Anthropol.* 78:327.

Yamashita, N. Functional Dental Correlates of Food Properties in Five Malagasy Lemur Species. *Am. J. Phys. Anthropol.* 106:169–188.

Young, A.L., Richard, A.F., Aiello, L.C. 1990. Female Dominance and Maternal Investment in Strepsirhine Primates. *Am. Natur.* 135:473–488.

CHAPTER 6

Tarsiiformes

INTRODUCTION

At present there is only one extant genus of Tarsiiformes, *Tarsius* (Figure 6-1) (but see Groves 1998). Many believe that modern tarsiers are derived from, or are the sister taxon of the Eocene omomyids, and the omomyids often are placed with tarsiers in the infraorder Tarsiiformes (e.g., Hill 1955, Szalay 1976, Rosenberger and Szalay 1980, Bown and Rose 1987, Beard et al. 1991, Covert and Williams 1992). Omomyids lack a number of the distinguishing features that characterize modern tarsiers and, therefore, many researchers now believe that they are not simply Eocene tarsiers (see Fleagle 1999). In fact, in the Eocene, a great variety of omomyids existed in North America and Europe. Body weight estimates range from 30 g animals, which are as small as the smallest extant primates, to as large as 2500 g, about the size of *Cebus*. Evidence for ecological diversification of these primates is found both in post-cranial morphology and in dentition (Simons 1972, 1974, Szalay 1975, 1976, Gingerich 1981, 1984, Conroy 1990).

There is widespread, but not universal acceptance that, among living prosimians, tarsiers are most closely related to the anthropoids (e.g., Luckett and Szalay 1975, Cartmill et al. 1981, MacPhee and Cartmill 1986, Pollock and Mullin 1987, Koop et al. 1989). This has led many taxomomists to divide the primates into the infraorders Strepsirhini (Lorisiformes and Lemuriformes) and Haplorhini (Tarsiiformes and Anthropoidea). Acceptance of this tarsier-anthropoid clade (close relationship) has led some to suggest that anthropoids arose from a tarsier-like, omomyid prosimian (e.g., Gazin 1958, Szalay et al. 1987, Beard et al. 1988). However, there is still a great deal of debate concerning the relationships of these taxa and their evolution (see Rose 1995). Summaries of alternative hypotheses can be found in Conroy 1990, Martin 1993, Rasmussen 1994, Fleagle 1999, and Gursky 1999.

Fossils of the family Tarsiidae are rare. There is a mandibular fragment of a tarsier-like primate from early Oligocene deposits in Africa (Simons and Bown 1985),

(a)   (b)   (c)

Figure 6-1 Different species of tarsier: a) *T. diana* [Photo by M. Shekelle]; b) *T. spectrum* [Photo by M. Shekelle]; c) *T. syrichta* [Photo by D. Haring].

and a single tarsier molar from the Miocene of Thailand (Ginsburg and Mein 1986). However, a recent discovery of a fossil *Tarsius* from the middle Eocene in China (Beard et al. 1994, Beard 1998) corroborates the great antiquity of tarsiers and indicates that Tarsiidae was part of the earliest radiation of primates (Rose 1995).

Today, tarsiers have a peculiar, discontinuous distribution on a scattering of islands of Southeast Asia which crosses Wallace's Line (Figure 6-2). There are five currently recognized species (but see Groves 1998 and Shekelle et al. 1997) (Table 6-1). The western tarsier, *T. bancanus,* is found on several islands of the Sunda Shelf, including Borneo, the eastern lowlands of Sumatra, and numerous small islands. The distribution of *T. bancanus* does not conform to areas of exposed land

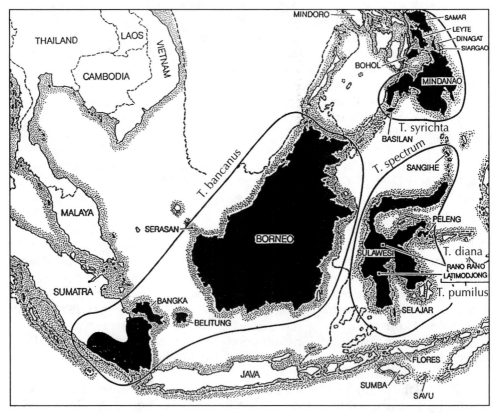

Figure 6-2 Summary of geographic distributions of five species of tarsiers. [Adapted from Musser and Dagpsto 1987].

Table 6-1 Species, Common Names, and Weights (Wild) of Tarsiers

Species	Common Name	Weight (gms)	References
Tarsius bancanus	Horsfield's, Bornean, or western tarsier	119–140 males 110–123 females 127.8 sd=6.18 (26 males) 116.9 sd=10.2 (16 females)	Niemitz 1979c[1] Crompton & Andau 1987 Niemitz 1979a
T. dianae	Diana's tarsier	110 (1 female) 105 (1 female) 104 (1 male) 100 (4 females) 119 (5 males)	Niemitz et al. 1991 Muskita pers. comm. Shekelle in prep.
T. pumilus	pygmy tarsier		
T. spectrum	spectral tarsier	113 (1 female) 110 (1 male) 104 (21 females) 115 (11 males) 107 (8 females) 126 (males)	Niemitz et al. 1991 Shekelle in prep. Gursky 1999
T. syrichta	Philippine tarsier		

[1] Niemitz states that two thirds of the adult females weighed between 110 and 120 g (n = 16), and 85% of the adult males weighed between 120 and 135 g (n = 21). A possibl;e new species *T. sangirensis* may be heavier than the other Sulawesi tarsiers (150g, 1male, 143g, 1 female) (Shekelle et al. 1997, in prep.).

during the ice ages. Reports of these tarsiers on Java, and the Lesser Sunda island of Savu are questionable. The Philippine tarsier, *T. syrichta* is from the Philippine islands of Samar, Leyte, Bohol, Mindanao, and a few smaller islands. Together, these islands make up Greater Mindanao, a landmass which was exposed during the ice ages. Tarsiers of the *T. spectrum* complex, *T. spectrum, T. pumilus,* and *T. dianae,* have a nearly continuous distribution on Sulawesi and surrounding islands. These tarsiers occur from sea level to over 2000 m, while other tarsier species are rare or non-existent above 300 m. As with other nocturnal primates (see Chapters 3 and 4), there may be more species of tarsiers yet to be recognized, especially in Sulawesi (MacKinnon and MacKinnon 1980, Niemitz et al. 1991, Nietsch and Niemitz 1993, Nietsch 1994, Shekelle et al. 1997, Nietsch and Kopp 1998).

Groves (1976) and Niemitz (1979a,b, 1984a) regard modern tarsiers as a recently evolved, specialized group derived from a form most similar to *T. spectrum,* with Sulawesi as the most probable dispersion point for all surviving forms. Crompton (1995), on the other hand, believes that *T. bancanus* is more similar to the ancestral tarsier. Little evidence exists as to the historical biogeography of tarsiers, however.

All living tarsiers are small, nocturnal, vertical clinging and leaping, faunivorous animals and, as such, they are anatomically and ecologically distinctive with regard to other primates. Among tarsiers, however, there exists previously under-emphasized variability on this theme. Many anatomical and behavioral traits show a clinal distribution with tarsiers of the *T. spectrum* complex being the most generalized, *T. bancanus* being the most specialized, and *T. syrichta* being intermediate. Other differences among taxa include such things as body weight, intramembral indices, finger pads, locomotor behavior, habitat selection, nesting sites, communication, and social and ranging behavior, to name only a few.

For example, until recently it was thought that body weights varied slightly across tarsier species. That remains true outside of Sulawesi, with adults generally weighing about 100–140 g. In Sulawesi, however, Shekelle (pers. comm.) found that among the *T. spectrum* complex, there is a large difference in body weight among various forms (Table 6-1). Males may be slightly heavier than females. Additionally, all tarsiers have relatively long legs. Statistical differences among tarsier species exist in these measurements, with *T. bancanus* having the relatively longest legs, feet and hands, and *T. spectrum* the shortest legs, feet and hands (Niemitz 1977). This led Niemitz (1979b) to predict that *T. spectrum* is less specialized for vertical clinging and leaping, comes to the ground more often, and has a more diversified habitat than do western or Philippine tarsiers. Each of these predictions was later supported by field studies of *T. spectrum* (MacKinnon and MacKinnon, 1980, Gursky 1997, Shekelle pers. comm.).

For most species of tarsier, fingers and toes end in large terminal pads which assist them in gripping smooth vertical surfaces, such as bamboo. For example, a tarsier is able to walk down a rain-slickened bamboo stalk head first (Shekelle pers. comm.) The relatively largest pads are found on *T. bancanus,* and the smallest on *T. pumilus.* In the latter species, the pad is almost non-existent, and long, keeled, claw-like nails protrude beyond the end of the fingers and toes. *T. pumilus* lives at very high altitudes in montane forests, where vertical surfaces are covered with moss. Musser and Dagosto (1987) speculate that the digits and toes of this species are adapted for clinging to the moss covered branches of the montane forest.

The body of a tarsier is only around 12–13 cm long, but the tail is twice that length. Besides being used for support on vertical substrates, the tail is also used to control momentum and direction during leaps (Preuschoft et al. 1979, Niemitz 1984b, (Figure 6-3). The tail of Sulawesi tarsiers is lightly haired along its length— as is the tail of all infant tarsiers—with a tuft of denser, longer hair along the distal third. *T. syrichta* and *T. bancanus* tails are nearly naked except for that tuft, which is more prominent in *T. bancanus.* The tail on these two species has developed a smooth sitting pad on the ventral surface, and the sitting pad of *T. bancanus* is further characterized by well-developed papillary ridges, similar to those found on fingers and toes (Hill 1955).

Figure 6-3 *Tarsius spectrum.* [Adapted from Oxnard, et al., 1980]

The huge eyes of tarsiers (Figure 6-4) lack a reflecting tapetum lucidum and, unlike most nocturnal animals whose eyes reflect light vividly, tarsiers eyes only glow a dull orange when light is shone on them. Western tarsiers have relatively the largest eyes among tarsiers, while Sulawesi tarsiers have the smallest. Owing to the limited eyeshine, their small size, and their ability to move through the understory in virtual silence, field studies of tarsiers have lagged behind those of other primates. Until recently, knowledge of tarsier behavior was limited to, and shaped by, numerous early reports from casual observations made on *Tarsius* in its natural habitat and in captivity (Cuming 1838, Le Gros Clark 1924, Lewis 1939, Cook 1939, Catchpole and Fulton 1939, Wharton 1950, Harrison 1962, 1963, Ulmer 1963).

Scientific study of *Tarsius* in its natural habitat is a relatively recent occurrence. The first field study, of *T. bancanus* (Fogden 1974), came about by accident when a number of tarsiers became trapped in mist nets, incidental to a study of birds. Niemitz (1979a, and papers in Niemitz 1984c) returned to Fogden's site near the city of Kuching in the Southwest corner of Sarawak, East Malaysia (Borneo) in 1972, but found tarsiers to be challenging research subjects. He ultimately was resigned to building enclosures at his field site and studying tarsier behavior in these semi-wild environments.

MacKinnon and MacKinnon (1980) conducted a 15 month study on *T. spectrum* incidental to other work they were pursuing at Tangkoko National Park on the extreme northern tip of Sulawesi, Indonesia. Tarsiers there proved more amenable to study and the MacKinnons were able to follow tarsiers for several hours at a time. Nietsch (1994), however, collected data on *T. spectrum* in a cage enclosure within the forest at Tangkoko over a six month period. A short term study of infant caretaking also was conducted on spectral tarsiers at this park by Gursky (1994), and she (1997) recently completed a 15 month study on this species focusing on infant transport.

Crompton and Andau (1986, 1987) took advantage of technological improvements in telemetry equipment which allowed for extremely small, 12 g transmitters. Their study site of *T. bancanus* is at Sepilok Forest Reserve in Sabah, East Malaysia, approximately 1000 km from the research site of Fogden and Niemitz. They were the first to conduct all-night follows of free-ranging tarsiers. Tremble et al. (1993) studied *T. dianae* at Kamarora in Central Sulawesi over a 15 month period. They followed a research design similar to that of Crompton and Andau, using telemetry equipment and all night follows. Gursky (1997) also mist-netted and radio-tracked animals to assist in all night follows. There has been a preliminary, two month field study of the Philippine tarsier, *T. syrichta* (Dagosto and Gebo 1997). The only field reports of *T. pumilus* since one was last trapped at 2200 m on Mount Latimojong, Central Sulawesi, in 1930, are those of Niemitz (1984d), and Nietsch and Niemitz (1993) of tarsiers in Central Sulawesi, but it is unclear whether these were indeed *T. pumilus* (Shekelle et al. 1997, Gursky 1999). Finally, Shekelle conducted population surveys of tarsiers from North and Central Sulawesi, capturing 102 tarsiers from 12 localities for taxonomic and behavioral studies (Shekelle et al. 1997, Shekelle in prep.)

THE ECOLOGY OF TARSIUS

In most recent studies of *Tarsius,* animals were caught in mist nets in a manner similar to that normally used to catch birds or bats. To facilitate recognition in earlier studies, aluminum or colored plastic bird rings often were placed on trapped animals. Thus, Niemitz and the MacKinnons were able to follow animals within their study sites and recognize a number of them individually, however they were unable to follow animals throughout the night.

T. spectrum sleeps at the same sleeping site each day, gives loud calls at dawn and is approachable, often within 2m. Thus, it was possible for the MacKinnons to maintain contact with these tarsiers for long periods. As mentioned above, to supplement his observations of free-ranging animals, Niemitz (1979b) kept a pair of *T. bancanus* in a cage (roughly 7.5 x 12 x 3.5 m) within the natural habitat of his study population. The cage was constructed around natural vegetation on a borderline between secondary and primary forest. Data collected by Fogden on *T. bancanus* were solely from casual observations and trapping of tarsiers during his two year study of birds. Crompton and Andau (1986, 1987), Tremble et al. (1993), and Gursky (1997, 1999) used radio collars and were able to follow individuals continuously throughout the night. Of the six major field studies of tarsiers which have been published, these three have the most comparable methodology. Dagosto and Gebo (1997) also used radio-tracking in their short term study of *T. syrichta*.

LOCOMOTION AND HABITAT PREFERENCES

Western tarsiers are found in a wide variety of lowland habitats. Niemitz (1979b) reports evidence of tarsiers in primary forest, shrub, and mangrove, in coastal areas, and even bordering plantations. Because of locations of capture, Fogden and earlier authors (e.g., LeGros Clark 1924, Davis 1962, Harrison 1963) considered primary forest vegetation to be a marginal habitat for *T. bancanus*. However, Niemitz (1979b) believes that Bornean tarsiers are common in primary forest, but that they may have larger home ranges and thus may be less densely populated in this vegetation. Crompton and Andau (1987) found no evidence that primary forest is marginal habitat for tarsiers in Sabah but they did find them to have much larger home ranges than did Fogden or Niemitz (Table 6-2). *T. spectrum* is found in nearly all habitats of Sulawesi, except those with dense human populations, agricultural areas where all potential nest sites have been cleared, or where insecticides or herbicides are used (Shekelle in prep.). The density of *T. spectrum* within various habitats in Sulawesi is given in Table 6-2. The habitat of Philippine tarsiers is less well known, though evidence suggests that it is more common in lowland, coastal regions (Catchpole and Fulton 1939, Wharton 1950).

The locomotor morphology of *Tarsius* is specialized for vertical clinging and leaping (Niemitz 1977, 1979b, 1984b; Preuschoft et al. 1979). All species that have been studied use leaping between vertical supports as the major means of locomotion, though there is some subtle variation in locomotion among species. Climbing is the

Table 6-2 Home Range Size and Population Densities of Tarsiers

Species	Home Range (hectares)	Population Density (per sq. km.)	Social Structure	Sources
Tarsius bancanus	0.9–1.6	80	pair bonds	Niemitz 1979b
T. bancanus	2.5–3.0	---	solitary but social	Fogden 1974
T. bancanus	mean=8.5 (4.5–11.25)	15–20	solitary but social	Crompton and Andau 1987
T. spectrum	1	70 study area[1] (30–100)[2]	pair bonds, some larger groups	MacKinnon & MacKinnon 1980
T. spectrum	2.32 ha(♀); 3.0(♂)	156	"groups"	Gursky 1997
T. dianae	0.5–0.8	---	"groups"	Tremble et al. 1993
T. syrichta	0.6–1.7 (2♂'s)	---	solitary	Dagasto & Gebo 1997

[1] Study area in Sulawesi contains a number of types of vegetation, including scrub, secondary forest, thicket, grassland, and some primary forest.

[2] Depending upon habitat, e.g., 30=montane; 100=thorn scrub.

second most common locomotor category used by tarsiers. For the most part, it consists of shinning up and sliding down vertical supports, rather than climbing in the usual sense (Oxnard et al. 1990). Some walking and frog-like or bipedal hopping is done, usually while tarsiers are on the ground searching for insects. A cantilever movement, where the animal stretches out while holding on with its hindlimbs, is used to grasp flying insects. Niemitz (1984b) describes these types of locomotion in detail.

Tarsiers leap around 60–70% of the time, but the amount of time they use other categories of locomotion differs between species (Figure 6-5). For example, *T. bancanus* relies on leaping and climbing almost exclusively (94% of locomotion), whereas *T. spectrum, T. dianae,* and *T. syrichta* appear to be much less specialized, with comparatively greater reliance on hopping, walking, and cantering (Crompton and Andau 1986, Tremble et al. 1993, Dagosto and Gebo 1997). Climbing in Western tarsiers is mainly used for short distance vertical displacements, associated with foraging (Crompton and Andau 1986). *T. dianae* and *T. spectrum* use quadrupedal climbing, walking and running in a wider variety of horizontal displacements, including somewhat rapid movement on the ground (Tremble et al. 1993, Shekelle in prep.).

T. bancanus and *T. syrichta* spend most of their time below 2 m in height, on small vertical supports of less than 5 cm in diameter. Although these tarsiers will range as high as to 6–10m, this is done rarely (Niemitz 1984b, Crompton and Andau 1986, Dagosto and Gebo 1998). Tarsiers from Sulawesi, however, have a much greater vertical ranging pattern, often nesting at heights of 20 meters. For example, *T. bancanus* in Sabah spent over 75% of its time below 1 m and only 5% above 3 m, whereas *T. dianae* spent 23% of the time above 3 m. *T. spectrum* also spent more time higher in the forest than did *T. bancanus.* This may be related to the fact that both Diana's and the spectral tarsier were studied in higher forest than was the western tarsier (Niemitz 1984e, Tremble et al. 1993) or it may be a real difference in habitat choice.

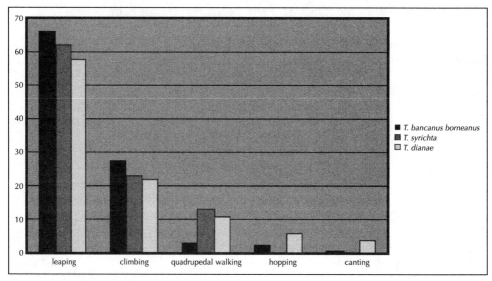

Figure 6-5 Categories of locomotor behavior among species.

Different activities are performed at different heights. For example, Crompton and Andau (1986) found that *T. bancanus* rests and grooms significantly higher (around 2 m) than it travels and that it travels significantly higher (around 1 m) than it forages (mean = 0.66 m). Feeding has the lowest mean height, 0.53 m (see also Dagosto and Gebo 1998 for *T. syrichta*). *T. spectrum* also rested at higher levels than it traveled and traveled higher than it foraged and fed (MacKinnon and MacKinnon 1980). In general, tarsiers use higher and larger supports when travelling between foraging sites than when foraging, and they use locomotor modes other than leaping less (Crompton and Andau 1986).

During resting, *T. dianae* sits more than does *T. bancanus,* and this may be related to its more frequent use of sloping and horizontal branches (Figure 6-6). The latter species uses the specialized sitting pad on its tail and presses tightly

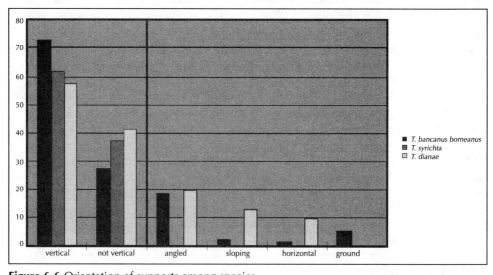

Figure 6-6 Orientation of supports among species.

against the vertical substrate while cling-ing (Figure 6-7). In this way, weight is dis-tributed to the sitting pad and the clinging tarsier actually sits on its tail (Neimitz 1984b). *T. dianae* and *T. spec-trum* also sit on their tails when using ver-tical supports. However, these species often sit on horizontal or oblique sup-ports, and their tails do not possess a sit-ting pad and will often loop over a branch or simply hang beneath the animal (Trem-ble et al. 1993, Shekelle pers. comm.).

T. bancanus also shows less variation in the angle of support it uses; 90% of the time it is found on vertical or only slight-ly angled supports, whereas *T. dianae* uses sloped and horizontal branches much more frequently (over 22% vs. 3.5%) (Figure 6-6). Both *T. bancanus* and *T. dianae* use small diameter supports in similar proportions, with about 70% of

Figure 6-7 *T. bancanus* using tail for support. [Photo by M. Shekelle]

their time being spent on supports smaller than 3 cm (Figure 6-8). The ground is used infrequently, mainly while capturing prey. On Sulawesi, tarsiers use a bipedal hop to move between trees in open habitat, such as in coconut groves or dry stream beds (Shekelle in prep.).

The average length of leaps of tarsiers are slightly over 1 m and only on rare occasions do they leap farther than 3 m, although they can do so with ease (Niemitz 1984b, Crompton and Andau 1986). Niemitz (1984b) notes that the mean leap is nearly double the mean separation of 3–4cm saplings in the pole forest where he studied *T. bancanus.* While being chased by conspecifics or primatologists, tarsiers

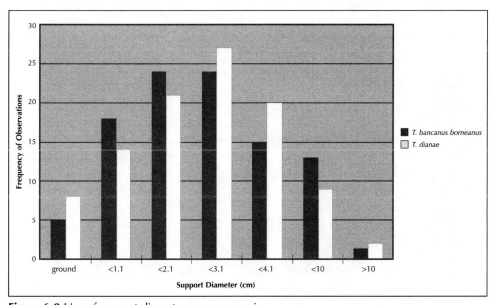

Figure 6-8 Use of support diameters among species.

can leap up to 5–6 m. *T. bancanus* has been observed to leap 30 times its head-to-body length (about 13 cm) without losing height (Niemitz 1979b).

ACTIVITY CYCLES AND SLEEPING SITES

Tarsiers are entirely nocturnal and begin their activity around sunset. Activity is continuous throughout the night with bouts of leaping occurring at intervals of about 1 hour, alternating with periods of rest (Niemitz 1979b). However, tarsiers have peaks of activity in the early evening and again just before sunrise. Niemitz (1984e) found that there was a minor peak in activity early and a major peak late in the evening in two *T. bancanus* individuals, whereas Crompton and Andau (1987) observed more travel in the early evening in the same species. However, different species spend differing amounts of time in various activities (Figure 6-9). In Sulawesi, the most energetic periods of *T. spectrum* occurred during the first half-hour of activity and, again, during the final hour of activity, when groups reformed and gave loud calls as they moved to their sleeping sites (MacKinnon and MacKinnon 1980).

There is a large amount of variation among tarsier species with regard to sleeping site preferences. This has important consequences for determination of social groups and mating systems, since tarsier social groups are often identified, in practice, by which animals sleep together.

Tarsiers from Sulawesi typically make nests in hollow trees to which they regularly return each morning (MacKinnon and MacKinnon 1980). Spectral tarsiers at Tangkoko seem to have a preference for nesting in large strangler figs (Gursky 1997). The nest site, therefore, is more like a huge arboreal cave, than simply a hole in a tree. Social groups as large as eight individuals (Gursky 1997), or with as many as six adults are known to sleep together in a nest (MacKinnon and MacKinnon 1980). Virtually nothing is known of tarsier behavior within the nest tree. Nest sites are highly stable and are used for several years without major interruption. At Tangkoko, rangers have been bringing tourists to the same nest trees since the study by the MacKinnons in the late 1970's (Shekelle pers. comm.).

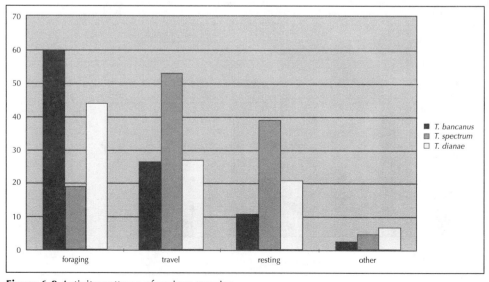

Figure 6-9 Activity patterns of various species.

Niemitz et al. (1991), in their description of *T. dianae,* state that this species does not return to the same nest each night. This description agrees with the MacKinnon's (1980) observation that tarsiers in the nearby Palu Valley do not use the same sleeping site each night. Tremble et al. (1993), however, found that their focal groups of Diana's tarsier did return to the same nest tree each night. Six years later the same nest sites were still being used (Shekelle pers. comm.). Tremble et al. studied tarsiers at the same locality as Niemitz et al. Gursky also found that *T. dianae* at her sight returned to the same nest each morning. Therefore, this aspect of nesting behavior remains ambiguous. Nest sites for this tarsier include liana vine tangles, or other dense vegetation, tree cavities, and fallen logs (Tremble et al. 1993, Shekelle, pers. comm.). As many as six tarsiers used individual nest sights and prominent duetting of *T. dianae* indicates that more than two adults are often present (Shekelle pers. comm.).

In the wild and in captivity, Western tarsiers sleep singly unless it is a mother and her infant (Niemitz 1979b, Crompton and Andau 1987, Roberts and Kohn 1993). Sleeping sites are likely chosen for shelter from the rain and for camouflage (Niemitz 1979b, Crompton and Andau 1987). Niemitz found his tarsiers to sleep clinging to nearly vertical branches, at elevation of 2 m and less, while Crompton and Andau (1987) found their tarsiers to typically sleep in vine tangles 4–5.5 m above the ground on 50–90 degree supports. They further note that sleeping sites tended to cluster at the edge of a tarsiers home range, in areas of overlap with neighbors of the opposite sex.

Until the recent report by Dagosto and Gebo 1997, the only reports of sleeping behavior of wild *T. syrichta* were anecdotal accounts from early naturalists. Cuming (1838) stated that *T. syrichta* sleeps under the roots of trees, particularly bamboo. Cook (1939) reported sleeping sites to include grass, vines, underbrush, hemp, with only one tarsier being found to nest in a tree. Philippine tarsiers have also been observed to sleep in stands of bamboo (Catchpole and Fulton 1939), and in hollow trees (Lewis 1939). Among captive *T. syrichta,* males and females often sleep huddled together (Wright et al. 1985). However, Dagosto and Gebo (1997) reported these tarsiers to be always solitary in the wild.

DIET AND FEEDING BEHAVIOR

The extant tarsiers have a highly specialized diet. They are the only primates whose diet is made up entirely of animal foods, although the slender loris is almost as faunivorous (Nekaris in prep., see Chapter 3). In over 1000 hours of observations of *T. bancanus* under seminatural conditions (i.e., the forest cage described above), no plant foods were eaten, even though natural vegetation was readily available (Niemitz 1979b, 1984f). Further, although many observations of feeding have been done (Davis 1962, Harrison 1963, Niemitz 1984f, Fogden 1974, MacKinnon and MacKinnon 1980, Crompton and Andau 1986), there are no reports of plant foods being eaten by free-ranging tarsiers.

The diet of wild *T. bancanus* consists mainly of insects, whereas to date wild *T. spectrum* has been observed eating only insects (Niemitz 1979, 1984f, MacKinnon and MacKinnon 1980, Crompton and Andau 1986, Jablonski and Crompton 1994). The following types of insects are among those eaten: beetles, grasshoppers, cockroaches, butterflies, moths, praying mantis, crickets, dragonflies, termites and occasionally ants, phasmids, spiders, and cicadas. Some of these prey are caught as they settle on foliage, but most are captured in the leaf litter on the ground. Tarsiers spend most of their night foraging, aurally and visually searching for prey from small, vertical perches just above the ground (Niemitz 1979b, 1984f, Crompton and Andau 1986). They have voracious appetites and eat many large prey items in one night (Jablonski and Crompton 1994). Tarsiers locate their prey primarily by sound and only secondarily by sight (see Chapter 2, and Crompton 1995). A small percentage of flying insects are captured in the air by cantilevering.

Besides insects, *T. bancanus* also has been observed feeding on snakes (poisonous ones at that! (Figure 6–10)), many species of birds, a few species of bats (Niemitz 1979b, 1984f, see also Harrisson 1963) and lizards (Fogden 1974), "flying frogs" and small crabs (Jablonski and Crompton 1994). Toads, snails, small rats and shrews were not eaten by either *T. bancanus* or *T. spectrum* though these potential prey were available.

The jaw muscles and teeth of tarsiers are highly specialized for acquisition and processing relatively large animal prey. However, Jablonski and Crompton (1994) believe that these specializations are not optimal for a single function but are compromises between a number of competing pressures. They must allow the tarsier to efficiently bite and chew tough, plywood-like insect exoskeletons and vertebrate bones, as well as soft, chewy insect intestines. Often, when tarsiers catch vertebrates, they will eat everything, including bones, bird beaks and feet, etc. These authors also argue that the digestive system entails compromise. Tarsiers have a simple gut (Hill 1955) but it takes a long time for the chitonous exoskeleton to pass through the digestive system. Jablonski and Crompton believe that tarsiers are able to digest the exoskeleton by a process of fermentation in their relatively large caecum.

Considering the foraging pattern and diet of *Tarsius*, Niemitz (1979b, 1984a, 1985) suggests that tarsiers occupy a niche essentially similar to that of owls. Tarsiers show striking convergence with owls both morphologically (e.g., large eyes and the ability to rotate their head through 180°) and behaviorally. Niemitz (1985) lists 33 parallelisms, convergences and analogies between tarsiers and owls, including the following four behavioral similarities: (1) owls and tarsiers feed on similar prey; (2) they locate their prey acoustically, rather than by smell or vision; (3) they are both completely noiseless during locomotion; and (3) they utilize ambush-type predation by moving about during the night just above ground level. Where the undergrowth is too dense for flying, Niemitz believes the tarsier replaces owls completely.

PREDATION AND COMPETITION

Both Niemitz (1979b, 1984f) and the MacKinnons (1980) comment on the fact that *Tarsius* is generally incautious to potential predators. Large owls, snakes, civets and slow lorises *(Nycticebus coucang)* elicited essentially no alarm response from tarsiers. In fact, as stated above, a *T. bancanus* has been observed catching and eating a 30 cm long poisonous snake *(Maticora intestinalis,* Figure 6-10). It is thought by some that the an unpleasant odor and, perhaps, taste of the unwary tarsier may be its best defense against predation (Niemitz 1979, MacKinnon and MacKinnon 1980). The quick, saltatory locomotion and dense habitat of this tarsier would make it a challenge to most predators. The MacKinnon's observed alarm calls given by *T. spectrum* toward feral and domestic cats and the MacKinnons' pet cat managed to catch and partially eat an adult male tarsier.

Gursky's (1997) observation differs from those described above. She found that tarsiers give alarm calls or show predator avoidance in several situations. Tarsiers gave alarm calls to the Malaysian palm civet. This species had been introduced to her site but another civet is endemic there. Gursky also believes that some snakes and birds of prey do prey upon tarsiers. She observed one close encounter between an unidentified snake and a tarsier, and upon hearing the calls of predatory birds, tarsiers freeze for up to several minutes. Gursky also believes that monitor lizards may prey upon tarsiers. One of the 18 individuals that she captured and marked was found partially eaten. She concludes that predation pressure is high on *T. spectrum* at Tangkoko. Shekelle (pers. comm.) concurs with Gursky. He reports one observation of a monitor lizard attacking a tarsier and two observations of attempted predation by snakes.

Figure 6-10 *T. bancanus* eating a neurotoxic snake. [From Niemetz, 1979b]

Humans represent the greatest threat to tarsiers. Animals are sometimes eaten but usually are caught and sold for pets, often dying within a few days. The habitat of tarsiers also is destroyed for agricultural purposes, timbering, and other, usually short term economic exploitation (Neimitz 1984f, MacKinnon 1987). This is especially so in the Philippines (Myers 1987). Currently, *T. syrichta* is considered as a threatened species, and the status of *T. pumilis* and T. *dianae* is unknown (Wright et al. 1987).

The most likely competitors of *T. bancanus,* and possibly other species of tarsiers, are owls, insectivorous bats, the slow loris *(Nycticebus coucang)* and tree shrews. Owls may be competitors in open areas but Niemitz does not believe they compete with tarsiers in dense vegetation. *Nycticebus* is most likely very much like *Perodicticus potto,* being highly frugivorous and probably feeding only on slow-moving, noxious prey (see Chapter 3). Ecological interactions of these tarsiers with bats and tree shrews have not been described (except that tarsiers eat small bats). Bats most likely feed on flying insects, and the species of tree shrews of the genus *Tupaia* are diurnal. Fogden (1974) states that the only nocturnal tree shrew, *Ptilocercus lowii,* occupies a niche quite distinct from that of the tarsier. Other sympatric primates, such as macaques, proboscis monkeys, gibbons and orang utans are not thought to compete with tarsiers (Niemitz 1984f).

Gursky (1997) believes that competitors of *T. spectrum* at Tangkoko include insectivorous bats. She observed eight encounters of tarsiers with these bats in which the tarsiers were challenged by bats over insect prey. In seven of the eight cases, the tarsier got the insect.

THE SOCIAL BEHAVIOR OF TARSIIFORMES

Studies of tarsiers have raised as many questions as they have answered regarding social organization and mating systems. From his data on trapping, Fogden

(1974) suggested that the social structure of *T. bancanus* was solitary but social, similar to that of *G. demidoff* or *M. murinus* (see Chapters 3 and 4). Niemitz (1979b, 1984a) believed *T. bancanus* to typically live in pairs. However, he (1984a:121) stated: "It is questionable whether Bornian tarsiers are strictly monogamous. Members of both sexes, however, live synterritorially for a long period developing strong pair bonds."

Crompton and Andau (1987) did not find this pattern in their radio-tracked animals. They found that individuals foraged and slept alone. In fact tarsiers were never seen together in 120 hour of following. Vocal (and probably olfactory) communication was common but always at a distance. Calling appeared to be associated with range boundaries and range overlaps. Varying numbers of individuals, sometimes as many as five, often engaged in these "calling concerts" simultaneously.

The range of one radio-collared female was visited by at least 4 males and male ranges overlapped several female ranges. Crompton and Andau (1987:67) believe that the social structure of *T. bancanus* "seems to be closer to the generalized 'noyau' pattern of many lorisines and galagines . . . than to any form of monogamy or pair bonding", though distinguished from these by the extreme degree of solitariness and the lack of contact at sleeping sites. In captivity, Wright et al. (1985) noted that the amount of contact between *T. bancanus* individuals is minimal compared to that for *T. syrichta*. Individuals of the latter species play, allogroom, urine mark, genital sniff, and call frequently throughout the night, while the Western tarsiers interact infrequently and rest alone for long periods.

MacKinnon and MacKinnon (1980) believed *T. spectrum* to be pair bonded. The bonded pairs they reported on were highly stable, with only minor changes (essentially demographic; i.e., due to births, deaths, age changes and immigration) occurring over the 15-month study period. Family members ranged close together, maintaining some auditory and visual contact throughout the night. The MacKinnon's reported that *T. spectrum* pair members relocated each other in the morning and gave a characteristic duet song, the male and female contributing different components to the song. Similar duets were given during intergroup encounters. Pair members frequently came together during the night, sitting together, sometimes intertwining tails, and often scent-marking each other (MacKinnon and MacKinnon 1980). During breeding season, copulations occured between pair bonded adults.

More recent studies, however, indicate that *T. spectrum* is able to live in groups containing more than one adult of the same sex. In fact, one group studied by the MacKinnons contained one adult female and 2 males, and another 3 adult females and 3 adult males. Nietsch and Niemitz (1993) and Gursky (1997, 1999) found that spectral tarsiers often live in groups with more than one adult female. Nietsch and Niemitz (1993) observed one group with three adult females and in 33 groups censused by Gursky (1997), group size ranged from 2-8 individuals and five groups contained more than one female. Other groups contained one or more subadults. Shekelle (in prep.) observed one group with one adult male and two adult females, each with an infant.

Niemitz (1984a) speculates that, in *T. spectrum,* juveniles may stay with their mothers, whereas in *T. bancanus* juveniles are not tolerated. From an examination of testes of captured *T. bancanus,* Fogden (1974) and Niemitz (1977, 1979b) regard about 20% of the adult males to be sexually inactive. Tremble et al. (1993) observed *T. dianae* in "groups" with more than one adult of the same sex living in the same home range (see also Niemitz 1979b, Niemitz 1984a, Shekelle in prep.).

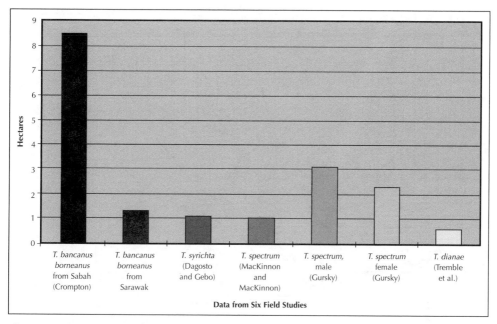

Figure 6-11 Home range estimates from different studies.

Home range estimates are quite variable among studies (Figure 6-11). Without following the animals throughout the night, Niemitz (1979b, 1984a) estimated *T. bancanus* to live in home ranges of 0.9 to 1.6 ha and to have a population density of approximately 80 tarsiers km^2 (Table 6-2). From trapping records, Fodgen (1974) gave an estimate of 2.5–3 ha for the same species. These are minimal estimates because they include mainly trapping locations or sleeping sites. Crompton and Andau (1987), following animals throughout the night, found home ranges between 4.5 and 11.25 ha (Table 6-2). Given their results, population densities would be much lower than previously estimated, approximately 15–20 per km^2.

T. spectrum home ranges have been estimated to be about 1 hectare (N=8) (MacKinnon and MacKinnon 1980), but these also are not from all-night follows. Using all-night follows, Gursky (1997) found female home ranges to average 2.32 ha and males to use ranges of approximately 3.0 ha. Population densities at her site were high, approximately 156 animals per km^2. Tremble et al. (1993), also using radio-tracking, found *T. dianae* to have small home ranges, 0.5–0.8 ha. In their short-term study, Dagosto and Gebo (1997) report home ranges of two male *T. syrichta* to be 0.6 and 1.7 ha. Thus, some species and populations may have small home ranges whereas in others ranges may be much larger. Some of these differences may be due to methodology, to species or population variability, to differences in habitat, or to a combination of these variables.

Home ranges appear to be stable over long periods of time. For example, Fogden (1974) trapped an adult male in the same area in 1965 and 1966. In 1972, the same animal (identified by the bird ring on its leg) was trapped by Niemitz (1979b) 30 m away from the original trap site. This animal was at least 8 years old. The longevity record for *Tarsius* in captivity is 12 years (Ulmer 1960, Roberts 1994), although that animal showed few signs of aging and a maximum life span of greater than 20 is predicted (Ulmer 1960).

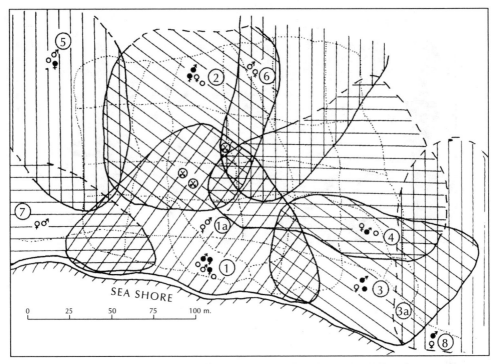

Figure 6-12 Map of known ranging limits, sleeping sites, group compositions, and sites of territorial clashes. Ringed animals are shaded. [Adapted from MacKinnon and MacKinnon, 1980]

From earlier studies, it was reported that tarsier groups came together, animals actively chasing each other, exchanging loud vocalizations (similar to duets, described above, in *T. spectrum*) and scent marked prominent branches. Because of these encounters, both Niemitz and the MacKinnons infer that tarsiers are territorial. However, the MacKinnons point out that there is considerable overlap between ranges (Figure 6-12). Crompton and Andau (1987) report that male day ranges at their site ranged between 8.75–11.25 ha, and females 4.5–9.5 ha. Most activity was concentrated around overlapping home range boundaries or near sleeping sites, which also were mostly near these boundaries. Further, activity was not concentrated in one area of the home range but in two to four well-dispersed regions. They believed that the relatively large home ranges and their clumped utilization made it unlikely that home ranges could be "patrolled" each night, despite long night ranges, especially by males. Thus, it appears that at least in some tarsiers agonistic interactions may be used to defend the integrity of the group and may not be territorial.

ACOUSTIC AND OLFACTORY COMMUNICATION

Tarsier species vary to the extent they employ acoustic and olfactory communication. Generally, those studies which find extensive use of olfactory communication find relatively less acoustic communication (e.g., Niemitz 1984d), while those that find extensive acoustic communication report relatively less olfactory communication (e.g., MacKinnon and MacKinnon 1980, Crompton and Andau 1986, 1987). All tarsiers, however, are known to use a variety acoustic and olfactory signals.

Niemitz found that olfactory communication among *T. bancanus* at his study site in Sarawak is quite complex. Tarsiers there urinate on vertical substrates, and some tree trunks are used exclusively by a single individual. Females in estrus mark branches with their vulva. The epigastric gland is more highly developed in tarsier males and secretions from this gland are rubbed on branches at Sarawak. The circumoral gland is also rubbed on branches, as well as an individual's own tail tuft. Rare examples of tactile communication, such as a female holding the tail of a male, have been observed in captive *T. bancanus* (Niemitz 1979b).

Western tarsiers have been described as typically silent (Le Gros Clark 1924). Harrison (1963) noted a distress call, and vocalizations associated with courtship and mating, and Niemitz (1984d) described four basic patterns for *T. bancanus* vocalizations including a short whistle, a "grate," a long call, and the clicks of infants. Male and female calls could not be distinguished.

Spectral tarsiers rely heavily on acoustic signals and are known to have at least 14 discrete calls (MacKinnon and MacKinnon 1980, Nietsch and Kopp 1998). The behavioral contexts have been established for several of these. Most prominent is the duet call described above. The call is typically sung in the morning, shortly before returning to the sleeping site. Parameters of the call have excellent characteristics for transmitting the location of the signaler (Niemitz, 1984d). Thus, the call is thought to function as a means for unifying the social group after an evening of solitary foraging (MacKinnon and MacKinnon 1980). The volume of the call, however, is far louder than is required for such a purpose, and the call is further thought to advertise to neighboring groups. Olfactory communication among spectral tarsiers includes marking each other with the circumoral gland, anogenital region, or with the epigastric gland. Branches are marked with the epigastric gland as well as urine. MacKinnon and MacKinnon (1980) further note that spectral tarsiers communicate through close contact facial expressions, body posture, and intimate contact. Communication among other species of tarsier are not well known but calls may be specific to particular species and not answered across species (MacKinnon and MacKinnon 1980, Niemitz et al. 1991, Nietsch 1994, Nietsch and Niemitz 1993, Nietsch and Kopp 1998, Shekelle in prep.).

REPRODUCTION AND INFANT DEVELOPMENT

Except for the study of Gursky (1997), little is known of reproductive and mating behavior among wild tarsiers. These behaviors have been observed in captive *T. bancanus, T. syrichta,* and *T. spectrum.* Nothing in the literature concerning captive tarsiers indicates major differences among tarsiers. The estrous cycle of each of these species is typically 24 days (Catchpole and Fulton 1939, Ulmer 1963, Wright et al. 1986a, Permadi et al. 1994). Seasonality in tarsier births remains an open issue. Neither studies of captive births in *T. bancanus* and *T. syrichta* (Haring and Wright 1989, Roberts and Kohn 1993), nor a survey of wild *T. bancanus* placentae (Zuckerman 1932) show evidence of seasonality. Nevertheless, seasonal birth peaks which follow local rainy seasons have been noted in *T. bancanus* near Kuching (Sarawak, East Malaysia) by Le Gros Clark (1924), Fogden (1974), and Niemitz (1977), as well as for *T. spectrum* at Tangkoko by MacKinnon and MacKinnon (1980) and Gursky (1997).

In captive *T. bancanus* housed in pairs, male and female calling and scent marking increased when the female was in estrus and proestrus. Males began initiating

Table 6-3 Reproductive Data for *T. bancanus*

Cycle length	Gestation	Litter Size	Age at Sex. Maturity	References
24±3.2 days	178–180days	1	@18 months	Wright et al. 1986a Roberts 1994

contact during proestrus, but were agonisticly countered by females. Once in estrus, females initiated contact. Courtship lasted for 60–90 minutes before a brief copulatory bout (60–90 seconds). Copulatory plugs were not observed in captive *T. bancanus* (Wright et al. 1986b), but have been observed in captive *T. syrichta* at the Cincinnati Zoo (R. Jones pers. comm.). Wright et al. (1986b) interpret the copulatory behavior of *T. bancanus* to be consistent with monogamy and inconsistent with the mating system proposed by Fogden (1984) or Crompton and Andau (1987). However, Roberts (1994) believes that the life history, reproductive and parental patterns are not consistent with monogamy in this species.

In *T. bancanus,* gestation is exceptionally long for such a small animal, 178–180 days (Izard et al. 1985, Roberts 1994, Gursky 1997). Normally, females give birth to one infant (Table 6-3). The infant is exceptionally large, attaining 25–33% of the mother's weight at birth (Roberts 1994, Gursky 1999) (Figure 6-13). Much of this weight is accounted for by the infant's brain, which is 60–70% of the adult brain size. In fact, compared to 25 other prosimians, tarsiers have the largest relative neonatal size, the longest gestation length, and the largest relative neonatal brain size (Roberts 1994).

At first, the infant is carried in the mother's mouth. When the mother is active she parks the neonate on small diameter branches, on which it clings with its body and tail pressed close to the substrate. During this time, infants of *T. spectrum* are often visited and played with by the family members of both sexes (MacKinnon and MacKinnon 1980, Gursky 1994, 1997). Gursky

Figure 6-13 *T. spectrum* mother with infant. [Photo by M. Shekelle]

(1997) observed subadult females retrieving fallen infants, allowing infants to take insect prey, and playing, grooming and "babysitting" with infants. Adult males were seen playing and grooming with them. This is not the case in *T. bancanus* and there is no evidence of paternal investment in this species (Roberts 1994).

Roberts (1994) found that *T. bancanus* had the slowest recorded fetal growth rate for any mammal, and that the relative postnatal growth rate to maturity was the lowest in a sample of 26 prosimians. However, he also found that the relative rate of behavioral development, especially in foraging and locomotion, was extremely rapid. After about 3 weeks neuromuscular coordination improves dramatically. By 6 weeks the mother no longer carries the infant, and within two months young *T. bancanus* begin to catch prey. Lactation is extremely short, lasting only 2–3 months (Roberts and Kohn, 1993, Roberts 1994, Permadi et al. 1994, Gursky 1997). The infant is nutritionally independent by 80 days and by 100 days it is able to capture prey with frequencies similar to those of adults. These skills are acheived with no aid from either parent (Roberts 1994). Roberts (1994) believes that there is an energetic/dietary basis for the slow physical growth rates, but that an extremely large neonatal brain size enables the rapid behavioral and neuromuscular development and coordination. This rapid behavioral development is needed for this specialized predator to attain early nutritional independence.

SUMMARY OF THE ECOLOGY AND BEHAVIOR OF THE TARSIIFORMES

Although there were a great variety of somewhat tarsier-like primates during the Eocene, only five known species of the genus *Tarsius* are extant today: *T. bancanus, T. dianae, T. pumilus, T. spectrum,* and *T. syrichta.* These forms are all found in Southeast Asia on many islands of the Malay Archipelago. From preliminary investigations, it appears that other species of tarsier may be recognized in the near future, especially in Sulawesi. Research has been conducted on naturally occurring populations of *T. bancanus, T. dianae, T. spectrum* and *T. syrichta.* These species have been followed throughout the night, but only the latter species has been the subject of a relatively long-term study and only at one site. Furthermore, only brief field studies have been done on the Philipine tarsier, *T. syrichta,* and none on *T. pumilus.* Thus, the range of behavioral variation that may occur in tarsiers over their entire geographical range is presently not known.

All extant tarsiers are active at night. They are vertical clingers and leapers and are active mainly on small vertical supports in the undergrowth, usually less than 3 meters above the ground. The great majority of travel and foraging of the studied populations takes place on supports of this type, though there is some variation between study populations in the amount of time spent higher in the forest, and in locomotor types other than clinging and leaping.

Tarsiers are the only primates whose diet consists entirely of animal foods (although the slender loris is a close second). Insects form the greatest proportion of the diet and, in fact, free ranging *T. spectrum* were observed to feed exclusively on insects. *T. bancanus* also feeds upon some small vertebrates, e.g., reptiles (snakes and lizards), birds and a few species of bats, and small frogs. Most prey of tarsiers are fast moving and are captured on the ground; hearing plays an important role in detecting prey. Tarsiers spend most of their time foraging from low, vertical sapling trunks. They ambush their prey on the ground, and move back to the trees to feed.

Niemitz suggests that *Tarsius* is convergent with owls in some aspects of morphology and behavior, and that it replaces these predatory birds completely in dense vegetation. Synecological relationships between tarsiers and other sympatric species (e.g., tree shrew, *Nycticebus coucang* and bats) indicate that they are relatively free from competition. Some believe that *T. bancanus* and *T. spectrum* appear to be incautious toward potential predators and may be "protected" by being odoriferous and bad tasting, as well as by their quick, saltatory locomotion and dense habitat. However, other researchers have provided evidence that tarsiers may be preyed upon by such predators as civets, snakes, monitor lizards and birds of prey.

The social behavior of tarsiers remains in question. Initially, *T. spectrum* was thought to live in family groups, containing an adult pair with their offspring. However, groups with more than one adult pair have been observed in this species and many groups especially contain more than one adult female. Group-living also may be typical of *T. dianae* and *T. syrichta*. In *T. spectrum,* group members remain in close contact throughout the night and these tarsiers use a rich array of communication (visual, tactile, olfactory and vocal) for inter- and intra-group spacing. *T. spectrum* give loud calls ("duets", in which male and female portions differ) both to relocate in the morning and during intergroup encounters. Groups usually sleep in the same site each night and have relatively small, but highly overlapping home ranges.

Although *T. bancanus* was thought to live in pair bonded groups, recent evidence suggests that it has a solitary but social type of social structure, similar to many lorisid species. In all night follows, these tarsiers were never seen together, there is no paternal involvement in offspring, and males and females do not participate in duets. Male and female *T. bancanus* do, however, have overlapping home ranges and exchange olfactory and vocal communications frequently, especially where range borders overlap. Thus, tarsiers seem to have multiple social systems, possibly including monogamy, multi-male/multi-female groups, one-male groups, and solitary but social systems. It is not known whether the variation seen is related to species differences, different environmental pressures, methodological differences between studies, or a combination of these factors. It is obvious that further research needs to be done on this genus.

Normally, a female gives birth to one offspring each year which is carried in her mouth or "parked" on branches during the first 3 weeks. After this, the infant becomes increasingly independent. The neonate is large in relation to the mother's body weight, and the brain of the infant accounts for 60–70% of its weight. Neonatal and postnatal growth is extremely slow, but behavioral development is very rapid. Infants move on their own within 6 weeks, and are as efficient as adults in capturing prey within 100 days. The combination of slow physical growth, large brain size, and rapid behavioral development is thought to be a compromise between the energetic/dietary requirements of the mother, and the necessity of the infant to become competent in its highly specialized locomotor and predatory skills, and thus attain independence.

BIBLIOGRAPHY

Beard, K.C. 1998. A New Genus of Tarsiidae (Mammalia: Primates) From the Middle Eocene of Shanxi Province, China, with Notes on the Historical Biogeography of Tarsiers. *Bull. Carn. Mus. Nat. Hist.* 34:260–277.

Beard, K.C., M. Dagosto, D.L. Gebo, and M. Godinot. 1988. Interrelationships Among Primate Higher Taxa. *Nature* 331:712–714.

Beard, K.C., L. Krishtalka, and R.K. Stucky. 1991. First Skulls of *Shoshonius cooperi* and the Anthropoid Tarsier Dichotomy. *Nature* 349:64–67.

Beard, K.C., T. Qi, M.R. Dawson, B. Wang, and C. Li. 1994. A Diverse New Primate Fauna From Middle Eocene Fissure Fillings in Southeastern China. *Nature* 368:604–609.

Bown, T. and K. Rose. 1987. Patterns of Dental Evolution in the Early Eocene Anaptomorphine Primates (Omomyidae) From the Bighorn Basin, Wyoming. *J. Paleo. Soc. Mem.* 23:1–162.

Cartmill, M. MacPhee, R.D.E., and Simons, E.L. 1981. Anatomy of the Temporal Bone in Early Anthropoids, With Remarks on the Problem of Anthropoid Origins. *Am J. Phys. Anthropol.* 56:3–21.

Catchpole, H.R. and J.F. Fulton. 1939. Tarsiers in Captivity. *Nature* 144:514.

Conroy, G.C. 1990. *Primate Evolution.* W.W. Norton, New York.

Cook, N. 1939. Notes on Captive *Tarsius carbonarius. J. Mammal.* 20173–20177.

Covert, H.H. and Williams, B.A. 1992. The Anterior Lower Dentition of *Washakius insignis* and Adapid-Anthropoidean Affinities. *J. Hum. Evol.* 21:463–467.

Crompton, R.H. 1995. "Visual Predation", Habitat Structure, and the Ancestral Primate Niche. Pp. 11–30 in *Creatures of the Dark.* L. Alterman; G.A. Doyle; K. Izard, (eds.). Plenum, New York.

Crompton, R.H. and P.M. Andau. 1986. Locomotion and Habitat Utilization in Free-Ranging *Tarsius bancanus:* a Preliminary Report. *Primates* 27:337–355.

Crompton, R.H. and P.M. Andau. 1987. Ranging, Activity, Rhythms, and Sociality in Free-Ranging *Tarsius bancanus.* a Preliminary Report." *Intl. J. Primatol.* 8:43–71.

Cuming, H. 1838. On the Habit of Some Species of Mammal From the Philippine Islands. *Proc. Zool. Soc. Lond.* 6:67–68.

Dagasto, D.L.; Gebo, M. 1997. A Preliminary Study of the Philippine Tarsier in Leyte. *Asian Primates* 6:5–8.

Dagasto, D.L.; Gebo, M. 1998. A Preliminary Study of the Philippine Tarsier *(Tarsius syrichta)* in Leyte. *Am. J. Phys. Anthropol.* Suppl. 26:73.

Davis, D.D. 1962. Mammals of the Lowland Rain-Forest of North Borneo. *Bull. Sing. Natl. Mus.* 31:

Fleagle, J.F. 1999. *Primate Adaptation and Evolution.* New York: Academic Press.

Fogden, M.P.L. 1974. A Preliminary Field-Study of the Western Tarsier, *Tarsius bancanus* Horsefield. Pp. 151–166. In *Prosimian Biology,* ed. R.D. Martin, G.A. Doyle, and A.C. Walker. Pittsburgh: University of Pittsburgh Press.

Gazin, C.L. 1958. A Review of the Middle and Upper Eocene Primates of North America. *Smith. Misc. Coll.* 136:1–112.

Gingerich, P.D. 1981. Early Cenozoic Omomyidae and the Evolutionary History of the Tasiiform Primates. *J. Hum. Evol.* 10:345–374.

Gingerich, P.D. 1984. Paleobiology of Tarsiiform Primates. Pp 33–44. In *The Biology of Tarsiers,* ed. C. Niemitz. New York: Gustav Fischer Verlag.

Ginsburg, L. and P. Mein. 1986. *Tarsius thailandica* nov. sp., Tarsiidae (Primates, Mammalia) fossile d'Asie. *C. R. Academie of Science (Paris)* t.304, ser. II:1213–1215.

Groves, C.P. 1976. The Origin of the Mammalian Fauna of Sulawesi (Celebes). Z. *Saugetierekunde* 41:201–206.

Groves, C.P. 1998. Systematics of Tarsiers and Lorises. *Primates* 39:13–27.

Gursky, S.L. 1994. Infant Care in Spectral Tarsier *(Tarsius spectrum)* Sulawesi, Indonesia. *Intl. J. Primatol.* 15:843–853. Anthropol. Supplement 16:100.

Gursky, S.L. 1997. *Modeling Maternal Time Budget; The Impact of Lactation and Infant Transport on the Time Budget of the Spectral Tarsier,* Tarsius spectrum. Ph.D Thesis. SUNY, Stony Brook, New York.

Gursky, S.L. 1999. The Tarsiidae: Taxonomy, Behavior And Conservation Status. Pp. 140–150 in *The Nonhuman Primates.* P. Dolhinow; A. Fuentes, eds., Mountain View, Mayfield.

Haring, D. and P. Wright. 1989. Hand-Raising an Infant Tarsier, *Tarsius syrichta. Zoo Biol.* 8:265–274.

Harrison, B. 1962. Getting to Know About *Tarsius. Mal. Nat. J.* 16:197–204.

Harrison, B. 1963. Trying to Breed *Tarsius. Mal. Nat. J.* 17:218–231.

Hill, W.C.O. 1955. *Primates: Comparative Anatomy and Taxonomy. II. Haplorhini: Tarsioidea.* Edinburgh: Edinburgh University Press.

Hill, W.C.O., Porter, A., Southwick, M.D. 1952. The Natural History, Ectoparasites and Pseudo-Parasites of the Tarsiers *(Tarsius carbonarius),* Recently Living in the Society's Menagerie. *Proc. Zool. Soc. Lond.* 122:79–199.

Izard, M.K., P.C. Wright, and E.L. Simons. 1985. Gestation Length in *Tarsius bancanus. Am. J. Primatol.* 9:327–331.

Jablonsky, N.G., Crompton, R.H. 1994. Feeding Behavior, Mastication, and Tooth Wear in the Western Tarsier *(Tarsius bancanus). Int. J. Primatol.* 15:29–59.

Koop, B.F., D. Siemienak, J.L. Slightom, M. Goodman, J. Dunbar, P.C. Wright, and E.L. Simons. 1989. *Tarsius* D- and B-Globin Genes: Conversions, Evolution, and Systematic Implications. *J. Biochem. Evol.* 264:68–79.

LeGros Clark, W. E. 1924. Notes on the Living Tarsier *(Tarsius spectrum). Proc. Zool. Soc. Lond.* 217–223.

Lewis, G.C. 1939. Notes on a Pair of Tarsiers from Mindanao. *J. Mammal.* 20:57–61.

Luckett, W.P. and Szalay, F.S. (eds.) 1975. *Phylogeny Of The Primates: A Multidisciplinary Approach.* Plenum, New York.

MacKinnon, J. and K. MacKinnon. 1977. The Formation of a New Gibbon Group. *Primates* 18:701–708.

MacKinnon, J. and K. MacKinnon. 1980. The Behavior of Wild Spectral Tarsiers. *Intl. J. Primatol.* 1:361–379.

MacKinnon, K. 1987. Conservation Status of the Primates of Malesia, with Special Reference to Indonesia. *Primate Conserv.* 8:175–183.

Martin, R.D. 1993. Primate Origins: Plugging the Gaps. *Nature* 363:223–234.

McPhee, R.D.E. and Cartmill, M. 1986. Basicranial Structures and Primate Systematics. In *Comparative Primate Biology, Vol. 1, Systematics, Evolution, and Anatomy.* pp. 219–276. D.R. Swindler and J. Erwin (eds.). Alan R. Liss, New York.

Musser, G.G. and M. Dagosto. 1987. The Identity of *Tarsius pumilus,* a Pygmy Species Endemic to the Montane Mossy Forests of Central Sulawesi. *American Museum Novitates* 2867:1–53.

Myers, N. 1987. Trends in the Destruction of Rain Forests. Pp. 3–22 in *Primate Conservation in the Tropical Rain Forest.* C.W. Marsh; R.A. Mittermeier, eds., New York, Alan R. Liss.

Niemitz, C. 1977. Zur Funktionsmophologie und Biometrie der Gattung *Tarsius* Storr, 1780 (Mammalia, Primates, Tarsiidae). *Cour. Forsch. Int. Senckenberg* 25:1–161.

Niemitz, C. 1979a. Results of a Field Study on the Western Tarsier *(Tarsius bancanus borneanus* Horsfeld, 1821) in Sarawak. *Sarawak Mus. J.* 27:171–228.

Niemitz, C. 1979b. Outline of the Behavior of *Tarsius bancanus.* Pp. 631–660. in *The Study of Prosimian Behavior,* ed. G.A. Doyle and R.D. Martin. . New York, Academic Press.

Niemitz, C. 1979c. Field Biology of Tarsius bancanus borneanus. Unpublished manuscript.

Niemitz, C. 1984a. An Investigation and Review of the Territorial Behaviour and Social Organization of the Genus *Tarsius.* Pp. 117–128. In *The Biology of Tarsiers,* ed. C. Niemitz. New York: Gustav Fischer Verlag.

Niemitz, C. 1984b. Locomotion and Posture of *Tarsius Bancanus.* Pp. 191–226 in *The Biology of Tarsiers,* ed. C. Niemitz. New York: Gustav Fischer Verlag.

Niemitz, C. 1984c. *The Biology of Tarsiers.* New York: Gustav Fischer Verlag.

Niemitz, C. 1984d. Vocal Communication of Two Tarsier Species *(Tarsius bancanus* and *Tarsius spectrum).* Pp. 129–142 in *The Biology of Tarsiers,* ed. C. Niemitz. New York: Gustav Fischer Verlag.

Niemitz, C. 1984e. Activity Rhythms and the Use of Space in Semi-Wild Bornean Tarsiers, With Remarks on Wild Spectral Tarsiers. Pp. 85–116 in *The Biology of Tarsiers,* ed. C. Niemitz. New York: Gustav Fischer Verlag.

Niemitz, C. 1984f. Synecological Relationships and Feeding Behaviour of the Genus *Tarsius.* In *The Biology of Tarsiers,* ed. C. Niemitz. pp. 59–76. New York: Gustav Fischer Verlag.

Niemitz, C. 1985. Can a Primate Be an Owl? - Convergences in the Same Ecological Niche. *Fortschritte der Zoologie* 30:667–670.

Niemitz, C., A. Nietsch, S. Warter, and Y. Rumpler. 1991. *Tarsius dianae:* A New Primate Species From Central Sulawesi (Indonesia). *Folia Primatol.* 56:105–116.

Nietsch, A. 1994. *A Comparative Study of Vocal Communication in Sulawesi Tarsiers.* Paper delivered at Congress of the International Primatological Society in Denpasar, Bali, Indonesia.

Nietsch, A and M.-L. Kopp. 1998. Role of Vocalization in Species Differentiation of Sulawesi Tarsiers. *Folia Primatol.* 69 Suppl. 1 :371–378.

Nietsch, A. and C. Niemitz. 1993. Diversity of Sulawesi Tarsiers. *Deutsche Gesellschaft Fur Saugetierkunde* 67:45–46.

Oxnard, C.E., Crompton, R.H., and Lieberman, S.S. 1990. *Animal Lifestyles and Anatomies: The Case of the Prosimian Primates.* Univ. Washington Press, Seattle.

Permadi, D., Tumbelaka L.I., Yusef, T.L. 1994. *Reproductive Pattern of* Tarsius *spp. in the Captive Breeding.* Paper delivered at Congress of Int. Soc. of Primatology, Bali, Indonesia.

Pollock, J.I. and Mullin, R.J. 1987. Vitamin C Biosynthesis in Prosimians: Evidence for the Anthropoid Affinity of *Tarsius. Am. J. Phys. Anthropol.* 73:65–70.

Preuschoft, H., Fritz, M. and Niemitz, C. 1979. Biomechanics of the Trunk in Primates and Problems of Leaping in *Tarsius.* Pp. 327–345. in *Environment, Behavior and Morphology: Dynamic Interactions in Primates.* M.E. Morbeck, H. Preuschoft, and N. Gomberg, (eds.) Gustav Fischer, New York.

Rasmussen, D.T. 1994. The Different Meanings of a Tarsioid-Anthropoid Clade and a New Model of Anthropoid Origin. Pp. 335–360 in *Anthropoid Origins,* ed. J.G. Fleagle and R.F. Kay. New York: Plenum Press.

Roberts, M. 1994. Growth , Development, and Parental Care in the Western Tarsier *(Tarsius bancanus)* in Captivity: Evidence for a "Slow" Life History and Non-monogamous Mating System. *Intl. J. Primatol.* 15:1–28.

Roberts, M. and Kohn, F. 1993. Habitat Use, Foraging Behavior and Activity Patterns in Reproducing Western Tarsiers, *Tarsius bancanus* in Captivity. *Zoo Biol.* 12:217–232.

Rose, K.D. 1995. The Earliest Primates. *Evol. Anthropol.* 3:159173.

Rosenberger, A. and Szalay, F.S. 1980. On the Tarsiiform Origins of Anthropoidea. Pp. 139–157 in *Evolutionary Biology of the New World Monkeys and Continental Drift.* R.L. Ciochon and A.B. Chiarelli (eds.). Plenum, New York.

Shekelle, M. In prep. Ph.D. Thesis. Washington University, St. Louis.

Shekelle, M., Mukti, S., Ichwan, L. and Masala, Y. 1997. The Natural History of the Tarsiers of North and Central Sulawesi. *Sulawesi Primate Newsletter* 2:4–11.

Simons, E.L. 1972. *Primate Evolution.* Macmillan, New York.

Simons, E.L. 1974. Notes on the Tertiary Prosimians. Pp. 415–433 in *Prosimian Biology,* R.D. Martin, G.A. Doyle, and A.C. Walker. Duckworth, London.

Simons, E.L. and T.M. Bown. 1985. *Afrotarsius chatrathi,* The First Tarsiform Primate (?Tarsiidae) From Africa. *Nature* 313:475–477.

Szalay, F.S. 1975. Phylogeny, Adaptations, and Dispersal of Tarsiiform Primates. Pp. 91–125 in *Phylogeny of Primates: A Multidisciplinary Approach.* W. Luckett and F.S. Szalay (eds.), Plenum, New York.

Szalaly, F.S. 1976. Systematics of the Omomyidae: Taxonomy, Phylogeny and Adaptations. *Bull. Am. Mus. Nat. Hist.* 156:157–450.

Szalay, F.S., Rosenberger, A.L., and Dagosto, M. 1987. Diagnosis and Differentiation of the Order Primates. *Yrbk. Phys. Anthropol.* 30:7–105.

Tremble, M., Y. Muskita, and J. Supriatna. 1993 Field Observations of *Tarsius dianae* at Lore Lindu Nation Park, Central Sulawesi, Indonesia. *Tropical Biodiversity* I:67–76.

Ulmer, F.A. 1960. A Longevity Record for the Mindanao Tarsier. *J. Mammol.* 41:512.

Ulmer, F.A. 1963. Observations on the Tarsier in Captivity. *Z. Gart.* 106–121.

Wharton, C.H. 1950. The Tarsier in Captivity. *J. Mammal.* 31:260–268.

Wright, P.C., D.M. Haring, and E.L. Simons. 1985. *Social Behavior of* Tarsius syrichta *and* Tarsius bancanus. Paper delivered at the annual meeting of American Association of Physical Anthropologists, Knoxville, TN.

Wright, P.C., M.K. Izard, and E.L. Simons. 1986a. Reproductive Cycles in *Tarsius bancanus. Am. J. Primatol.* 11:207–215.

Wright, P.C., L.M. Toyama, and E.L. Simons. 1986b. Courtship and Copulation in *Tarsius bancanus. Folia Primatol.* 46:142–148.

Wright, P.C., D. Haring, E.L. Simons, and P. Andau. 1987. Tarsiers: A Conservation Perspective. *Primate Conserv.* 8:51–54.

Zuckerman, S. 1932. *The Social Life Of Monkeys And Apes.* Kegan Paul, London.

CHAPTER 7

Summary Chapter: Patterns of Variability, Diversity and Conservation Among Prosimians

During the Eocene, prosimians occupied tropical regions all over the world. By the Oligocene, however, they had given rise to the monkeys and apes and most of the prosimians had disappeared. Currently only a relatively few species survive outside of Madagascar. In Africa, there are only two species of loris and twelve species of galagos recognized at this time. In Asia, three species of loris and five tarsiers have been identified to date. All of these forms are nocturnal and occupy fairly specialized niches in the forests they inhabit. In Madagascar, there are at least 25 species of lemurs, approximately half of which are nocturnal. In the recent past, at least 17 lemur species have become extinct, and most of these are larger than living forms and are thought to have been active during the day.

It is likely that more species of nocturnal prosimians will be identified in future years because of the difficulty researchers often have in identifying the subtle differences that exist among nocturnal forms. Also, in the future, it is hoped that more long-term ecological and behavioral studies will be conducted on these fascinating animals. Until recently, very few studies had been conducted on any of these species for over 12 months. Within the last 10 years, there has been a resurgence of research on prosimians, especially the diurnal lemurs. However, there are still some species that have been studied little or not at all and very little research exists on these long-lived animals over a number of generations. Furthermore, very few species have been studied at more than one field site.

In reviewing the various sections of Chapters 3–6, I was struck mainly by the diversity and variation that exists among the living species of prosimians. In many ways, this variability is even greater than that found among monkeys and apes. It also is likely that there is a great deal of yet undiscovered diversity and variation in behavior and ecology within different populations of the same species living in different localities. The ecosystems these primate populations occupy offer ideal laboratories in which to test hypotheses concerning the relationships between morphology, physiology, ecology, social structure and organization, and the evolution of social behavior. Much remains to be done. Let us hope that primate populations and the forests in which they live can be conserved and that these laboratories are not shut down in the near future.

In the following pages I will summarize some of this diversity, covering the major topics that I have dealt with in each of the last four chapters, including activity cycle, habitat and locomotion, diet, predation, and social structure and organization. I will also discuss the conservation status of these animals and some problems with current conservation policy.

ACTIVITY CYCLES

Diversity among prosimians is exemplified by variations in activity cycle. African and Asian Lorisiformes and the Southeast Asian Tarsiiformes are all nocturnal. Differences among these forms in their patterns of activity throughout the night may be important in relation to specific adaptations and to niche separation. These differences should be related to differences in size, diet, environmental factors, social organization and to predator pressure. However, specific nocturnal activity cycles have not been the subject of much scrutiny.

The Malagasy Lemuriformes exhibit a wide array of activity patterns. These include exclusively diurnal, exclusively nocturnal, crepuscular and, uniquely for primates, cathemeral activity (active both during the day and at night). In fact, the mongoose lemur shows all of these patterns depending on location and/or season. A number of reasons have been proposed for cathemeral activity cycles, including seasonal and hourly differences in resource availability (e.g., floral nectar), competition for resources between closely related species, thermoregulation, and avoidance of predators. Finally, there are unusual yearly patterns of hibernation (in dwarf lemurs) and torpor (in mouse lemurs). Less dramatic is the seasonal decrease in activity among some galagos. These yearly patterns of activity directly relate to diet and to the availability of resources (or lack thereof) for these diminutive creatures during seasons of resource scarcity in Madagascar and Africa.

Differences in activity cycle are related to a number of morphological and behavioral differences. For example, diurnal primates, including the diurnal lemurs, are larger than nocturnal species. Interrelated with differences in activity cycle and size are differences in diet, predation pressure, and ranging behavior. Nocturnal primates also use smell and vocal communication, whereas diurnal species use vocal and visual signals more frequently. Since most nocturnal species sleep in well-protected nests, tree holes, or dense vegetation during the day and most do not carry their young when they are active, they can afford to have litters of 2–3 offspring. The larger, diurnal species (except for the ruffed lemur) do not have nests and carry their infants while active. This limits the ability of females to care for more than one infant at a time and births are typically single. Finally, except for a few secondarily nocturnal species (species who are likely to be derived from a diurnal ancestor), those active during the night do not travel together in cohesive groups. Essentially, all diurnal primates including those in Madagascar (with the exception of the Asian orang utan) live in relatively permanent social groups.

HABITAT AND LOCOMOTION

The Lorisiformes of Africa are found in a wide array of habitat types, including primary and secondary rain forests, clearings along secondary forest, and along the margins of savannahs. Where they coexist, the five species studied by Charles-Dominique in Gabon are habitually found in quite specific substrates, forest levels, and biotopes. One species, *G. elegantulus* has claw-like nails which enables it to exploit all levels of the forest including large highway branches. The differences in habitat preferences enable these species to coexist.

East African galagos are very widely distributed, living in many different habitat types. These include woodland/savannah, bush and forest fringes, riverine forests, highland forests, and relic primary forests. Subtle habitat distinctions must exist between sympatric forms, and those with overlapping geographical ranges. For example, two coexisting forms of galagos, *G. garnettii* and *G. zanzibaricus,* are normally found at different heights with the larger species, *G. garnettii,* higher in the forest. Where *G. garnettii* occurs with *G. crassicaudatus,* the former prefers riverine and highland forest, whereas the latter is found most frequently in woodland habitat. The Somali galago *(G. gallarum)* is the most xerically (dry forest) adapted of all bushbabies.

The Asian loris, *Nycticebus,* is found in primary and secondary rain forest, and in bamboo and scrub forest. The genus *Loris* occurs in primary and secondary forest, as well as scrub, riverine, and high altitude forest. It moves mainly in dense vegetation and also does not hesitate to come to the ground when necessary.

The major distinction between the lorises and the galagos is their form of locomotion. Lorises are slow-moving and deliberate. They do not leap nor have an in-air phase during locomotion, unless they are falling. The galagos, on the other hand, are fast-moving. They mainly are quadrupedal runners and leapers. Some forms of galagos, such as *G. moholi* and *G. alleni,* are considered vertical clingers and leapers (VCLs) because they spend most of their time clinging to and leaping between vertical branches and trunks.

Some nocturnal Malagasy lemurs have adaptations that enable them to live in very seasonal environments. These include, as mentioned above, hibernation and seasonal torpor. As in the lorisiformes, as many as five nocturnal species can coexist in some dry western forests, each using separate forest substrates and strata. Two genera *Phaner* and *Allocebus* have claw-like nails similar to those of *G. elegantulus.* There are virtually no field data available for *Allocebus.* Various forms of nocturnal lemurs can be found in highland and lowland rain forest, riverine forest, forest edge, brush and scrub forest, on the borders of savannah, and even in the desert-like forests of the southern Madagascar.

The nocturnal family of Cheirogaleidae includes quadrupedal running and leaping animals that spend most of their time grasping branches smaller than their body size. The claw-like nails of *Phaner,* and probably those of *Allocebus,* allow these forms to use larger supports. *Lepilemur* is a vertical clinger and leaper. In fact, this locomotor adaptation allows the sportive lemur to be one of the few primates able to exploit desert-like vegetation, since most of the plants found here have only vertical supports to grasp.

Daubentonia, the oddest of all primates, also is found in many habitats. Once thought to have a restricted distribution, it now has been observed in many different localities including disturbed and undisturbed rain forest, dry deciduous forest, coastal forest, and mangrove. It is a slow, deliberate, quadruped and is found at all levels of the forest up to 35 meters high. In some localities, ayes-ayes use the ground for up to a quarter of the time (more than any other nocturnal primate). This species also has claw-like nails enabling it to use large branches, but this does not limit its locomotion on thin branches. While on the ground, the aye-aye usually moves slowly with the fingers raised not touching the ground. A strange set of adaptations, indeed.

As with the nocturnal forms, the diurnal lemurs exist in an extremely diverse set of habitats and exhibit an enormous array of locomotor adaptations, especially if one considers the subfossil forms. These lemurs are found in all major forest types in Madagascar including savannah, primary and secondary rain forest, desert, riverine and deciduous forests, mongroves, limestone and brush and scrub forests, bamboo forests, and reed bed marshes. They live in both high and lowland forests. The ringtailed lemur is the only extant prosimian that does most of its traveling on the ground and it is quite adept in forest edges. Populations of ring-tailed lemur even occupy rocky outcrops in the highlands of the central plateau. Although there are only a few areas of the central plateau where lemurs exist today, in the recent past many lemurs could be found there and the habitat of the plateau probably was composed of a mosaic of forest types until recently. Furthermore, one subfossil form was probably very much like the totally ground-living gelada baboon of Africa.

Sympatric diurnal species have different habitat choices within their shared forests. Differences in habitat choice between ringtailed and brown lemurs in Southwest are not subtle. However, sympatric pairs of Lemuridae coexisting in the North and East of Madagascar do display subtle distinctions in choices of substrate and forest strata. These habitat choices, along with subtle dietary differences, are the major factors enabling them to coexist. For example, *Eulemur fulvus* and *Eulemur coronatus* exhibit significant differences in the use of forest strata in certain seasons. Similar distinctions were found in the use of forest strata and height between *Eulemur fulvus* and *Varecia variegata*. Differences between *Eulemur fulvus* and *Eulemur rubriventer* were more related to diet than to locomotion and substrate preference, and included the proportion of different plant parts eaten and methods of flower feeding.

Locomotor behavior is extremely variable among the diurnal lemurs. Most of the Lemuridae are fast moving quadrupeds. *Hapalemur* is an exception, being a vertical clinger and leaper like the Indriidae. Vertical clinging and leaping in large species (like the indri and the sifaka) is a sight to behold. This means of locomotion necessitates that major travel be done where vertical branches are available but it does not limit exploitation of terminal branches for foraging or feeding. It also allows *Propithecus,* like *Lepilemur,* to exploit desert-like vegetation and enables *Hapalemur* to move about in bamboo forest and in reed beds.

The ringtailed lemur is adept at locomoting both on the ground and in the trees and has morphological adaptations that reflect this adaptability. As stated above, one of the subfossils, *Hadropithecus,* was a purely terrestrial quadruped. Other subfossil species had a variety of locomotor adaptations not found in extant forms, including sloth-like suspensory locomotion, koala-like slow locomotion, and orang utan-like hanging and climbing. Thus, the diurnal lemurs, extinct and extant, display a greater variety of locomotor behavior than do the living monkeys and apes.

Just as in the other prosimians, tarsiers are found in a great variety of habitats. These include scrub, secondary forest, thicket, grassland, primary forest, coconut groves, and along dry stream beds, and in montane and lowland forests. They are specialized VCLs, leaping 60–70% of the time. The amount of time they use other categories of locomotion differs between species. There is some evidence that certain species of tarsier may be more adept in higher forest strata (e.g., *T. dianae, T. spectrum)* and others in lower strata (e.g., *T. bancanus, T. syrichta),* but further study is needed to confirm this conclusion.

DIET

Just as for activity cycles, habitat preferences, and locomotion, prosimians display a wide array of dietary choices. In fact, in this aspect of their behavior there is remarkable variability, including dietary specialties found in no other primate and in some cases, in few other mammals. As discussed in Chapter 2, primates and squirrels are the major mammalian taxa to exploit the forest and canopy during the day—birds are their main competitors. During the night, there are few birds and no squirrels active in the canopy. Bats and the night-living prosimians are the main nocturnal canopy dwellers.

In the rain forests of Gabon, differences in the diets of the Lorisiformes are related to both body size and locomotion. The smallest lorisis and galagos are highly insectivorous whereas the larger species include more plant material in their diet. The majority of primates are omnivorous and this relationship between size and major dietary choice, generally, is predictable, with smaller species able to obtain more of their nutritional needs from animal prey and larger primates needing to eat more plant parts. The fast-moving galagos can catch fast-moving insects whereas the slower lorises must rely on slow, easy-to-catch insects. These slow insects protect themselves from most predators by being toxic or, in some way, noxious. Many of the differences between the galagos and lorises can be related to this ability to utilize these normally avoided insects.

The smallest of these species, such as *G. demidoff, G. moholi, G. zanzibaricus, A. calabarensis,* and *Loris tardigradus,* are highly insectivorous. In fact, the Indian slender loris is among the most faunivorous of all primates. However, the majority of the diet of most lorises and galagos consists of plant material, mainly fruit and gums. The needle-clawed galago *(G. elegantulus)* is morphologically specialized to feed on gums. However, other galagos living in dry habitats (e.g., *G. crassicaudatus, G. senegalensis, G. moholi,* and probably *G. matschiei)* also are highly dependent on gums during part of the year. The distribution and density of some of these galagos is dependent upon certain species of gum producing plants.

In forests in which a number of Lorisiformes coexist, resources are divided by a number of factors. These include: the types of insects consumed; the biotope in which they are caught; the types and placement in the canopy of the fruit eaten; and the dependence, by the mainly plant-consuming species, on either fruit or gums during seasons of scarcity. Thus, besides differences in basic habitat choice, subtle dietary differences are important in niche separation between these species, and these differences can be related to specific morphological and physiological characteristics.

The nocturnal lemurs display many dietary adaptations similar to those of the lorises and galagos. However, there is even more diversity in the array of specializations seen in these creatures of the night. In the five species of western Madagascar which were studied in sympatry, the ability to survive the very resource-poor, dry season has put extreme pressure on these animals so that each species is quite unique in its dietary adaptations.

Cheirogaleus medius (the fat-tailed dwarf lemur) eats mainly plant foods and increases its fruit intake to correspond with a pattern of fattening that precedes hibernation. *Microcebus* (the mouse lemur) is the smallest of all primates and, therefore, was assumed to be mainly insectivorous. In fact, it eats a mix of insects and fruits, and over half of its diet is fruit. In some forests, it relies heavily on highly nutritious plants like mistletoe, a plant often eaten by frugivorous birds. However, resources are scarce in the dryer season and mouse lemurs go through a period of torpor. But not all individuals enter torpor at the same time, and females do so more and for longer periods than do males.

Mirza coquereli is highly frugivorous but exploits insect secretion during the dry season in some forests. *Phaner,* like *G. elegantulus,* is a gum-feeder and it has morphological adaptations for this dietary specialty which parallel those of the needle-clawed galago. Though it has not been studied, the eastern rain forest species, *Allocebus,* shares this suite of adaptations for gum-feeding.

Lepilemur has a specialized, folivorous diet. This is unique for a small, nocturnal primate. *Lepilemur* is almost entirely folivorous and its diet is very monotonous, with a few plant species accounting for an inordinate proportion of the diet.

The nocturnal *Daubentonia* has one of the most specialized diets of the nocturnal prosimians. The aye-aye's diet is unique among primates. It uses its teeth to bore into hard seeds and nuts, eating the softer material inside, and into tree bark to obtain wood-boring insects. Then it uses its long middle finger and incisors to extract this material. In doing this, the aye-aye fills a niche similar to that of squirrels and woodpeckers, both of which are absent from the forests of Madagascar.

The diurnal Malagasy lemurs display a similar diversity of dietary preferences. The Lemuridae, except for *Hapalemur,* are mainly frugivorous with eastern Malagasy *Eulemur* and especially *Varecia* eating very high (80–90%) proportions of fruit. Western *Eulemur fulvus* is seasonally highly folivorous, eating the highly nutritious leaves of the tamarin tree. Nectar is an important seasonal component of the diet of some forms, especially *E. mongoz, E. rubriventer* and *Varecia,* and there may exist a long-term co-evolutionary relationship between these lemurs and the plants they feed upon. *L. catta* has a diverse diet made up mainly of fruit but supplemented by leaves, mainly from ground-dwelling herbs and tamarin. The ringtailed lemur also includes some insects in its diet, as do some species of *Eulemur.*

The Indriidae are mainly folivorous and have morphological, physiological, and behavioral adaptations that enable them to exist on low nutrient diets. Over 70% of the diet of *Indri* consists of leaves. The western species of sifaka, *Propithecus verreauxi,* eats mainly leaves, whereas eastern *P. diadema* feeds on leaves and/or seeds depending on the season and the site. The third genus of Indriidae, the secondarily nocturnal *Avahi,* is also folivorous.

The various species of *Hapalemur* have very specialized diets, eating mainly bamboo and grasses. Some of the bamboo that these species eat contains very high levels of deadly cyanide and it is not yet known how the animals avoid being poisoned. Few other mammals exploit this potentially abundant food resource and yet, in some forests of Madagascar, three sympatric species of gentle bamboo lemur coexist, dividing this resource among them.

As in many other ways, tarsiers are highly specialized in their feeding behavior. In fact, they are the only living primates whose diet is entirely faunivorous. *T. bancanus* eats mainly insects but also has been observed eating birds, bats, lizards, frogs, small crabs, and snakes. *T. spectrum* has been seen eating only insects. The jaw muscles and teeth of tarsiers are highly specialized for acquisition and processing of relatively large animal prey.

PREDATION

All living prosimians are relatively small and thus are vulnerable to a variety of predators. However, some of these primate species are much smaller than others. As with other aspects of behavior and ecology, size makes a big difference. The nocturnal primates, being generally smaller than diurnal ones, are potentially more vulnerable to predation. However, being nocturnal is probably one important component of their protection. During the night, these species are difficult to find and keep up with. Furthermore, the small nocturnal species do not travel together in groups and thus are less visible individually. During the day, these animals must find concealed sleeping spots, either nests, tree holes and/or very dense vegetation. Infants are often left in their nests or "parked" and remain immobile during the night.

Galagos are fast-moving animals and can escape from predators by leaping away. They also give alarm calls and sometimes mob predators after spotting them. Because of their slow locomotion, lorises have a unique predator protection system, including morphological and physiological components (e.g., a specialized circulation system, bad odor, toxicity, scapular shield) and behavioral components (e.g., cryptic locomotion, defense postures). Loris newborns are not parked during the night but are able to cling to the mother's fur as she moves about. Potential predators of Lorisiforms are small, nocturnal felids, snakes and birds of prey. Galagos also have been observed being preyed upon by chimpanzees.

Some of the most detailed studies of predation on primates have been done on the Malagasy lemurs. Potential predators are viverrids, cats, domestic dogs, snakes, and birds of prey. There are many observations of these predators dining on both nocturnal and diurnal lemurs. The nocturnal lemurs, like their Asian and African counterparts, are mainly cryptic and quick, and sleep in hidden places during the day. However, they are still quite vulnerable. In fact, it is estimated that the yearly predation rate on mouse lemurs at one forest, Beza Mahafaly, is approximately 25% (520–580 lemurs eaten), the highest known for any primate species.

Unlike the nocturnal species, diurnal lemurs travel in groups and this adds eyes, ears and voices to their predator defense system. All of these species have specific predator alarm calls and conspecifics, other lemur species, and even nonpredatory birds react to one another's predator calls. Some species, such as *L. catta* for example, have different calls and different reactions to different classes of predators such as cats, birds and snakes. Even given these defense behaviors, a large number of diurnal lemurs of many species and of all ages have been observed being successfully preyed upon by predatory birds, viverrids, and snakes. Furthermore, in the recent past, there was at least one quite large, subfossil eagle that was surely an important predator of lemurs.

Some investigators have not seen tarsiers react to predators and believe that these primates may have a repulsive odor or perhaps are toxic to prey. Others, however, have seen reactions to predators by tarsiers and there have been observations of attempted attacks on tarsiers by monitor lizards and by snakes. Predatory birds and civets also are thought to prey on tarsiers.

Humans prey on most of these species, either for food, for medicinal purposes, for the pet trade, for sport, or, in some regions, because the lemurs are considered to be pests.

SOCIAL ORGANIZATION

As I have described so far in this summary chapter, the behavior and ecology of prosimians is extremely diverse and variable. Yet, the social behavior of these forms is probably the most diverse and variable of all of their behavioral characteristics. Here, I only briefly will outline some of these differences.

For the most part, the nocturnal prosimians live in solitary but social types of communities. Basically, they are "social" in relation to their nesting patterns and to the amount, quality and patterns of social interactions throughout their nocturnal activity cycle. They are "solitary" in that, during the night when active, they usually move about and forage alone, and generally do not travel in cohesive groups. However, there is a great deal of variation in nesting patterns, in patterns of social interaction between individuals, and in tolerance between and among the sexes.

In most species, nesting groups consist of a various number of females with their young and sometimes an adult male. In most cases, adult males are not tolerant of one another, but this is one of the more variable characteristics among and between species. Generally, the male home range is larger than that of the females and may overlap more than one female nesting group. However, during mating season, there is evidence that both males and females mate with many partners in most species.

A few nocturnal species have a pattern quite different from the general one described above. In *Phaner, Lepilemur,* and *Galago zanzibaricus* a small number of individuals share home ranges and might be considered range-mates. These "groups" generally include one male with one or two females, but sometimes they consist of two females without a male (e.g., in *Lepilemur).* In these species "range-mates" interact physically and/or communicate throughout the night. Social organization and structure also is variable among tarsiers. *T. spectrum* and *T. dianae* appear to stay in groups similar to, or even larger than those described for *Phaner, Lepilemur* and *G. zanzibaricus. T. bancanus* may have a solitary-but-social social structure similar to that described for most other nocturnal prosimians. For *T. syrichta* the data are equivocal. In captivity they are quite social but in a short-term field study individuals of this species were always observed alone.

The diurnal prosimians of Madagascar, like all monkeys and apes except the orang utan, live in relatively permanent, more or less cohesive social groups. Among these diurnal forms there is a great deal of variation in social organization and social structure. Most species in the genus *Eulemur* live in multi-male/multi-female groups averaging less than 10 individuals. These groups do not have noticeable dominance hierarchies and do not defend territories, though they do maintain group integrity during inter-group encounters. In many lemur species females are dominant to males but this *is not* the case in most species of *Eulemur. E. macaco* is an exception.

Mating is generally promiscuous in that both males and females mate with more than one partner and with individuals both within and outside of their group. Mating season among all lemurs (except *Hapalemur griseus* and *Daubentonia)* is extremely short, usually lasting less than three weeks per year. *Eulemur mongoz* and *Eulemur rubriventer* differ from other species of the genus in that they live in pair bonded groups containing a female, a male and their offspring. However, in the mongoose lemur, I have observed mating occur between animals of different groups. *Eulemur mongoz* is not territorial whereas *Eulemur rubriventer* is. Although it has been suggested that all species of *Eulemur* form pair bonded subgroups within the larger group, there is no evidence from any of the long-term studies that this is the case.

Among extant prosimians, the ringtailed lemur is found in the largest groups, containing up to 27 animals. These groups are multi-male/multi-female. Males regularly migrate out of their natal group whereas females are philopatric (remain with their mother's group). Females are dominant to males. Home ranges are not exclusive and mating is promiscuous. In all of these species, as in all diurnal primates except *Varecia,* the young actively grasp the mother's fur and are carried whenever and wherever she travels. The great majority of births are singletons.

Varecia has a number of features that are unique to diurnal prosimians. They live in multi-male/multi-female communities with defended ranges but do not move in cohesive groups. Individuals interact with members of the community in a fission-fusion manner. Mating is promiscuous. Infants are kept in nests or stashed and there are normally 2–3 infants per birth. They are cared for communally by both male and female community members. The young grow very quickly and are independent at about 70 days after birth. Females are dominant to males.

Among the Indriidae, *Indri,* and most likely *Avahi,* live in pair bonded groups that defend territories. *Propithecus* lives in small multi-male/multi-female groups of about 4–6 animals though there can be as many as 14 in a group. Natal males migrate. Mating is promiscuous, occurring both within and between groups. Some groups in some areas are territorial whereas others are not. Again, females are dominant to males.

Finally, *Hapalemur* lives in small groups (2–9 individuals). Some groups contain one male and one to two adult females other groups are larger but have not been observed to have more than two adult females and three adult males. Approximately half of the groups contain two adult females. At some sites they are territorial and both males and females migrate. As mentioned earlier, the Alaotran gentle lemur is only one of two species of lemur that does not have a highly restricted birth season.

Thus, almost all types of social structure found among primates occur among the diurnal lemurs. By far the most common group structure among them is the multi-male/multi-female pattern, also the most common social structure among monkeys and apes. A few species occur in pair bonded groups and at least *Varecia* lives in fission-fusion groups. In some *Hapalemur* populations animals commonly have a one-male two-female group structure. Neither the one-male, multi-female groups found among some Old World monkeys nor the polyandrous group structure of the marmosets and tamarins are not found among extant prosimians. A unique feature of many species of diurnal lemurs is female dominance.

Over the years there have been many theories developed attempting to explain the relationships between social structure, social organization, and social behavior in group-living primates, and the relationships between ecology and social systems. Currently, one of the major theories relating individual spacing to that of resource distribution is based on sexual selection. Proponents of this theory claim that, since females must invest a great deal of time and energy in successfully reproducing offspring and males invest a small amount, the distribution of females is related to that of resources and that of males to the distribution of females. Thus, females compete for energy resources and males compete for females.

This may or may not be true. However, to date we still have not been able to relate patterns of resource distribution to patterns of spacing among primates. We know very little about the factors that underlie the enormous amount of variability in primate social systems and we still need much more data about this variability, both between and among species. So far these theories have not led to a better understanding of the relationships that exist between ecology and social structure. Let us hope that the primates are here long enough to collect sufficient data to test these theories, and to generate new theories based on past and future field data.

CONSERVATION

Conservation of primates is a problem all over the world. However, it is not just a problem of biology and the natural sciences. It also is a problem that includes the human population and its resources and the distribution of those resources. Thus, conservation policies and programs must include the people of the countries in which these programs are focused. Furthermore, it is often rural, relatively isolated human populations that are effected by conservation policy.

These people, not just the central government, need to have input into programs that have a direct input into their daily lives. However, there are often conflicting goals and aspirations among the scientific community, international conservation organizations, global agencies, national and local governments, and the local people. These conflicts are often extremely complex, with each group having real and legitimate problems related to conservation. Because of this, conservation often becomes as much, or more, a human social problem as it is a problem of purely the biological and natural sciences. Thus, although often left out of the formula of policy-making, the social sciences, and anthropology in particular, must begin to play a central role in worldwide conservation policy.

It is possible that the ecological carrying capacity of humans already may be exceeded in several regions of tropical Africa (e.g., Rurundi, Rwanda, Uganda, and parts of the Central African Republic) (Butynski 1996/97). This has led to malnutrition and conflict among many populations. The rate of population growth is still close to 3% per annum in most of Africa and at this rate the population will double in around 25 years. Thus, attempts to reach a sustainable resource base are and will remain extremely difficult.

The spiral of population growth coupled with resource decline has led to what Myers (1993) has called the "demographic gap", where population growth and the concomitant environmental decline blocks the possibility of economic growth that would allow for reduction of population growth. Any attempt to conserve the remaining forests must take into account increasing population, increasing poverty, reducing resources, and the problems related to the fair and equitable distribution of those resources.

At this point in time, the greatest threat to the African primates, generally, is unsustainable hunting and habitat destruction, mainly for logging and clearing for agriculture. With habitat destruction, primate populations become smaller and more isolated thus making them more susceptible to extinction. At present, tropical moist forests cover only approximately 10% of Africa and deforestation is exceeding reafforestation by at least 13-fold and is estimated to be higher than for any other continent (Butynski 1996/97). Countries of the European Union are the most important market of tropical timber from Africa, accounting for almost 90% of the market (Ammann and Pearce 1995). Logging and hunting usually go hand in hand, with hunters supplying bushmeat to logging company workers as well as to growing agricultural communities and to people in the distant larger towns and cities. In some areas over 40% of the bushmeat comes from primates.

Only one species of Lorisiform *(Galago gallarum)* is not found in a protected area, and none are listed as endangered at this time (Butynski 1996/97). Although the small nocturnal prosimians of Africa are not the major target of hunting, habitat destruction is certainly affecting the populations of these species. Concerning conservation projects in Africa, Butynski (1996/97:96–97) states:

> *. . . no less than 40 million dollars have been spent...over the past 10 years towards the direct and indirect conservation of the 14 eastern Africa priority sites. . . . Some of this money was put to good use. . . . Unfortunately, a number of the primate and forest "conservation" projects . . . have been driven and misguided by politics and economics (both nationally and internationally), while research findings and recommendations of trained, experienced, and independent professionals have often been ignored....The unwise use of conservation funds is one reason why countries in tropical Africa continue to lose their moist forests, and why many primate populations continue to be threatened.*

In Asia, the combination of factors leading to deforestation is different from that in Africa. Human population density and increase is definitely a problem in Asia. However, although over half of the world's population live in Asia (3.3 million people), and poverty is certainly a major problem, it is predicted that within the next 20–25 years 2 billion people will be raised out of poverty. Thus, in Asia it is population numbers and economic growth and development, and not poverty, that are the main factors threatening the environment and the wildlife (Eudey 1996/97).

The cause of deforestation in Asia is mainly legal and illegal timbering, conversion to agriculture, and hydroelectric projects. This has led to the direct loss of primate populations and their increased vulnerability to hunting and capture for illegal trade (Eudey 1996/97). Although the larger diurnal primates are most susceptible to hunting, lorises are used for medicinal and other purposes and regularly sold in the markets of big cities (Nekaris pers. comm.).

Currently only two Asian Lorisiformes, *Loris tardigradus tardigradus* and *Nycticebus pygmaeus,* are listed as vulnerable in the *IUCN Red List of Threatened Animals* (1996). Data are deficient for most of the tarsiers to even determine their current conservation status. However, due to continued habitat destruction, even in protected areas, Gursky (1999, in press) believes that at least the Sulawesi spectral tarsier should be considered as vulnerable and possibly even threatened. Tarsiers are extremely difficult to rear in captivity. Habitat destruction is not selective as to which primate species are affected and, as in Africa, loss of forest poses the greatest threat to the Asian prosimians.

Since the arrival of humans on Madagascar approximately 2000 years ago, at least 17 lemur species have become extinct, including all species over 7 kg. Because of the relatively small geographic region in which they live and their small overall range, all lemurs are considered to be in danger of extinction. Even though hunting of certain lemurs is traditionally taboo in many regions of Madagascar, and currently is against the law, this still remains a problem. The distribution of the larger species away from villages, even where suitable habitat exists, is testimonial to this fact (Ganzhorn et al. 1996/97). However, the small nocturnal forms are not usually hunted and, as in Africa and Asia, the main threat to all species is habitat destruction. Forest is cut mainly for local subsistence purposes, though recently commercial mining has become a problem in some regions. As in Africa, poverty is widespread in Madagascar and the rate of population increase here is even higher than in Africa, at 3.2% per year.

We did a study of the deforestation of the eastern rain forest using aerial photographs and satellite images, covering a thirty-five year period (1950–1985)

(Green and Sussman 1990). During this time, 50% of the rain forest had been lost, mainly due to subsistence agriculture, increasing population, and the need for more fertile land. Furthermore, as population increases, forest on steeper and steeper slopes is being used.

In the eastern rain forests of Madagascar, conservation efforts have been directed toward natural reserves and already protected areas, but most of these reserves were established in remote, isolated, and steep areas so they have not impinged upon the current need for land. From our data, we predict that, because most of the reserves are located on steep slopes, they will remain untouched by the year 2020. However, 50% of the remaining rain forests will be gone and only small isolated forest on the steepest slopes will remain (Sussman et al. 1994). In other regions of Madagascar, the need for firewood or charcoal for fuel, or to create grassland for cattle leads to forest cutting. In fact, the dry forests of the west and south are probably in more danger than are the eastern rain forests (Ganzhorn et al. 1996/97, Smith 1997, Sussman et al. in prep.).

However, these are not simply problems of nonhuman primate conservation. They also are related to human primate conservation. Human as well as nonhuman primates must be protected, and the needs of both must be met, or conservation efforts will never succeed. As stated above, conservation has become a problem of human existence and co-existence with the natural habitat. Therefore, the solutions must come not only from the natural and biological sciences but also from the social sciences. Since anthropologists are the social scientists working most closely with people of different cultures, they must become major players in this effort. Furthermore, many different scientists and agencies must get involved in conservation.

Preserving natural habitat is probably the most urgent problem that we face in the near future. There is little time left to save the remaining large tracts of natural forest. If this task is left solely in the hands of the large conservation and development organizations, without monitoring and outside input, and if the economic and social needs of local, rural people are not met, deforestation will continue until there is little forest left to cut. If this happens, of course, most of the prosimians will be gone and our natural laboratories will be closed forever.

BIBLIOGRAPHY

Ammann, K. and J. Pearce. 1995. *Slaughter of the Apes: How the Tropical Timber Industry Is Devouring Africa's Great Apes.* World Society for the Protection of Animals, London.

Butynski, T.M. 1996/1997. African Primate Conservation—The Species and the IUCN/SSC Primate Specialist Group Network. *Primate Conserv.* 17:87–100.

Eudey, A.A. 1996/1997. Asian Primate Conservation—The Species and the IUCN/SSC Primate Specialist Group Network. *Primate Conserv.* 17:101–110.

Ganzhorn, J.U., O. Langrand, P.C. Wright, S. O'Conner, B. Rakotosamimanana, A.T.C. Feistner, and Y. Rumpler. 1996/1997. The State of Lemur Conservation in Madagascar. *Primate Conserv.* 17:70–86.

Green, G.M. and R.W. Sussman. 1990. Deforestation History of the Eastern Rainforest of Madagascar from Satellite Images. *Science* 248:212–215.

Gursky, S. 1999. The Tarsiidae: Taxonomy, Behavior, and Conservation Status. Pp. 140–145 in *The Nonhuman Primates.* P. Dolhinow and A. Fuentes (eds.). Mayfield, Mountain View, California.

Gursky, S. In press. Conservation Status of the Spectral Tarsier, *Tarsius spectrum:* Data on Population Density and Home Range. *Folia Primatol.*

IUCN. 1996. *IUCN Red List of Threatened Animals.* IUCN, Gland.

Myers. 1993. Population, Environment, and Development. *Environ. Conserv.* 20:205–216.

Smith, A.P. 1997. Deforestation, Fragmentation, and Reserve Design in Western Madagascar. Pp. 415–441 in *Tropical Forest Remnants, Ecology, Management and Conservation of Fragmented Communities.* W.F. Lawrence and R.O. Bierregaard Jr. (eds.). University of Chicago Press, Chicago.

Sussman, R.W., Green, G.M. and Sussman, L.K. 1994. Satellite Imagery, Human Ecology, Anthropology, and Deforestation in Madagascar. *Hum. Ecol.* 22:333–354.

INDEX

271